SpringerWienNewYork

CISM COURSES AND LECTURES

The series presents lecture notes, monographs, edited works and proceedings in the field of Mechanics, Engineering, Computer Science and Applied Mathematics.
Purpose of the series is to make known in the international scientific and technical community results obtained in some of the activities organized by CISM, the International Centre for Mechanical Sciences.

INTERNATIONAL CENTRE FOR MECHANICAL SCIENCES

COURSES AND LECTURES - No. 520

NEW TRENDS IN VIBRATION BASED STRUCTURAL HEALTH MONITORING

EDITED BY

ARNAUD DERAEMAEKER
UNIVERSITE LIBRE DE BRUXELLES, BRUXELLES, BELGIUM

KEITH WORDEN
UNIVERSITY OF SHEFFIELD, SHEFFIELD, GREAT BRITAIN

SpringerWienNewYork

This volume contains 149 illustrations

SPIN 80021260

All contributions have been typeset by the authors.

ISBN 978-3-7091-0398-2 SpringerWienNewYork

PREFACE

This book is the outcome of a European project sponsored by the European Science Foundation(ESF) under the EUROCORES S3T scheme, called S3HM (Smart Sensing for Structural Health Monitoring). It all began in July 2004 at the European Workshop on Structural Monitoring in Munchen, Germany, where we had the first discussions for the preparation of the project. It was clear from the beginning that we wanted a collaborative project, where the partners had complimentary expertise. This is essential in vibration-based structural health monitoring (SHM) which is a truly multidisciplinary topic. It was also clear that we wanted to look ahead, to explore new paths in vibration-based SHM. Apart from the numerous conference and journal papers produced during the project, we decided to share our views on vibration-based SHM through a course at the CISM in Udine, in September 2009. The course was entitled "New trends in vibration based SHM". This book, with the same name, is the next logical step.

In constructing this book, we wanted it to be a reference book for students, scientists and engineers starting in the field of SHM. The book, of course is not exhaustive in describing all the methods available in SHM, but it illustrates, through simple examples, the necessary steps for the successful implementation of an SHM system: (i) sensor networks, and data acquisition, (ii) signal processing, data reduction and feature extraction, (iii) statistical analysis of the data. Emphasis is put on important issues for the practical implementation of SHM systems such as data reduction in large sensor networks, sensor failure detection, extraction of modal parameters from operational data, removal of the variability due to changing environmental or operational conditions, statistical analysis of the data through supervised (neural networks) or unsupervised (outlier analysis, control charts) learning techniques. The last chapter is dedicated to wave-based methods, which are not, as such vibration-based SHM techniques, but are seen as a very promising complementary technique to the vibration-based methods presented in this book. All the methods and techniques presented are illustrated through numerical examples, experimental data from small scale laboratory demonstrators, or field measurements made on real structures.

It is our hope that the reader will share the enthusiasm of the many scientists who have contributed to all the research work presented in this book.

Arnaud Deraemaeker
Keith Worden

CONTENTS

An Introduction to Structural Health Monitoring

Charles R. Farrar[1] & Keith Worden[2]
[1]The Engineering Institute,
Los Alamos National Laboratory Los Alamos, New Mexico, USA.
[2]Dynamics Research Group, Department of Mechanical Engineering,
University of Sheffield, Mappin Street, Sheffield S1 3JD, UK.

Abstract This introduction begins with a brief history of SHM technology development. Recent research has begun to recognise that a productive approach to the Structural Health Monitoring (SHM) problem is to regard it as one of statistical pattern recognition (SPR); a paradigm addressing the problem in such a way is described in detail herein as it forms the basis for the organisation of this book. In the process of providing the historical overview and summarising the SPR paradigm, the subsequent chapters in this book are cited in an effort to show how they fit into this overview of SHM. In the conclusions are stated a number of technical challenges that the authors believe must be addressed if SHM is to gain wider acceptance.

1 Introduction

The process of implementing a damage identification strategy for aerospace, civil and mechanical engineering infrastructure is referred to as *Structural Health Monitoring* (SHM). A wide variety of highly-effective local Non-Destructive Evaluation (NDE) tools are traditionally available for such monitoring. However, the majority of SHM research conducted over the last thirty years has attempted to identify damage in structures on a more global basis using permanently installed sensors. The past ten years has seen a rapid increase in the amount of research related to SHM as quantified by the significant escalation in papers published on this subject. The increased interest in SHM and its associated potential for significant life-safety and economic benefits has motivated the need for this book.

In the most general terms, damage is usually understood as changes introduced into a system that adversely affect its current or future performance. Implicit in this definition is the idea that damage is not meaningful without a comparison between two different states of the system, one of which is assumed to represent the initial, and often undamaged, state. This book is focused on the study of damage identification in structural and mechanical systems. Therefore, the definition of damage will be limited to changes to the material and/or geometric properties of these systems, including changes to the boundary conditions and system connectivity, which adversely affect the current or future performance of these systems.

In terms of length-scales, all damage begins at the material level. Although not necessarily universally accepted terminology, such damage is referred to as a defect or flaw and is present to some degree in all materials. Under appropriate loading scenarios the defects or flaws grow and coalesce at various rates to cause component, and then system-level, damage. The term damage does not necessarily imply total loss of system functionality, but rather that the system is no longer operating in its optimal manner. As the damage grows it will reach a point where it affects the system operation to a point that is no longer acceptable to the user. This point is referred to as failure. In terms of time-scales, damage can accumulate incrementally over long periods of time such as that associated with fatigue or corrosion damage evolution. On relatively shorter time-scales, damage can also result from scheduled discrete events such as aircraft landings and from unscheduled discrete events such as enemy fire on a military vehicle or natural hazards such as earthquakes.

The SHM process involves the observation of a structure or mechanical system over time using periodically-spaced measurements, the extraction of damage-sensitive features from these measurements, and the statistical analysis of these features to determine the current state of system health. For long-term SHM, the output of this process is periodically updated information regarding the ability of the structure to continue to perform its intended function in the light of the inevitable aging and damage accumulation resulting from the operational environments. Under an extreme event, such as an earthquake or unanticipated blast loading, SHM could be used for rapid condition screening. This screening is intended to provide, in near real-time, reliable information about system performance during such extreme events and the subsequent integrity of the system. A more detailed description of SHM can be found in Worden and Dulieu-Barton (2004).

Damage identification is carried out in conjunction with five closely related disciplines that include SHM, Condition Monitoring (CM, see Bently and Hatch (2003)), Non-Destructive Evaluation (NDE, see Shull (2002)), Statistical Process Control (SPC, See Montgomery (1997)) and Damage Prognosis (DP, see Farrar et al. (2001, 2003)). Typically, SHM is associated with on-line, global damage identification in structural systems such as aircraft and buildings. CM is analogous to SHM, but addresses damage identification in rotating and reciprocating machinery, such as used in manufacturing and power generation. NDE is usually carried out off-line in a local manner after the damage has been located, and requires access to the component or structure of interest. There are exceptions to this rule, as NDE is also used as a monitoring tool for *in situ* structures such as pressure vessels and rails. NDE is therefore primarily used for damage characterisation and as a severity check when there is *a priori* knowledge of the damage location. SPC is process-based rather than structure-based and uses a variety of sensors to monitor changes in a process, one cause of which can result from structural damage. Once damage has been detected, DP is used to predict the remaining useful life of a system.

1.1 Motivation for SHM Technology Development

Almost all private industries and government organisations want to detect damage in their products as well as in their manufacturing infrastructure at the earliest possible time. Such detection requires these industries to perform some form of SHM and is motivated by the potential life-safety and economic impact of this technology. As an example, the semiconductor manufacturing industry is adopting this technology to help minimise the need for redundant machinery necessary to prevent inadvertent downtime in their fabrication plants. Such downtime can cost these companies on the order of millions of dollars per hour. Aerospace companies in the US along with government agencies are investigating SHM technology for identification of damage to the space shuttle control surfaces hidden by heat shields. Clearly, such damage identification has significant life-safety implications. Also, there are currently no quantifiable methods to determine if buildings are safe for reoccupation after a significant earthquake. SHM may one day provide the technology to significantly reduce the uncertainty associated with such post-earthquake damage assessments. The prompt reoccupation of buildings, particularly those associated with manufacturing, can significantly mitigate economic losses associated with major seismic events. Finally, many portions of our technical infrastructure are approaching or exceeding their initial design life. As a result of economic issues, these civil, mechanical, and aerospace structures are being used in spite of aging and the associated damage accumulation. Therefore, the ability to monitor the health of these structures is becoming increasingly important.

Most current structural and mechanical system maintenance is done in a time-based mode. As an example missiles are retired after a set amount of captive-carry hours on the wing of an aircraft. SHM represents the group of technologies that will allow the current time-based maintenance philosophies to evolve into potentially more cost effective condition-based maintenance philosophies. The concept of condition-based maintenance is that a sensing system on the structure will monitor the system response and notify the operator that damage has been detected. Life-safety and economic benefits associated with such a philosophy will only be realised if the monitoring system provides sufficient warning such that corrective action can be taken before the damage evolves to a failure level. The trade-off associated with implementing such a philosophy is that it requires a more sophisticated monitoring hardware to be deployed on the system and it requires a sophisticated data analysis procedure that can be used to interrogate the measured data. It is also critical that any monitoring system installed should be at least as reliable as the structure or system of interest.

Finally, many companies that produce high-capital-expenditure products such as airframes, jet engines, and large construction equipment would like to move to a business model where they lease this equipment as opposed to selling it. With these models the company that manufactures the equipment would take on the responsibilities for it's maintenance. SHM has the potential to extend the intervals between scheduled maintenance and, hence, keep the equipment out in the field where it can continue to generate revenue for the owner. Also, the equipment owners would like to base their lease fees on the amount of system life used up during the lease time rather than on the current simple time-based lease fee arrangements. Such a business model will not be realised without the ability to monitor the damage initiation and evolution in the rental hardware.

1.2 Motivation for this Book

Directly reflecting the increased interest in this emerging technology, there have been several conference series initiated in the last fifteen years that focus directly on SHM; (the most recent examples in these series being[1,2,3,4].) Focussed meetings and conferences related to the condition mon-

[1] The 7th International Workshop on Structural Health Monitoring, Palo Alto, CA, 2009.

[2] The 8th International Conference on Damage Assessment of Structures, Beijing, China, 2009.

[3] 14th International Symposium on Nondestructive Evaluation and Health Monitoring of Aging Infrastructure, San Diego, CA, 2009.

[4] The 4th European Workshop on Structural Health Monitoring, Krakow, Poland, 2008

itoring of rotating machinery are much older[5,6]. These conferences have shown that the topic of SHM is of interest to a wide range of industries and government agencies; They have also shown that many technical disciplines need to be integrated to properly address the SHM problem. In addition, the first refereed journal devoted specifically to SHM has recently been initiated[7], and others have followed. The proceedings of the specialised conferences as well as the extensive number of refereed journal articles devoted to various aspects of SHM show that significant knowledge and experience has been gained through the reported studies. Finally, the emergence of a number of specialised courses on SHM technologies and methodologies is further testimony to the interest expressed by industry. Despite the clear interest, there is a limited number of published textbooks and monographs on the subject of SHM (recent exceptions of note are Adams (2007); Giurgiutiu (2007); Staszewski et al. (2003)). A theme issue of the Transactions of the Royal Society of London was also devoted to the topic (Farrar and Worden (2007)), and makes a useful first port-of-call for an overview. Most notably, a comprehensive reference work has also recently appeared, Boller et al. (2009); although the focus of this work is not pedagogical. All of this means that it is timely to devote a new book in an effort to provide the engineering community with an up-to-date overview of SHM technology focussed on vibration-based methods and statistical pattern recognition - aspects of the subject which are arguably neglected in the coverage of SHM to date.

2 Brief Historical Overview

The current authors believe that damage identification - as determined by changes in the dynamic response of systems - has been practiced in a qualitative manner, using acoustic techniques (e.g tap tests on train wheels), since modern man has used tools. More recently, the development of quantifiable SHM approaches has been closely coupled with the evolution, miniaturisation and cost reductions of digital computing hardware. In conjunction with these developments SHM has received considerable attention in the technical literature and a brief summary of the developments in this technology over the last thirty years is presented below. Specific references are not cited; instead the reader is referred to a number of comprehensive sur-

[5]The 22nd Conference on Condition Monitoring and Diagnostic Engineering Management - COMADEM, San Sebastian, Spain, 2009.

[6]The 63rd Meeting of the Society for Machinery Failure and Prevention Technology, Dayton, OH, 2009.

[7]*Structural Health Monitoring, An International Journal*, Sage Publications, London.

veys (Doebling et al. (1996); Sohn et al. (2003); Randall (2004a,b), for more detailed summaries of this subject.

To date, the most successful applications of SHM technology have been for the monitoring of rotating machinery. The rotating machinery applications have taken an almost exclusively data-based (as opposed to model-based) approach to damage identification. The identification process is usually based on pattern recognition methods applied to displacement, velocity or acceleration time-histories (or spectra), generally measured at a single point on the housing or shafts of the machinery during normal operating conditions or start-up or shut-down transients. Often this pattern recognition is performed only in a qualitative manner based on a visual comparison of the spectra obtained from the system at different times; this is nonetheless pattern recognition. Databases have been developed that allow specific types of damage to be identified from particular features of the vibration signature. For rotating machinery systems the approximate damage location is generally known, making a single-channel fast-Fourier-transform (FFT) analyser sufficient for most periodic monitoring activities. Typical damage that can be identified includes loose or damaged bearings, misaligned shafts, and chipped gear teeth. Today, commercial software integrated with measurement hardware is marketed to help the user systematically apply this technology to the operating equipment.

The success of CM is due in part to:

1. Minimal operational and environmental variability associated with this type of monitoring,
2. Well-defined damage types that occur at known locations,
3. Large databases that include data from damaged systems,
4. Well-established correlation between damage and features extracted form the measured data, and
5. Clear and quantifiable economic benefits that this technology can provide.

These factors have allowed this application of SHM to make the transition from a research topic to industry practice several decades ago resulting in comprehensive condition management systems such as the U.S. Navy's Integrated Condition Assessment System. Condition monitoring is not discussed in any further detail here, the curious reader can find many interesting texts and reviews; a good recent review is by Randall (2004a,b).

During the 1970s and 1980s, global oil industry made considerable efforts to develop vibration-based damage identification methods for offshore platforms. This damage identification problem is fundamentally different from that of rotating machinery because the damage location is not known *a priori* and because the majority of the structure is not readily accessible for measurement. To circumvent these difficulties, a common methodology adopted by this industry was to simulate candidate damage scenarios with numerical models, examine the changes in resonance frequencies that were produced by these simulated changes, and correlate these changes with those measured on a platform. A number of very practical problems were encountered including measurement difficulties caused by platform machine noise, instrumentation difficulties in hostile environments, changing mass caused by marine growth and varying fluid storage levels, temporal variability of foundation conditions, and the inability of wave motion to excite higher vibration modes. These issues prevented adoptation of this technology, and efforts at further developing SHM technology for offshore platforms were largely abandoned in the early 1980s.

The aerospace community began to study the use of vibration-based damage identification during the late 1970s and early 1980s in conjunction with the development of the space shuttle. This work has continued with current applications being investigated for the National Aeronautics and Space Administration's space station and future reusable launch vehicle designs. The Shuttle Modal Inspection System (SMIS) was developed to identify fatigue damage in components such as control surfaces, fuselage panels and lifting surfaces. These areas were covered with a thermal protection system making them inaccessible and, hence, impractical for conventional local non-destructive examination methods. The Shuttle Modal Inspection System has been successful in locating damaged components that are covered by the thermal protection system. All orbiter vehicles have been periodically subjected to SMIS testing since 1987. Space station applications have primarily driven the development of experimental/analytical methods aimed at identifying damage to truss elements caused by space debris impact. These approaches are based on correlating analytical models of the undamaged structure with measured modal properties from both the undamaged and damaged structure. Changes in stiffness indices as assessed from the two model updates are used to locate and quantify the damage. Since the mid-1990s, studies of damage identification for composite materials have been motivated by the development of a composite fuel tank for a reusable launch vehicle. The failure mechanisms, such as delamination caused by debris impacts, and corresponding material response for composite fuel tanks are significantly different than those associated with metallic structures. Also, the composite fuel tank problem presents challenges because the sensing systems must not provide a spark source. This challenge has lead to the development of SHM methodologies based on fibre-optic sensing systems. The overview Boller and Buderath (2007) provides a more detailed discussion of SHM applied to aerospace structures for the interested reader.

The civil engineering community has studied vibration-based damage assessment of bridge structures and buildings since the early 1980s. Modal properties and quantities derived from these properties such as modeshape curvature and dynamic flexibility matrix indices have been the primary features used to identify damage in bridge structures. Environmental and operating condition variability presents significant challenges in the bridge monitoring applications. The physical size of the structure also presents many practical challenges for vibration-based damage assessment. Regulatory requirements in Asian countries, which mandate that the companies that construct the bridges periodically certify their structural health, are driving current research and commercial development of bridge SHM systems. Good references on these specific issues are Brownjohn (2007); Lynch (2007) and a useful very recent collection of articles is by Karbhari and Ansari (2009). The International Society for Structural Health Monitoring of Intelligent Infrastructures (ISHMII) has emerged recently and has periodic conferences on SHM issues in civil engineering[8]. Some of the concerns with respect to civil infrastructure are highlighted in the chapter by Deraemaeker later in this volume.

In summary, the comprehensive reviews of the technical literature presented in Doebling et al. (1996); Sohn et al. (2003), show an increasing number of research studies related to damage identification. These studies identify many technical challenges to the adaptation of SHM that are common to all applications of this technology. These challenges include the development of methods to optimally define the number and location of the sensors, identification of the features sensitive to small damage levels, the ability to discriminate changes in these features caused by damage from those caused by changing environmental and/or test conditions, the development of statistical methods to discriminate features from undamaged and damaged structures, and performance of comparative studies of different damage identification methods applied to common data sets. These topics are currently the focus of various research efforts by many industries including defence, civil infrastructure, automotive, and semiconductor manufacturing where multi-disciplinary approaches are being used to advance the current capabilities of SHM and CM.

[8]SHMII - 4 Conference, ETH Zurich, Switzerland, 2009.

3 The Statistical Pattern Recognition Paradigm

There are many ways by which one can organise a discussion of SHM. The authors have chosen to follow the one described in the article Farrar et al. (2001), that defines the SHM process in terms of a four-step statistical pattern recognition paradigm. This process includes:

1. Operational evaluation,
2. Data acquisition, normalisation and cleansing,
3. Feature selection and information condensation, and,
4. Statistical model development for feature discrimination.

Almost all papers published in the fields of SHM and CM arguably address some parts of this paradigm, but the number of studies that address all portions of the paradigm is much more limited. An alternative approach to SHM which is often pursued is based on the solution of inverse problems using linear-algebraic methods; this is not discussed in any detail here, the reader can refer to Doebling et al. (1996); Sohn et al. (2003); Friswell (2007) for the background and further references.

3.1 Operational Evaluation

Operational evaluation attempts to answer four questions regarding the implementation of a damage identification capability:

1. What are the life-safety and/or economic justifications for performing the SHM?
2. How is damage defined for the system being investigated and, for multiple damage possibilities, which cases are of the most concern?
3. What are the conditions, both operational and environmental, under which the system to be monitored functions?
4. What are the limitations on acquiring data in the operational environment?

Operational evaluation begins to set the limitations on what will be monitored and how the monitoring will be accomplished. This evaluation starts to tailor the damage identification process to features that are unique to the system being monitored and tries to take advantage of unique features of the damage that is to be detected.

3.2 Data Acquisition, Normalisation and Cleansing

The data acquisition portion of the SHM process involves selecting the excitation methods, the sensor types, number and locations, and the data acquisition/storage/transmittal hardware. Again, this process will be application-specific. Economic considerations will play a major role in making these decisions. The intervals at which data should be collected is another consideration that must be addressed.

Because data can be measured under varying conditions, the ability to normalise the data becomes very important to the damage identification process. As it applies to SHM, data normalisation is the process of separating changes in sensor readings caused by damage from those caused by varying operational and environmental conditions. One of the most common procedures is to normalise the measured responses by the measured inputs. When environmental or operational variability is an issue, the need can arise to normalise the data in some temporal fashion to facilitate the comparison of data measured at similar times of an environmental or operational cycle. Sources of variability in the data acquisition process and with the system being monitored need to be identified and minimised to the extent possible. In general, not all sources of variability can be eliminated. Therefore, it is necessary to make the appropriate measurements such that these sources can be statistically quantified. Variability can arise from changing environmental and test conditions, changes in the data reduction process, and unit-to-unit inconsistencies. These issues are discussed in some detail in the chapter by Kullaa later in this volume; a recent survey on environmental variations in SHM which is of value is given in Sohn (2007).

Data cleansing is the process of selectively choosing data to pass on to or reject from the feature selection process. The data cleansing process is usually based on knowledge gained by individuals directly involved with the data acquisition. As an example, an inspection of the test setup may reveal that a sensor was loosely mounted and, hence, based on the judgment of the individuals performing the measurement, this set of data or the data from that particular sensor may be selectively deleted from the feature selection process. Signal processing techniques such as filtering and re-sampling can also be thought of as data cleansing procedures.

Finally, it should be noted that the data acquisition, normalisation, and cleansing portion of the structural health-monitoring process should not be static. Insight gained from the feature selection and statistical model development processes will invariably provide information regarding changes that can improve the data acquisition process. Issues relating to data acquisition and processing will be discussed in all of the later chapters in this book.

3.3 Feature Extraction and Information Condensation

The area of the structural health-monitoring process that receives the most attention in the technical literature is the identification of data features that allows one to distinguish between the undamaged and damaged structure. As such, the chapters in this book will devote considerable space to the feature extraction portion of SHM; in particular, the pattern recognition context of feature selection is the major focus of sections in the chapters by Kullaa and Worden. Inherent in the feature selection process is the condensation of the data. The best features for damage identification are, again, application-specific. In the context of vibration-based SHM, the features are usually those measurements associated with structural dynamic (or *modal*) testing. The extraction of dynamic parameters: frequencies, dampings, modeshapes etc., is an art in itself; the chapter by Reynders and De Roeck in this volume discusses an algorithm for this purpose which is state-of-the-art.

One of the most common feature extraction methods is based on correlating measured system response quantities, such as vibration amplitude or frequency, with first-hand observations of the degrading system. Another method of developing features for damage identification is to apply engineered damage, similar to that expected in actual operating conditions, to systems and develop an initial understanding of the parameters that are sensitive to the expected damage. The damaged system can also be used to establish that the diagnostic measurements are sensitive enough to distinguish between features identified from the undamaged and damaged systems. The use of analytical tools such as experimentally-validated finite element models can be a great asset in this process. In many cases the analytical tools are used to perform numerical experiments where the damage is introduced through computer simulation. Damage accumulation testing, during which significant structural components of the system under study are degraded by subjecting them to realistic loading conditions, can also be used to identify appropriate features. This process may involve induced-damage testing, fatigue testing, corrosion growth, or temperature cycling to accumulate certain types of damage in an accelerated fashion. Insight into the appropriate features can be gained from several types of analytical and experimental studies as described above and is usually the result of information obtained from some combination of these studies.

One of the main issues faced in using statistical classifiers in a SHM context is that the amount of *training data* - the *a priori* data needed in order to establish the diagnostic - grows explosively with the dimension of the feature vector. Because data sets acquired by engineering experimentation are typically small, it becomes crucial to reduce the dimension of feature vectors without compromising their information content. Many statistical (and other) methods are available for these purposes, including principal component analysis and factor analysis, techniques which are discussed in some detail in the later chapters by Kullaa and Worden.

The operational implementations of the diagnostic measurement technologies needed to perform SHM invariably produce more data than is traditional in the use of structural dynamics information. A condensation of the data is usually advantageous and can be essential when comparisons of many feature sets obtained over the lifetime of the structure are envisioned. Also, because data will be acquired from a structure over an extended period of time and in potentially many operational environments, robust data reduction techniques must be developed to retain feature sensitivity to the structural changes of interest in the presence of environmental and operational variability (again, the reader can consult Sohn (2007) for a survey). To further aid in the extraction and recording of the high-quality data needed to perform SHM, the statistical significance of the features should be characterised and used in the condensation process. The discipline of *data-mining* has emerged recently as a means of bringing together methods for the extraction of information from large data sets; however, although there are projects successfully applying data-mining in a SHM context Liang and Austin (2004), they are rather rare.

3.4 Statistical Model Development

The portion of the SHM process that has arguably received least attention in the technical literature is the development of statistical models for discrimination between features from the undamaged and damaged structures. Statistical model development is concerned with the implementation of algorithms that operate on the extracted features to quantify the damage state of the structure. The algorithms used in statistical model development usually fall into three categories. When data are available from both the undamaged and damaged structure, the statistical pattern recognition algorithms fall into the group concerned with *supervised learning*; Group classification and regression analysis are examples of learning algorithms which fall into this category. The term *unsupervised learning* refers to those algorithms that are applied to data *not* containing examples from the damaged structure. As engineering structures are typically produced at very high cost; unsupervised learning is often the only course of action as it is not economically viable to damage structures in order to produce data for supervised learning. The group of algorithms based around the idea of *outlier* or *novelty* detection is the primary one applied in the unsupervised learning context. All of the algorithms analyse statistical distributions of the measured or derived features to enhance the damage identification process.

The damage identification process for a system or structure can be summarised in terms of a hierarchical structure along the lines discussed in Rytter (1993); where the objective is to answer the following questions:

Existence: Is there damage in the system?;
Location: Where is the damage in the system?;
Type: What kind of damage is present?;
Extent: How severe is the damage?; and
Prognosis: How much useful life remains?

Answers to these questions in the order presented, represent increasing knowledge of the damage state. When applied in an unsupervised learning mode, statistical models are typically used to answer questions regarding the existence (and sometimes the location) of damage. When applied in a supervised learning mode and coupled with analytical models, the statistical procedures can be used to better determine the type of damage and the extent of damage. Prognosis of remaining useful life is more difficult and will usually require detailed physical models of the damage processes of interest and good predictions of the future loading regime of the structure of interest.

The statistical models are also required to minimise false indications of damage. False indications of damage fall into two categories: (1) False-positive damage indication (indication of damage when none is present), and (2) False-negative damage indication (no indication of damage when damage is present). Errors of the first type are undesirable as they will cause unnecessary downtime and consequent loss of revenue as well as loss of confidence in the monitoring system. More importantly, there are clear safety issues if misclassifications of the second type occur. Many pattern recognition algorithms allow one to weigh one type of error above the other, this weighting may be one of the factors decided at the operational evaluation stage.

The chapter by Worden later in this volume discusses pattern recognition approaches to SHM in detail and the chapter by Kullaa discusses some powerful statistical algorithms in detail.

4 Challenges for SHM

The basic premise of vibration-based SHM feature selection is that damage will significantly alter the stiffness, mass or energy dissipation properties of a system, which, in turn, alter the measured dynamic response of that system.

Although the basis for feature selection appears intuitive, its actual application poses many significant technical challenges. The most fundamental challenge is the fact that damage is typically a local phenomenon and may not significantly influence the lower-frequency global response of structures that is normally measured during system operation. (As an adjunct or alternative to vibration-based approaches to SHM, a number of strategies based on the use of high-frequency waves have developed as a means of detecting small damage; the last chapter of this book by Kudela and Ostachowicz, is concerned with one such approach.) Stated another way, this fundamental challenge is similar to that in many engineering fields where the ability to capture the system response on widely varying length and time scales, as is needed to model turbulence or to develop phenomenological models of energy dissipation, has proven difficult.

Another fundamental challenge is that in many situations feature selection and damage identification must be performed in an unsupervised learning mode; that is, data from damaged systems are not available. Damage can accumulate over widely varying time scales, which poses significant challenges for the SHM sensing system. This challenge is supplemented by many practical issues associated with making accurate and repeatable measurements over long periods of time at a limited number of locations on complex structures often operating in adverse environments.

Finally, a significant challenge for SHM is to develop the capability to define the required sensing system properties before field deployment and, if possible, to demonstrate that the sensor system itself will not be damaged when deployed in the field. If the possibility of sensor damage exists, it will be necessary to monitor the sensors themselves. This monitoring can be accomplished either by developing appropriate self-validating sensors or by using the sensors to report on each other's condition. Sensor networks should also be 'fail-safe'. If a sensor fails, the damage identification algorithms must be able to adapt to the new network. This adaptive capability implies that a certain amount of redundancy must be built into the sensor network.

In addition to the challenges described above, there are other non-technical issues that must be addressed before SHM technology can make the transition from a research topic to actual practice. These issues include convincing structural system owners that the SHM technology provides an economic benefit over their current maintenance approaches and convincing regulatory agencies that this technology provides a significant life-safety benefit. All these challenges lead to the current state of SHM technology, where outside of condition monitoring for rotating machinery applications, SHM remains a research topic that is still making the transition to field demonstrations and subsequent field deployment. There are lots of ongoing and new structural monitoring activities, but these systems have been put in place without a pre-defined damage to be detected and without the corresponding data interrogation procedure. As such, these monitoring activities do not represent a fully integrated hardware/software SHM system with pre-defined damage identification goals. A final non-technical challenge is concerned with providing the educational materials and opportunities for engineers to learn the (rapidly-developing) state-of-the-art in SHM technologies and analysis.

5 Concluding Remarks

The development of robust SHM technologies has many elements that make it a potential "Grand Challenge" for the engineering community. First, almost every industry wants to detect damage in its structural and mechanical infrastructure at the earliest possible time. Industries' desire to perform such monitoring is based on the tremendous economic and life-safety benefits that this technology has the potential to offer. However, as previously mentioned with the exception of rotating machinery condition monitoring, there are few examples of where this technology has made the transition from research to practice.

Significant future developments of this technology will, in all likelihood, come by way of multi-disciplinary research efforts encompassing fields such as structural dynamics, signal processing, motion and environmental sensing hardware, computational hardware, data telemetry, smart materials, and statistical pattern recognition, as well as other fields yet to be defined. These topics are the focus of significant discipline-specific research efforts, and it is the authors' speculation that to date not all technologies from these fields that are relevant to the SHM problem have been explored by the SHM research community. Furthermore, there are few efforts that try to advance and integrate these technologies with the specific focus of developing SHM solutions. Without such a focus in mind, these technologies may well evolve in manner that is not optimal for solving the SHM problem. Finally, the problem of global SHM is so significantly complex and diverse that it will not be solved in the immediate future. Like so many other technology fields, advancements in SHM will most likely come in small increments requiring diligent, focused and coordinated research efforts over long periods of time.

Bibliography

D.E. Adams. *Health Monitoring of Structural Materials and Components: Methods and Applications.* Wiley Blackwell, 2007.

D.E. Bently and C.T. Hatch. *Fundamentals of Rotating Machinery Diagnostics.* ASME Press, 2003.

C. Boller and M. Buderath. Fatigue in aerostructures where structural health monitoring can contribute to a complex subject. *Philosophical Transactions of the Royal Society, Series A: Mathematical, Physical and Engineering Sciences*, 365:561–587, 2007.

C. Boller, F.-K. Change, and Y. Fujino (Eds.). *Encyclopedia of Structural Health Monitoring.* Wiley Blackwell, 2009.

J.M.W. Brownjohn. Structural health monitoring of civil infrastructure. *Philosophical Transactions of the Royal Society, Series A: Mathematical, Physical and Engineering Sciences*, 365:589–622, 2007.

S.W. Doebling, C.R. Farrar, M.B. Prime, and D. Shevitz. Damage identification and health monitoring of structural and mechanical systems from changes in their vibration characteristics. Technical Report LA-13070, Los Alamos National Laboratory, Los Alamos, New Mexico, USA, 1996.

C.R. Farrar and K. Worden. *Theme Issue on Structural Health Monitoring; Philosophical Transactions of the Royal Society, Series A; Mathematical, Physical and Engineering Sciences.* Royal Society of London, 2007.

C.R. Farrar, S.W. Doebling, and D.A. Nix. Vibration-based structural damage identification. *Philosophical Transactions of the Royal Society, Series A: Mathematical, Physical and Engineering Sciences*, 359:131–149, 2001.

C.R. Farrar, H. Sohn, F.M. Hemez, M.C. Anderson, M.T. Bement, P.J. Cornwell, S.W. Doebling, J.F. Schultze, N.A.J. Lieven, and A. N. Robertson. Damage prognosis: current status and future needs. Technical Report LA-14051-MS, Los Alamos National Laboratory, Los Alamos, New Mexico, USA, 2003.

M.I. Friswell. Damage identification using inverse methods. *Philosophical Transactions of the Royal Society, Series A: Mathematical, Physical and Engineering Sciences*, 365:393–410, 2007.

V. Giurgiutiu. *Structural Health Monitoring with Piezoelectric Wafer Active Sensors: Fundamentals and Applications.* Academic Press, 2007.

V.M. Karbhari and F. Ansari. *Structural Health Monitoring of Civil Infrastructure Systems.* Woodhead Publishing Ltd., 2009.

B. Liang and J. Austin. Mining large engineering data set on the grid using aura. *Lecture Notes in Computer Science*, 3177:430–436, 2004.

J.P. Lynch. An overview of wireless structural health monitoring for civil structures. *Philosophical Transactions of the Royal Society, Series A: Mathematical, Physical and Engineering Sciences*, 365:345–372, 2007.

D.C. Montgomery. *Introduction to statistical quality control.* John Wiley & Sons, New York, 3rd edition, 1997.

B.R. Randall. State of the art in monitoring rotating machinery - part 1. *Sound and Vibration*, March:14–21, 2004a.

B.R. Randall. State of the art in monitoring rotating machinery - part 2. *Sound and Vibration*, May:10–17, 2004b.

A. Rytter. *Vibration Based Inspection of Civil Engineering Structures.* Phd thesis, University of Aalborg, Denmark, Department of Building Technology, 1993.

P.J. Shull. *Nondestructive Evaluation: Theory, Techniques and Applications.* Marcel Dekker, Inc., New York, 2002.

H. Sohn. Effects of environmental and operational variability on structural health monitoring. *Philosophical Transactions of the Royal Society, Series A: Mathematical, Physical and Engineering Sciences*, 365:539–560, 2007.

H. Sohn, C.R. Farrar, F.M. Hemez, J.J. Czarnecki, D.D. Shunk, D.W. Stinemates, and B.R. Nadler. A review of structural health monitoring literature: 1996-2001. Technical Report LA-13976-MS, Los Alamos National Laboratory, Los Alamos, New Mexico, USA, 2003.

W.J. Staszewski, C. Boller, and G.R. Tomlinson (Eds.). *Health Monitoring of Aerospace Structures: Smart Sensor Technologies and Signal Processing.* Wiley Blackwell, 2003.

K. Worden and J.M. Dulieu-Barton. An overview of intelligent fault detection in systems and structures. *International Journal of Structural Health Monitoring*, 3:85–98, 2004.

Vibration Based Structural Health Monitoring Using Large Sensor Arrays: Overview of Instrumentation and Feature Extraction Based on Modal Filters

Arnaud Deraemaeker

Université Libre de Bruxelles, Active Structures Laboratory
50 av. F.D. Roosevelt, CP 165/42, B-1050 Brussels, Belgium

Abstract Vibration based Structural Health Monitoring (SHM) techniques are based on the measurement of vibration signals on structures. The first part of this chapter is dedicated to an overview of the existing sensors and acquisition units which are available for such measurements, with an emphasis on recent advances in this area. With these recent advances in sensors and instrumentation, structures can be equipped with very large sensor arrays which monitor the vibration in real time. The difficulty is to be able to extract meaningful information from the huge amount of data generated by these large arrays. The second part of the chapter presents a data reduction technique based on modal filters. It is shown how features can be extracted from the output of these filters and subsequently used for damage detection.

1 Sensors and Instrumentation for Vibration Based SHM

Measuring vibrations has been possible for many years, so what has changed during the last decade? Important advances have been made in sensors and instrumentation. For existing sensors, the quality has been improved. New types of sensors (such as fibre optics) have appeared, and also new manufacturing techniques (MEMS technology) have allowed lowering of the price of sensors considerably. Acquisition units have also benefited from the great advances in electronics and computing. Sampling rates, filtering and storage rates have all improved; prices have also considerably decreased. Where in the past, a structure was instrumented only for the short time of testing, it is now possible to monitor continuously, and in real-time, its vibrations.

1.1 Most Common Transducers for SHM Applications

There are so many types of sensors on the market that it is not possible to give a comprehensive list of them. The goal of this section is to introduce the most common types of sensors as well as show the new trends in this area.

Accelerometers By far the most common sensor for measuring vibrations is the accelerometer. The basic working principle of such a device is presented in Figure 1(a). It consists of a moving mass on a spring and dashpot, attached to a moving solid. The acceleration of the moving solid results in a differential movement x between the mass M and the solid. The governing equation is given by,

$$M\ddot{x} + c\dot{x} + kx = -M\ddot{x_0} \tag{1}$$

In the frequency domain $x/\ddot{x_0}$ is given by,

$$\frac{x}{\ddot{x_0}} = \frac{-1}{-\omega^2 + \omega_n^2 + 2j\xi\omega\omega_n} \tag{2}$$

and for frequencies $\omega << \omega_n$, one has,

$$\frac{x}{\ddot{x_0}} \simeq \frac{-1}{\omega_n^2} \tag{3}$$

showing that at low frequencies compared to the natural frequency of the mass-spring system, x is proportional to the acceleration $\ddot{x_0}$. Note that since the proportionality factor is $\frac{1}{\omega_n^2}$, the sensitivity of the sensor is increased as ω_n^2 is decreased. At the same time, the frequency band in which the accelerometer response is proportional to $\ddot{x_0}$ is reduced.

The relative displacement x can be measured in different ways which are presented in Figure 2: using piezoelectric material (longitudinal or shear mode), in which case the strain results in an electric charge that can be measured with a charge amplifier, or using piezoresistive material, for which the strain results in a change in the resistance. Strain gauges are made of piezoresistive material and can be placed at different locations on the material playing the role of the spring. The advantage of the piezoresistive accelerometer is the possibility to measure acceleration down to 0 Hz, which cannot be achieved with piezoelectric materials. There exists a very wide family of accelerometers, ranging from very heavy seismic accelerometers able to measure very small accelerations at low frequency, to tiny light accelerometers, with a lower sensitivity but a much wider frequency bandwidth.

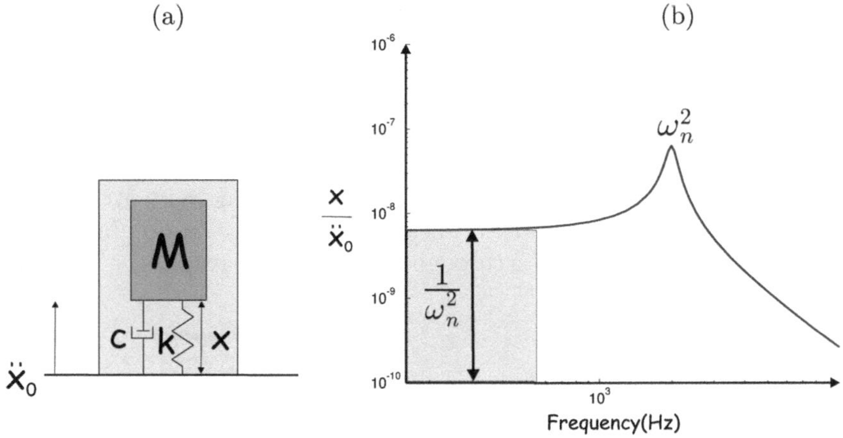

Figure 1. Working principle of an accelerometer

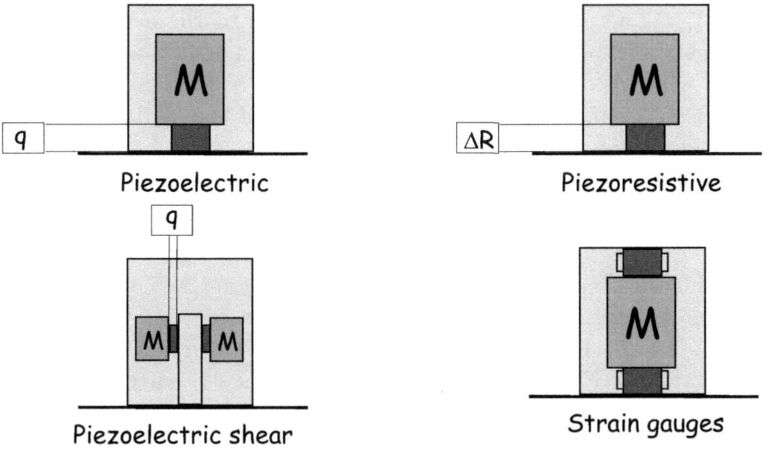

Figure 2. Different sensing principles for standard accelerometers

In the last decade, MEMS (Micro-Electro-Mechanical-Systems) sensors have started to appear on the market (Stein (2001)). These small devices come in very different shapes and characteristics. Their main advantage is the very low cost (in the range of 50 Euros compared to traditional accelerometers which are in the range of several hundreds of Euros). MEMS are flat devices (due to the silicone "layer" manufacturing technology). As in traditional accelerometers, they consist of a moving mass attached to a spring. The most common way to measure the relative motion between the casing and the moving mass is through the change of capacitance (Figure 3).

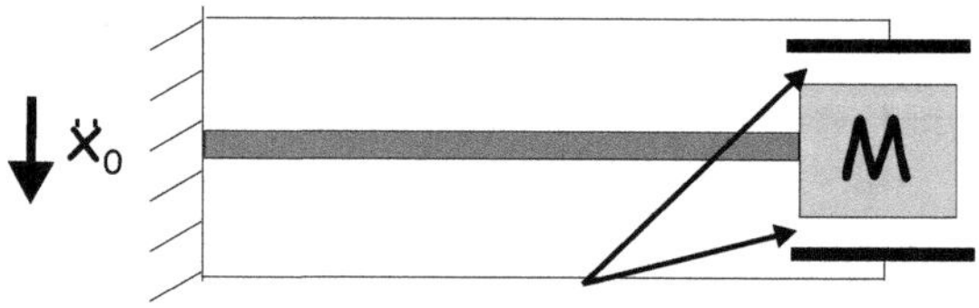

Figure 3. Working principle of a MEMS capacitive accelerometer

A more recent kind of MEMS accelerometer is the resonant MEMS (Seshia (2002)). There, the longitudinal strain due to the acceleration of the moving mass is responsible for a change in the first lateral natural frequency of a beam which is excited electrostatically (Figure 4). The main advantage is a better signal-to-noise ratio (the frequency of the signal transmitted is less sensitive to noise than the amplitude).

Strain Sensors

Piezoresistive Strain Gauges The strain gauge is the fundamental sensing element for many type of sensors (accelerometers, pressure sensors, load cells etc.). It consists of a pattern of metallic conductor mounted on a backing material (Figure 5). As the conductor is stretched or compressed, the electrical resistance changes. This change of resistance can be measured using a Wheatstone bridge. Many applications with strain gauges are at low frequencies, but it is possible to measure vibrations, even up to several kHz.

Piezoelectric Strain Sensors Piezoelectric materials can produce an electric charge when deformed. In order to increase this effect, the materials are poled in a preferential direction (the poling direction). Charge can be

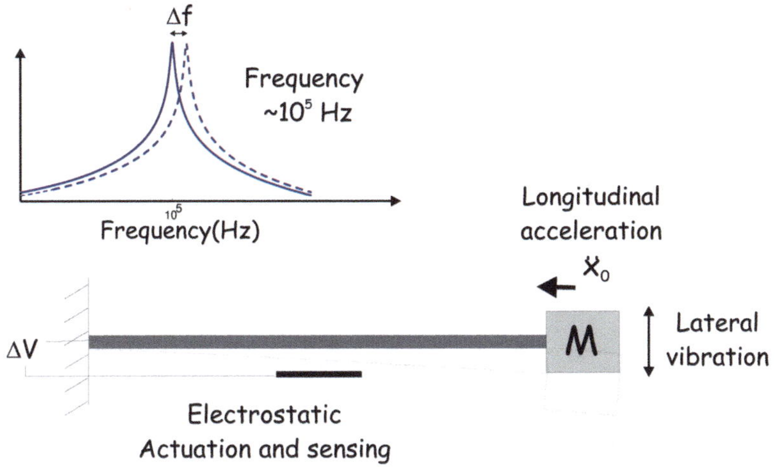

Figure 4. Working principle of a MEMS resonant cantilever beam

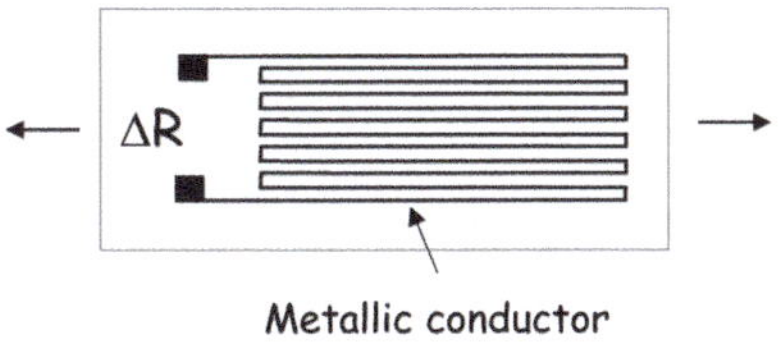

Figure 5. Working principle of a piezoresistive strain gauge

produced for different deformation mechanisms as illustrated in Figure 6. Because of the isotropic nature of most piezoelectric materials in the plane perpendicular to the poling direction, it is difficult to produce a strain sensor which reacts to only one of the strain components in the plane, so that lateral sensitivity is a problem. Electric charges are measured using charge amplifiers which are not suitable for static measurements, but can work up to several MHz. In the last decade, piezocomposite strain sensors have been developed (Deraemaeker et al. (2009)). For certain volume fractions, they exhibit a strong orthotropy, making them potential candidates for uni-directional piezoelectric strain sensors.

Fibre Bragg Grating Sensors As an alternative to the traditional strain gauges, new sensors based on fibre optics have appeared on the market in the last ten years (Glisic and Inaudi (2007)). Fibre Bragg Grating Sensors

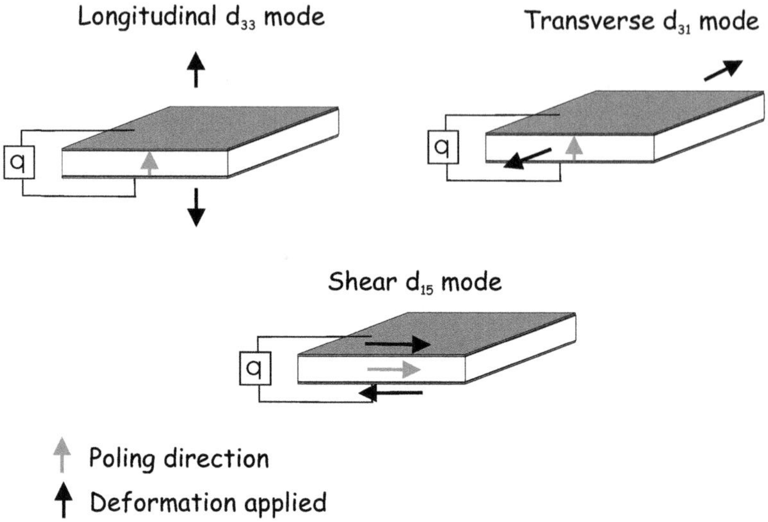

Figure 6. Three modes of deformation generating a charge on a piezoelectric material

(FBGS) consist of a grating written on an optical fibre. When a broadband light source is directed into the optical fibre, a specific wavelength, proportional to the gratings's size Λ (distance between the lines) is reflected (Figure 7). When strain is induced into the grating, the grating's size changes, therefore changing the frequency of the reflected wavelength and allowing the measurement of the strain applied to the grating. The main advantages of the technique are:

- immunity to electromagnetic interferences;
- the possibility to introduce several sensors on one long fibre optic cable. Two techniques can be used: (i) using different wavelengths for each sensor, and (ii) sending pulses instead of broadband light and calculating each sensor's response based on the time of flight.

The main disadvantage of fibre optic sensors is the brittleness of the fibres and the rather high cost associated with the interrogation units. Typical interrogation units scan at low rates (100 Hz) but new systems can reach up to 20kHz.

Brillouin Scattering Sensors Brillouin scattering is a quite complex phenomenon (Glisic and Inaudi (2007)). When light travels through a medium, a small portion of it is back-scattered, due to inhomogeneities

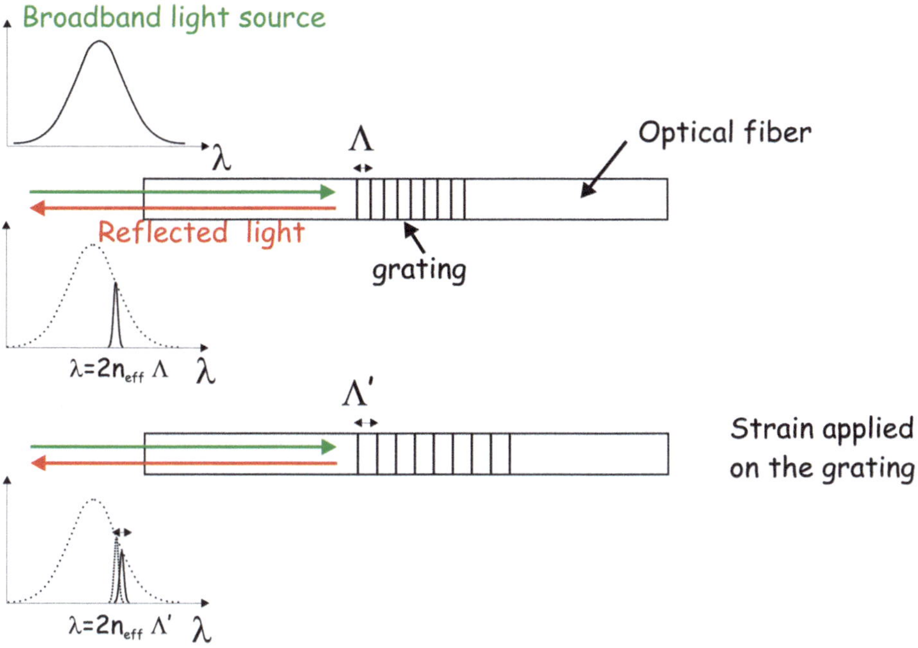

Figure 7. Working principle of a fibre Bragg Grating Sensor (FBGS)

of the refractive index. Brillouin scattering is one of the components of the back-scattered light, the frequency of which is dependent on the strain and temperature. The interrogation unit sends a light pulse at a given frequency and records the back-scattered spectrum as a function of time. The operation is repeated for a wide range of frequencies. The time information is then used to determine the position along the fibre, and a three-dimensional plot can be drawn, giving the frequency of the back-scattered light as a function of the distance along the fibre. In this manner, strain can be measured with a very high spatial distribution (but not simultaneously at all the points). Strains can be monitored over very large distances (10km or more), but the scanning time is high so that only static measurements can be made. No special treatment (such as writing a grating) is necessary, so the sensors are very cheap, but the acquisition unit is rather expensive.

Vibrating Wires Vibrating wires are commonly used for quasi-static measurements of strains in civil engineering structures. The principle is based on the relationship between the first eigenfrequency f of a taut string and the applied tension T : $f = 1/2L\sqrt{T/\rho}$, where L is the length and ρ is the mass per unit length of the cable (Yu and Gupta (2005)). The string in this case is ferromagnetic and both ends are fixed to the structure (Figure 8). By sending a pulse to the electromagnet, the string is attracted and released, resulting in a vibration of the string in its first mode. The motion of the string is then measured with the same coil (the movement of the string, which acts as a moving magnet, induces a voltage difference in the wires). The post-processing consists of estimating the frequency (and therefore the tension in the wire) by "counting" the periods. The tension in the wire can be directly transformed into a strain. The main advantage of the system is the low sensitivity of the identified frequency to the noise in the signal, so that the signal can be sent over large distances for post-processing at a remote distance. Note that this is not truly a vibration sensor, since the sampling rate is rather low, but it operates on a vibration principle.

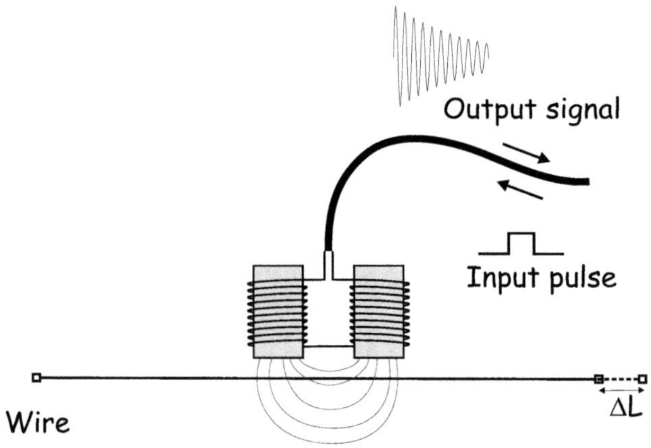

Figure 8. Working principle of a vibrating wire

Cable Force Sensor Load cells or strain gauges can be used for the measurement of forces in cables. For strain gauges, the disadvantage is that they only give a very local value of the strain which may not be representative of the average force in a section. The disadvantage of load cells is that they have to be inserted in the path of the force, and modify the structure of the cable attachment.

One nice alternative is to use electromagnetic sensors (Wang et al. (2005), Figure 9). The underlying principle is that the magnetic permeability μ of the cable changes with the applied stress. Two windings are installed around the cable. A pulse input current is applied to the first coil, resulting in a magnetic field H. The change in the magnetic field causes a difference of potential V in the second coil, which is proportional to B. The permeability $\mu = B/H$ is measured and is proportional to the applied stress.

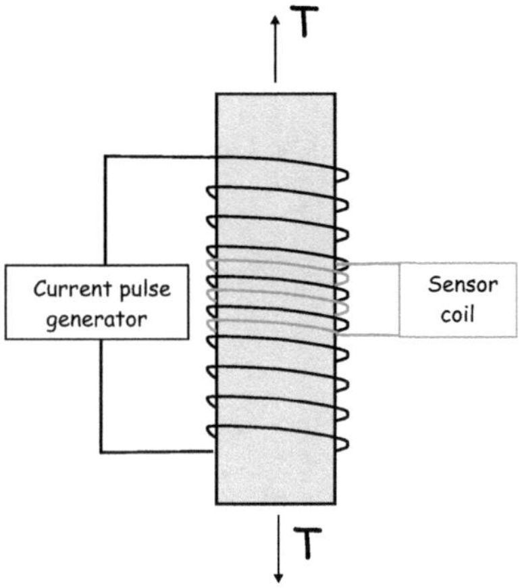

Figure 9. Working principle of an electromagnetic cable force sensor

Another possibility is to measure the vibrations of the cable in order to determine the tension. The principle is similar to that of the vibrating wire. For a string (without sag), the eigen frequencies are given by,

$$f_k = \frac{k}{2L}\sqrt{\frac{T}{\rho}} \tag{4}$$

where f_k is the kth eigenfrequency of the cable, L is the length and ρ is mass per unit length. By measuring the first or the lowest eigenfrequencies, it is possible to determine the tension in the cable (Figure 10). The eigenfrequencies are traditionally measured with an accelerometer attached to the cable. More advanced techniques take into account the flexural rigidity of the cable as well as the effect of the attachments in order to obtain a more accurate measure of cable tension (Geier et al. (2005)).

Position Sensors: GPS The GPS (Global Positioning System) allows one to determine the position of a receiver, based on a set of satellites deployed in space. The receivers have become cheaper and cheaper and the possibility of measuring the vibrations of large structures with such

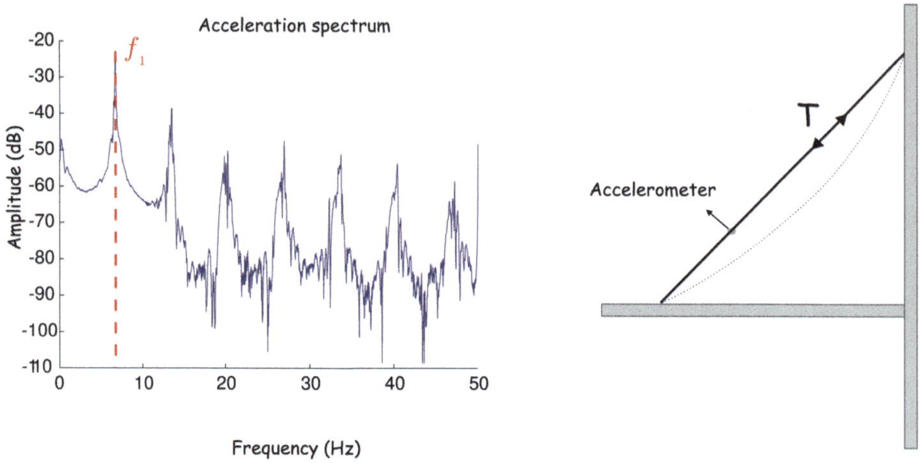

Figure 10. Measuring the force in a cable via measurement of the acceleration spectrum

sensors has been studied by many research groups (Meng (2002)). In order to get a good accuracy, a typical setup is made of one or several reference receivers positioned outside of the structure and an array of receivers on the structure (Figure 11). The technique used is referred to as differential GPS. The sampling frequency can reach up to 20Hz with a centimetre accuracy. If time averaging is performed, mm accuracy can be reached, at the cost of a much lower sampling rate.

Other Sensors Although these sensors do not measure vibrations, they are often used as a complement to the mechanical sensors.

- Humidity sensors: highly accurate humidity sensors are available nowadays at very economical prices. Different types exist: resistive, capacitive and thermal conductivity. In most cases, the environmental conditions dictate the sensor choice.
- Temperature sensors: temperature sensors can be classified into two types: contact and non-contact. For contact sensors, it is assumed that the temperature of the sensor is in thermal equilibrium with the material to which it is in contact. Non-contact sensors measure the radiant power of the infrared or optical radiation.
- Tilt meters: a tilt meter is used to measure the static angle with respect to gravity. The most common type of tilt meter is the "bubble"

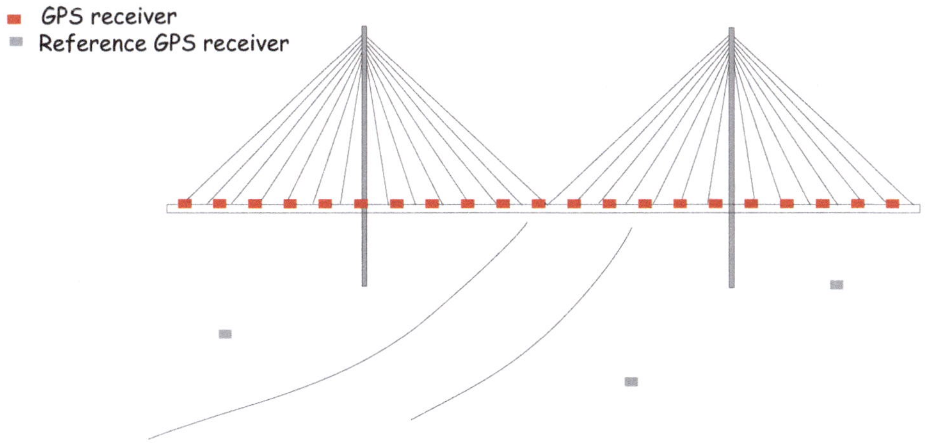

Figure 11. Principle of a differential GPS system for the measurement of vibrations on a bridge

tilt meter. Modern types include electrodes which enable the measurement of the position of the bubble in an electrolyte solution. This measurement has a much higher precision.

- Corrosion sensors: corrosion produces areas in concrete where there is a large concentration of negative ions, which creates a small electric voltage potential. By mapping the electric potential along a rebar, it is possible to detect corrosion. This is done usually using a half-cell potential method, which measures the potential difference between the rebar and the half-cell.
- Crack sensors: typically, these are strain sensors which measure the relative displacement between the two sides of a crack. The simplest crack sensor is a crack-width gauge which allows one to measure visually and by hand the size of a crack. Alternatively, the reading can be performed with a more precise dial. More advanced types of measurement techniques include inductive non-contact measurement, optical fibres, strain gauges, etc. Another interesting idea is the CVM (crack vacuum monitoring, http://www.smsystems.com.au/). A small volume of vacuum is kept in a series of interconnected tubes. As the crack propagates, the tubes break causing a leak in the vacuum; The size of the leak is proportional to the number of broken tubes, hence the propagation of a crack along the sensor can be monitored. Modern crack monitoring techniques can rely on 2D image correlation techniques, allowing one to detect automatically the position and evolution of a

crack in the range of vision of a camera.

1.2 Acquisition Units

Many different types of acquisition units exist on the market. The most sophisticated type is made of an independent platform equipped with ADC (analog to digital converters) for acquisition and DAC (digital to analog converters) for actuation purposes (Figure 12). The platform can include on-board storage as well as a processor and an operating system. This platform is linked to a regular or laptop computer. The computer is used to program the processor and send and receive information to and from the platform. The link is a fast link such as fibre optic or ethernet. This type of platform is usually modular, and can be configured for a given application.

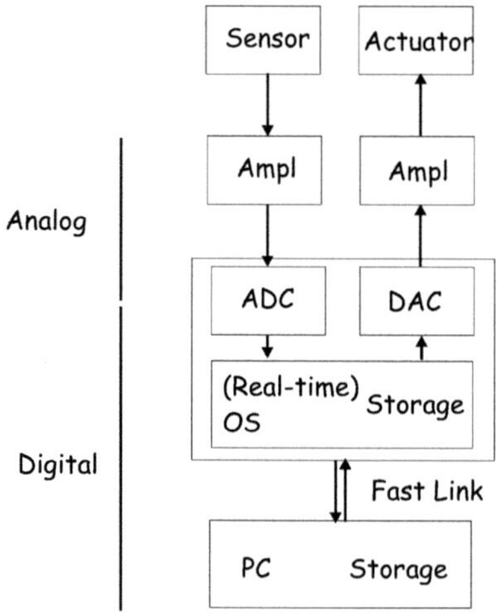

Figure 12. Block diagram of an acquisition unit with on-board storage and computation capabilities

Cheaper alternatives exist; the first one is a simpler platform without on-board computing and storage capabilities, which is linked to the computer via a standard USB link (Figure 13). This type of platform can also be modular.

For dedicated applications, it is also possible to use PCI (Peripheral Component Interconnect) acquisition cards, directly inserted into the computer chassis (Figure 14).

The important factors to take into account when choosing an acquisition unit are:

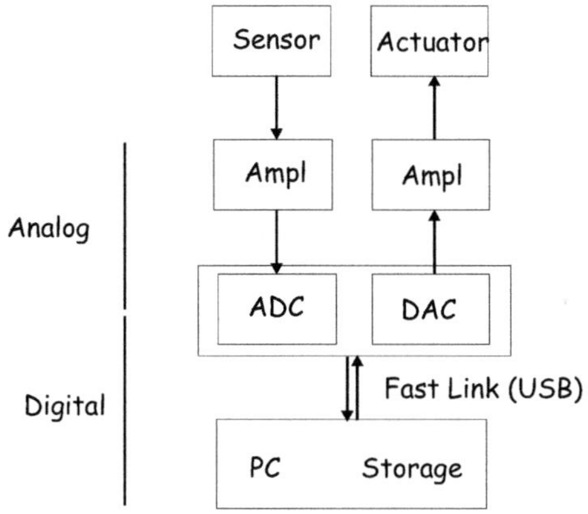

Figure 13. Block diagram of a USB based acquisition unit with ADC and DAC

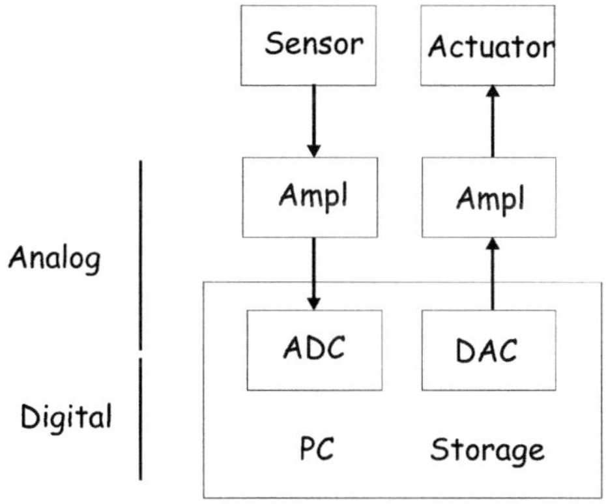

Figure 14. Block diagram of a PCI acquisition card directly inserted into the PC

- the signal-to-noise ratio (linked to the type of ADC/DAC),
- the sampling frequency,
- the time delays,
- the presence and type of anti-aliasing filters,
- the transfer rate to the storage (hard disk),
- the number of channels (or possibility to extend it),
- the possibility to use preamplified accelerometers (IEPE - Integral Electronics Piezoelectric, or ICP - Integrated circuit piezoelectric').

Wifi Networks With the development of wireless networks for home internet, there have been a lot of developments in this area. The main advantages of such systems are:

- low cost,
- ease of installation (this is probably the most important point),
- ease of maintenance.

There are however some major drawbacks in current systems for applications to vibration measurements:

- packet loss (loss of data transmitted from the sensor to the base station),
- high power needed for transmission (this is a problem since wireless sensors are usually battery powered),
- problems of synchronization of the sensors,
- reliability and stability of the network.

Examples of applications of wireless sensor networks for structural health monitoring can be found in Lynch and Loh (2006).

1.3 Examples of Instrumented Bridges

Figure 15 shows a plot of the evolution of the instrumentation put on bridges over the years, in China and elsewhere. It is clear that China has been the first country to put massive instrumentation on bridges, back in 1997 with the Tsing Ma and the Ting Kau bridges (more than 400 sensors on each). The Stonecutters bridge, currently under construction, is going to become the most instrumented bridge in the world; more than 1100 sensors are foreseen on the bridge. The interested reader can find a description of many case studies of instrumented bridges on the following websites,

- http://www.samco.org: website of the thematic network SAMCO (Structural Assessment, Monitoring and Control).
- http://www.ishmii.org/: website of the International Society for Structural Health Monitoring of Intelligent Infrastructure.

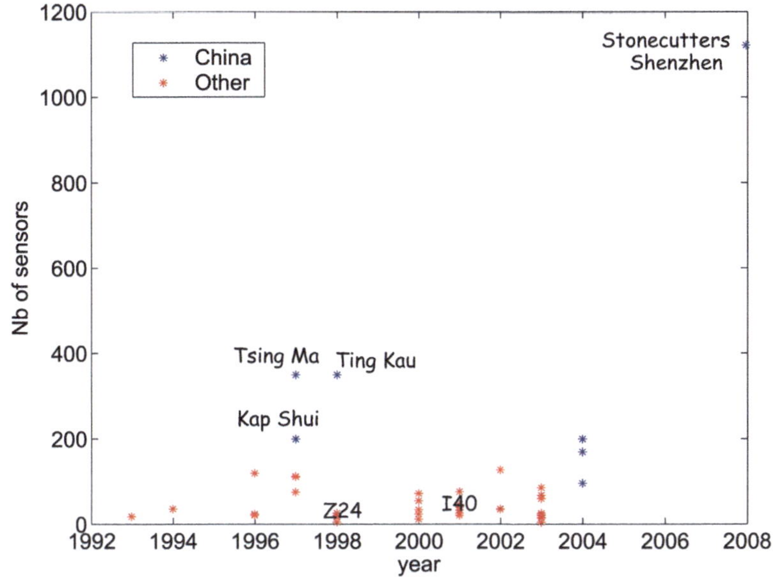

Figure 15. Examples of instrumented bridges : evolution of the number of sensors with time

- http://nomotida.net: this website developed by EMPA in Switzerland allows one to follow in real time a wireless sensor network monitoring the frequencies of the cables of a bridge.

Although many of the case studies are very interesting, the data collected on the structures cannot always be obtained by other research teams around the world, and is sometimes not adequate in order to test a given SHM strategy. Small-scale laboratory experiments have advantages over real structures: they are always available and induce much smaller operating costs; the instrumentation is less costly and testing campaigns are much easier to organize. There are many examples of such structures in the SHM community. Described below is the experimental setup developed at the Université Libre de Bruxelles (ULB) in the Active Structures Laboratory (ASL).

The test structure is a small-scale mock-up of a cable-stayed bridge in the construction phase. The bridge is made of a central steel pillar, a deck made of two U-shaped aluminum beams and steel rectangular stiffeners, forty additional masses and 8 steel cables (Figure 16). The total length is 3m, the height is 1.4 m and the width of the deck is 18cm. The central pillar is resting on a concrete block. Two wooden blocks are placed between the concrete base and the floor; this introduces flexibility between the concrete and the floor which is necessary for the seismic excitation described hereafter.

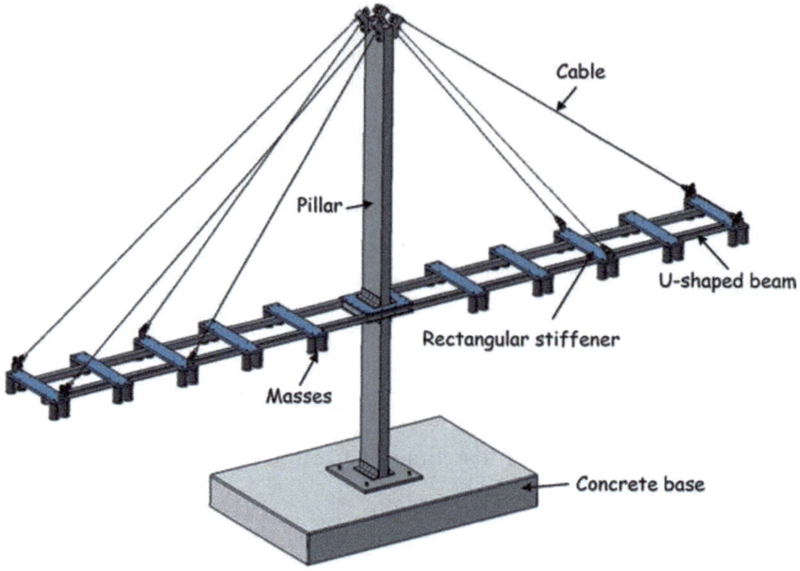

Figure 16. The bridge demonstrator : description of the main components

At the pillar side, each cable is attached to a screw which can move relatively to the pillar, allowing the tension to be adjusted manually (Figure 17).

Each cable is equipped with a collocated actuator-sensor pair consisting of a piezoelectric elliptic actuator (Cedrat APA 100M for the long cables and APA 50S for the short cables), and a piezoelectric force sensor (B&K 8200) (Figure 18).

The actuators can be used to induce vibrations in the bridge as well as for active vibration control. In addition, a shaker is attached to the concrete base in order to induce seismic type excitation (Figure 19).

The deck and the pillar are equipped with traditional accelerometers. The left part of a U-shaped beam has also been equipped with five dynamic fibre optic strain sensors distributed on a single fibre (fibre Bragg grating sensors - FBGS). Each grating is 8 mm long and measures the average strain applied to it. The position of the five strain measurements is shown

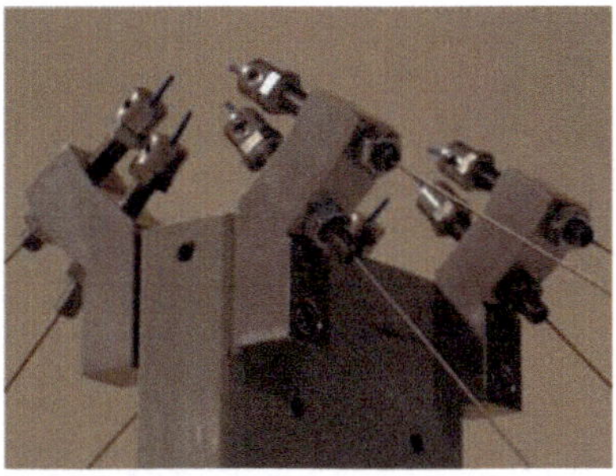

Figure 17. Cable tension tuning system installed at the top of the pillar

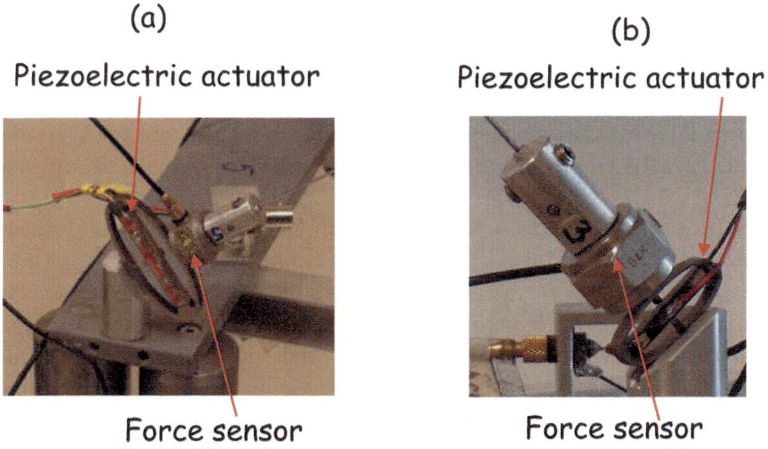

Figure 18. Cables equipped with a piezoelectric actuator and a force sensor: (a) long cable with large piezoelectric actuator, (b) short cable with small piezoelectric actuator

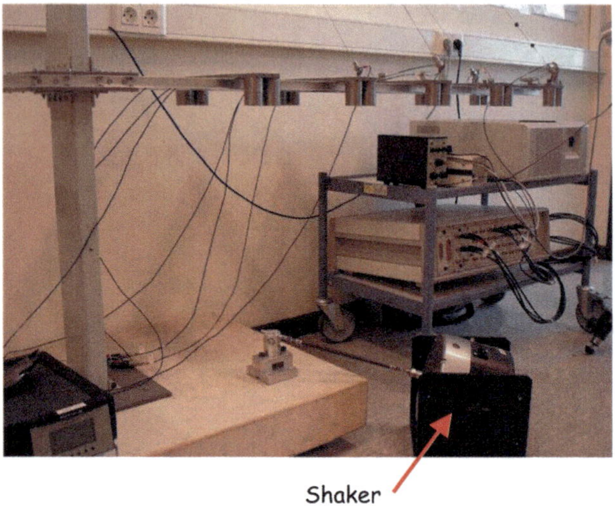

Figure 19. Shaker attached to the concrete base in order to induce seismic excitation

on Figure 20. The measurements are made using the hi-speed Dynosense 300 from *FOS&S* which allows one to measure on all sensors simultaneously at 4.3 kHz.

The data acquisition system consists of a DSpace DS1005 modular board equipped with 64 ADC used for the sensors and 6 DAC used for the actuators (all 16 bit converters). Control or signal processing routines can be implemented on the DSP using the Matlab/Simulink environment. DSpace Control Desk Software is used to interact with the DSP board in real time to adjust settings and retrieve measurements from the sensors and actuators (usually in the form of time domain signals).

For the FBGS sensors, the signal processing is not implemented fully in hardware, so that a special software for data acquisition is needed.

Currently, experimental as well as numerical studies are carried out on the small-scale bridge. The main focus is on cable vibrations. Point masses have been added to one of the cables of the bridge in order to increase the linear mass and introduce some sag (Figure 21).

1.4 Summary

An overview of instrumentation for vibration-based SHM has been presented, together with links to real case studies and the description of a small

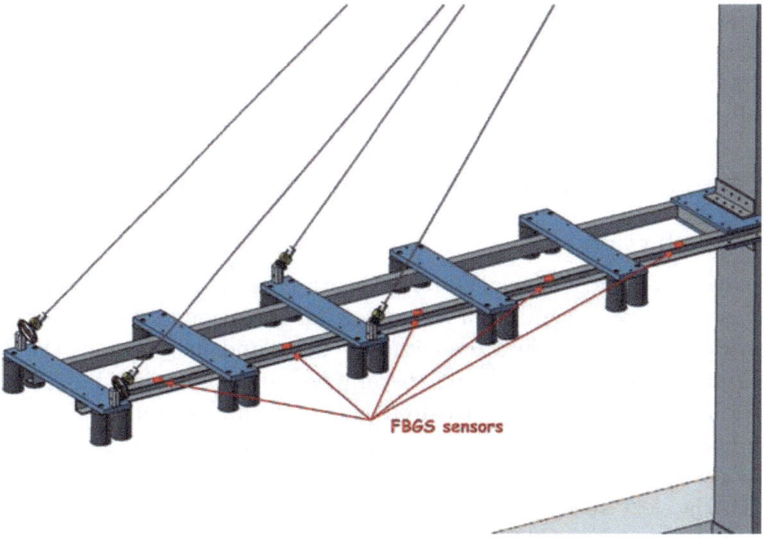

Figure 20. Position of the five FBGS sensors on the U-shaped beam

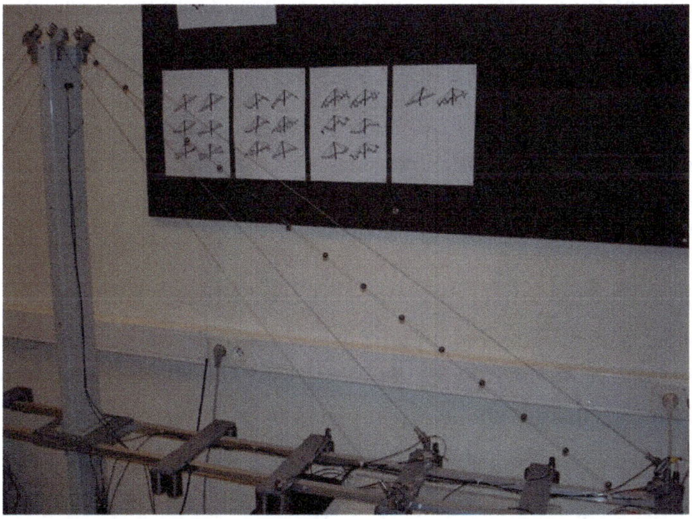

Figure 21. Cable with additional point masses. The additional weight is responsible for a small sag of the cable

scale test setup developed at ULB-ASL. For the reader interested in recent advances in instrumentation, valuable information can be found on the website *http://www.sensorsportal.com/*, where one can also find the open access journal "The Sensors & Transducers Journal (S&T e-Digest)"

2 Vibration-Based SHM Using Modal Filters

2.1 Introduction

The first, and most crucial step in vibration based structural health monitoring is to extract meaningful features from vibration measurements. The most widely used features are eigenfrequencies and modeshapes (Doebling et al. (1998)). Even in the case where only sensor outputs are known and the excitation forces cannot be measured, adequate modal analysis techniques can be used to extract those features from ambient vibrations (Peeters and Roeck (1999)).

With the very recent advances in sensors and instrumentation, massive instrumentation can be implemented on large civil engineering structures such as bridges. As explained in Section 1.1, hundreds, even thousands of sensors can monitor the vibration of such structures in real time. The difficulty is that most algorithms are not suited to treat such a large amount of data in real-time. Moreover, modal analysis techniques still rely today on some user interactions, although some efforts have been made to make this task as automatic as possible (Deraemaeker et al. (2008)).

The motivation for the present study is therefore to develop a very simple and robust methodology to detect changes in modeshapes of structures under ambient vibrations without any user interaction. The idea is based on the principle of spatial and modal filtering which is complemented with signal processing techniques.

2.2 Data Reduction Using Spatial and Modal Filters

Consider a structure equipped with a large array of n sensors. The array is linked to a linear combiner such that the sensor outputs y_1 ... y_n are combined to form a single output $y(t) = \sum_{k=1}^{n} \alpha_k y_k(t)$.

Depending on the selection of α_k, various meaningful outputs may be constructed, for example, modal filters. The idea behind modal filtering is to select the coefficients α_k such that they are orthogonal to all the modes of a structure in a frequency band of interest, except mode l. The modal filter is then tuned to mode l and all the contributions from the other modes are removed from the signal. Figure 23 represents the FRF of a modal filter

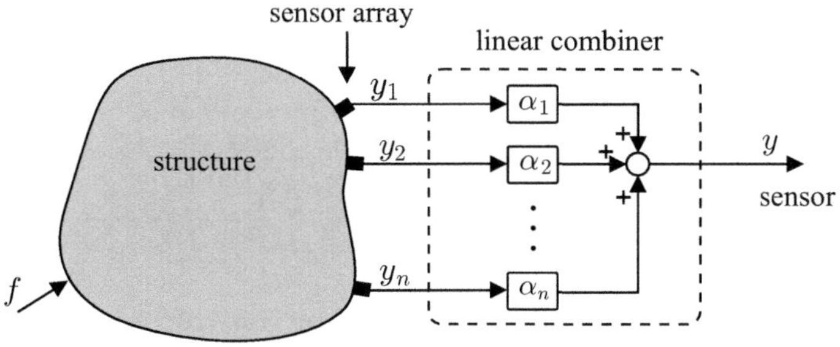

Figure 22. Principle of spatial filtering based on a network of n sensors

tuned to mode l.

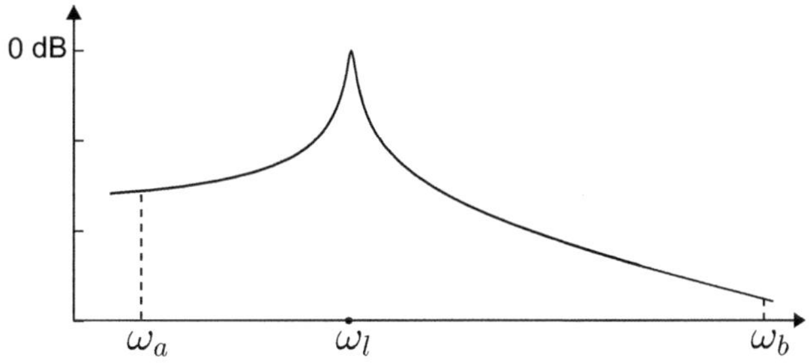

Figure 23. Perfect modal filter tuned to mode l within the bandwidth $[\omega_a, \omega_b]$.

If one assumes that the m modeshapes, eigenfrequencies and modal damping coefficients are known in the frequency band of interest $[\omega_a, \omega_b]$; for a given input, the modal expansion of the FRF of the sensor array reads,

$$Y_k(\omega) = \sum_{i=1}^{m} \frac{c_{ki} b_i}{(\omega_i^2 - \omega^2 + 2j\xi_i\omega_i\omega)} \qquad k = 1...n \tag{5}$$

where b_i is the modal input gain (at the actuator) and c_{ki} is the modal output gain of sensor k in the array. If the n sensors in the array are connected to a linear combiner with gain α_k for sensor k, the output of the linear combiner is $y(t) = \sum_{k=1}^{n} \alpha_k y_k(t)$ and the global frequency response is,

$$G(\omega) = \sum_{k=1}^{n} \alpha_k Y_k(\omega) = \sum_{i=1}^{N} \frac{\{\sum_{k=1}^{n} \alpha_k c_{ki}\} b_i}{(\omega_i^2 - \omega^2 + 2j\xi_i\omega_i\omega)} \tag{6}$$

A modal filter which isolates mode l can be constructed by selecting the weighing coefficients α_k of the linear combiner in such a way that,

$$\sum_{k=1}^{n} \alpha_k c_{ki}(\omega) = \delta_{li} \tag{7}$$

or

$$\{\alpha\}^T [C] = \{e_l\}^T \tag{8}$$

where $\{\alpha\}$ is the vector of the linear combiner coefficients, $[C]$ is the matrix of modal output gain (column i is the sensor array output when the structure vibrates according to mode i) and $\{e_l\}^T = (00...1...0)^T$ is the vector with all entries equal to 0 except entry l which is equal to 1. Assuming that matrix $[C]$ is known accurately, the modal filter coefficients for mode l can be found by solving the rectangular system of equations,

$$[C]^T \{\alpha_l\} = \{e_l\} \tag{9}$$

The number of columns of $[C]^T$ is equal to the number of sensors n and the number of lines is equal to the number of modes in the frequency band of interest m. In order to satisfy equation (9), the rank of matrix $[C]$ must be $\geq m$. There are two situations in which $rank(C) \leq m$: (i) the number of sensors is smaller than the number of modes, (ii) some columns of $[C]$ are linearly dependent, which leads to a spatial aliasing effect. In both of these situations, it is not possible to obtain a perfect modal filter in the frequency band of interest.

When $rank([C]) \geq m$, some care must be taken in order to solve the over-determined system of equations (9). The most common approach is to use the Moore-Penrose pseudo-inverse of $[C]^T$ defined as

$$([C]^T)^\dagger = \left([C][C]^T\right)^{-1} [C] \tag{10}$$

which gives the best approximation in the least-squares sense. This approach is however not well-suited when some of the singular values of the matrix are very small. In such a case, the computation of the inverse leads to very irregular coefficients. This problem can be overcome using a singular value decomposition of the matrix and a truncation,

$$[C]^T = [U]\,[S]\,[V]^T \tag{11}$$

where $[U]$ and $[V]$ are unitary matrices and $[S]$ is a rectangular matrix of dimension $[m, n]$ with the singular values σ_i on the diagonal. If u_i are the column vectors of $[U]$ and v_i the column vectors of $[V]$, equation (11) can be written,

$$[C]^T = \sum_{i=1}^{r} \sigma_i u_i v_i^T \tag{12}$$

where r is the rank of the matrix $[C]$. The solution of equation (9) is,

$$\{\alpha_l\} = \left(\sum_{i=1}^{r} \frac{1}{\sigma_i} v_i u_i^T \right) \{e_l\} \tag{13}$$

This shows that the solution is dominated by the small singular values. The solution to this problem is to truncate the expression of the inverse, neglecting the contribution of the small singular values. For more details on the subject of spatial filtering, the interested reader may refer to Preumont et al. (2003).

2.3 Effect of Damage and Environment on Modal Filters

If one now assumes that the structure is damaged, the FRF of the output of the linear combiner is given by,

$$G(\omega) = \sum_{k=1}^{n} \alpha_k \tilde{Y_k}(\omega) = \sum_{i=1}^{N} \frac{\{\sum_{k=1}^{n} \alpha_k \tilde{c_{ki}}\}\tilde{b_i}}{(\tilde{\omega_i}^2 - \omega^2 + 2j\tilde{\xi_i}\tilde{\omega_i}\omega)} \tag{14}$$

where $\tilde{}$ refers to the damaged structure.

The impact of damage can be decomposed into the three effects on:

- $\tilde{b_i}$: the change in the modeshapes of the structure will affect the modal input gain which changes the amplitude of the modal filter;
- $(\tilde{\omega_i}, \tilde{\xi_i})$: the change in the eigenfrequencies and modal damping will affect respectively the position and peak amplitude of the modal filters;
- $\sum_{k=1}^{n} \alpha_k \tilde{c_{ki}}$: because of the change in the modeshapes, equation(7) may not be satisfied. In this case, the output of the modal filter does

not isolate mode l perfectly and the other modes may appear in the response.

The third effect is interesting because it is a clear indicator that the shapes of the eigenmodes have changed. As a result, peaks will appear at some of the resonant frequencies of the damaged system (Figure 24a). The appearance of these peaks can be used to detect damage. When the stiffness change is global and proportional to the stiffness matrix, the modeshapes are not altered and equation (7) still holds. In this case, no peaks will appear, but the existing peak will shift due to a change in the eigenfrequencies of the system (Figure 24 b).

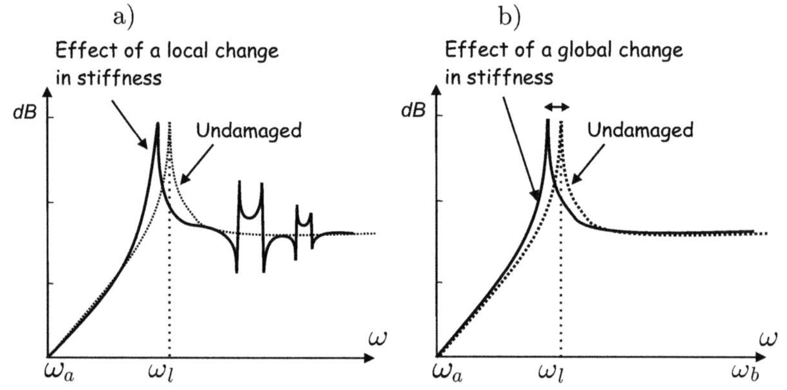

Figure 24. Example of changes in a modal filter tuned to mode 1: a) Effect of local changes (damage), b) effect of global changes (i.e. environment)

From the above, one can conclude that the appearance of new peaks in the FRF of the modal filters can be used to detect local changes due to damage. In addition, if the environment induces global changes to the stiffness and mass of the structure, the modeshapes will be very slightly altered, so that only very small peaks will appear, and the damage detection will be robust to such environmental changes. This idea was first proposed and developed in Deraemaeker and Preumont (2006).

One major drawback of using FRF measurements is the fact that input forces are not always available when monitoring a structure in its environment. For large structures such as bridges, ambient measurements are generally used for damage detection (Wenzel and Pichler (2005)). Vibrations are induced by natural excitations (wind, traffic etc.) and output-only measurements are recorded. Since spatial filtering is applied in the time domain, modal filters can also be used with output-only measurements. The idea is to compute the power spectral density $S_{yy}(\omega)$ of the time domain output of the modal filter defined as,

$$S_{yy}(\omega) = \int_{-\infty}^{\infty} R_{yy}(\tau)\, e^{-j\omega\tau}\, d\tau \tag{15}$$

where $R_{yy}(\tau)$ is the auto-correlation of $y(t)$ defined by,

$$R_{yy}(\tau) = E\left[y(t)\, y(t-\tau)\right] \tag{16}$$

$E[.]$ is the mathematical expectation. In addition, for a linear system with transfer function $H(j\omega)$ the following relationship holds (Preumont (1994)),

$$S_{yy}(\omega) = |H(j\omega)|^2 S_{xx}(\omega) \tag{17}$$

in which $S_{xx}(\omega)$ is the power spectral density of the input to the system. If $S_{xx}(\omega)$ is constant, one easily sees that,

$$\sqrt{S_{yy}(\omega)} = C\,|H(j\omega)| \tag{18}$$

where C is a constant, so that $\sqrt{S_{yy}(\omega)}$ is proportional to the amplitude of the transfer function. If $S_{xx}(\omega)$ is not constant but smooth (no sharp peaks), the peaks in $\sqrt{S_{yy}(\omega)}$ will correspond to the eigenfrequencies of the system. Even if strong components exist at specific frequencies in the input spectrum $S_{xx}(\omega)$, this is not a problem as long as that frequency does not correspond to an eigenfrequency of the system, since the peak detection will be performed close to the known eigenfrequencies of the undamaged system.

In conclusion, in the general case for a perfect modal filter, only one sharp peak will appear in $\sqrt{S_{yy}(\omega)}$, and the same conclusions with respect to the appearance of new peaks can be drawn. The next step for autonomous damage detection is to define a procedure to detect the appearance of the new peaks automatically.

2.4 Feature Extraction Based on Modal Filter Outputs

Each frequency point of $\sqrt{S_{yy}(\omega)}$ of a modal filter can be used as a feature for damage detection. This is not a very good approach, since, as discussed before, not all the frequencies contain information about the damage to be detected, and this would lead to a very large number of features. Moreover, one is interested in detecting the appearance of peaks around the natural frequencies of the undamaged system. The approach followed is to define a *peak indicator* in each of these frequency bands (Figure 25). For n_f

modal filters, the total number of peak indicators extracted is $n_f\,(n_e - 1)$ where n_e is the number of eigenfrequencies in the frequency band for which the modal filters are computed.

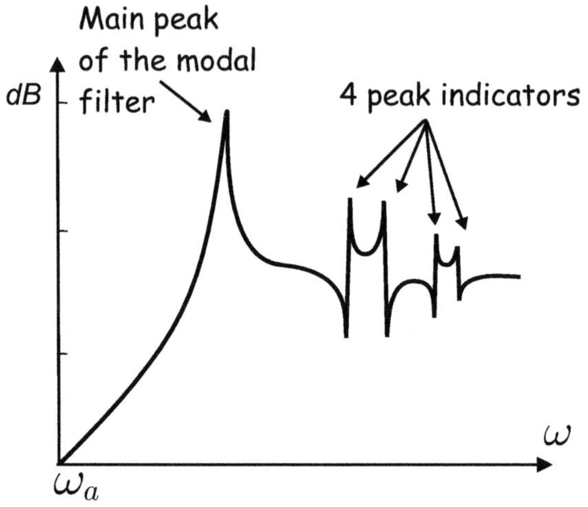

Figure 25. Feature extraction from the output of the modal filter in the frequency domain : computation of peak indicators in frequency bands around the natural frequencies of the undamaged system

Let us assume that the entire frequency bandwidth is divided into frequency bands $[\omega_1, \omega_2]$ around each natural frequency of the structure (the bandwidth is given in % of the natural frequency, typical values are 10% or 20 %). One notes $s(\omega) = \sqrt{S_{yy}(\omega)}$ and computes the following quantities (Figure 26, Jarman et al. (2003)),

- The frequency center (FC) $= \dfrac{\int_{\omega_1}^{\omega_2} \omega s(\omega) d\omega}{\int_{\omega_1}^{\omega_2} s(\omega) d\omega}$

- The root variance frequency (RVF) $= \left[\dfrac{\int_{\omega_1}^{\omega_2} (\omega - FC)^2 \, s(\omega) d\omega}{\int_{\omega_1}^{\omega_2} s(\omega) d\omega} \right]^{\frac{1}{2}}$

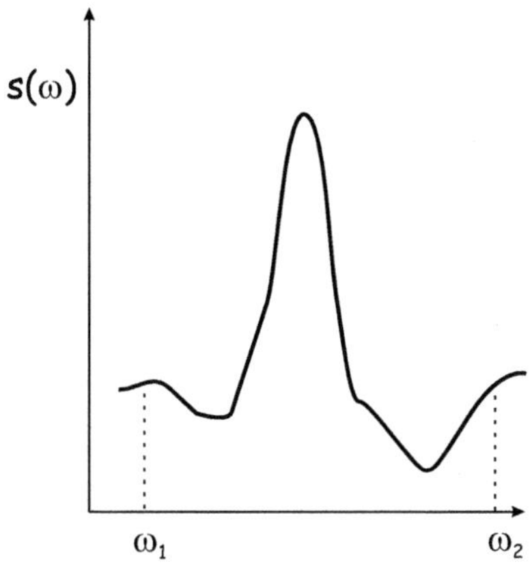

Figure 26. Restriction of $s(\omega) = \sqrt{S_{yy}(\omega)}$ to the frequency band $[\omega_1, \omega_2]$ for the computation of the peak indicator

The peak indicator is given by,

$$I_{peak} = \frac{2\sqrt{3}\,(RVF)}{(\omega_2 - \omega_1)} \tag{19}$$

It has the following properties,

- if $s(\omega)$ is a Dirac function, $I_{peak} = 0$
- if $s(\omega)$ is constant, $I_{peak} = 1$
- A drop of I_{peak} below 1 corresponds to the appearance of a peak. In practice, a threshold value of 0.8 should be used for robust peak detection

Illustration In Figure 27, $s(\omega)$ is plotted for a modal filter tuned to mode 1 of an undamaged structure (a), and of a damaged structure (b). This example is based on the numerical model of a three span bridge developed in Section 2.5. The frequency intervals used for the computation of the peak indicators are shown in Figure 27b), and the values of the peak indicators are given in Figure 28. For the undamaged structure, the peak indicators are all close to unity, except for interval 1 which corresponds to the main peak of the modal filter. For the damaged structure, the peak indicators for intervals 3 and 4 decrease clearly below 1, but for intervals 6 and 7, there is only a very small drop in the peak indicators, although the peaks can be clearly seen on Figure 27b). As illustrated in the next section, it is possible to improve the detection of smaller peaks using some signal processing techniques.

a)

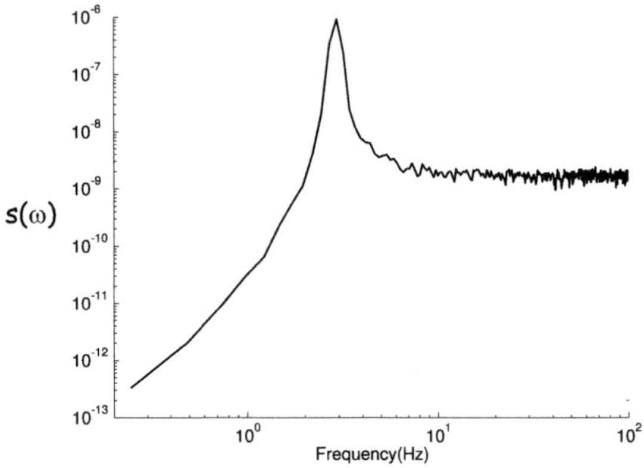

b)

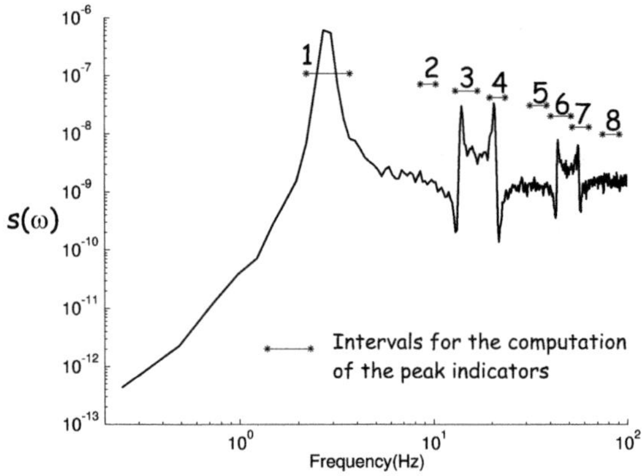

Figure 27. $s(\omega)$ computed from the output of a modal filter tuned to mode 1: a) undamaged structure; b) damaged structure and definition of the intervals for the computation of the peak indicators

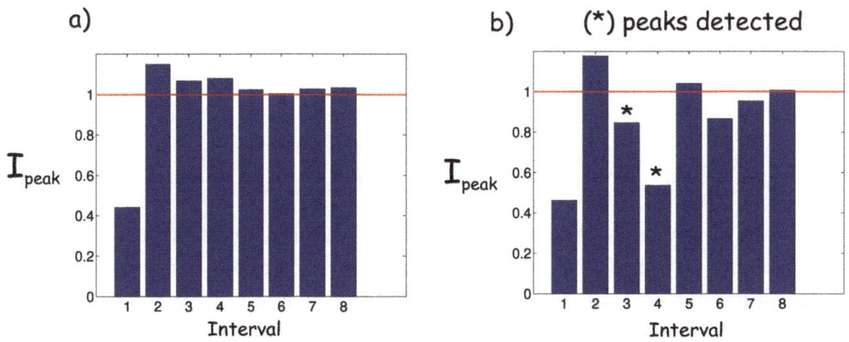

Figure 28. Value of the peak indicator in each interval for: a) the undamaged structure; b) the damaged structure

Improved Peak Indicator Using Signal Processing Techniques In order for the peak indicator to be more sensitive (increase of signal-to-noise ratio), a technique called "second-derivative matched filtering" (Danielsson et al. (2002)) is proposed. Let $s(\omega)$ be the signal to be filtered and $f(\Omega)$ be the filtering function. Simple filtering consists of computing the following convolution integral,

$$s_f(\omega) = \int_{-\infty}^{\infty} f(\Omega)s(\omega + \Omega)\, d\Omega \tag{20}$$

In order to remove background noise, the second derivative is computed,

$$s''_f(\omega) = \int_{-\infty}^{\infty} f(\Omega)s''(\omega + \Omega)\, d\Omega = \int_{-\infty}^{\infty} f''(\Omega - \omega)s(\Omega)\, d\Omega \tag{21}$$

This expression shows that it is only necessary to differentiate the filtering function, which is much less sensitive to the noise than differentiating the signal itself. On the other hand, the filtering function needs to be twice-differentiable. Although the philosophy of matched filtering is to have a filtering function equal to the equation of the peak to be detected, this choice is not adopted here (due to the complicated expression for the second derivative). Instead, a simpler and more typical choice for such a function is a Gaussian distribution. There is in fact an analogy of this method with wavelet analysis where the so-called "Mexican-hat" corresponds to the second derivative of the Gaussian (Daubechies (1992)). Both the Gaussian distribution and its second derivative are represented in Figure 29. Their analytical expressions are given by,

$$f(x) = e^{\frac{-x^2}{\sigma^2}} \tag{22}$$

$$f''(x) = \frac{1}{\sigma^2}\left(1 - \frac{x^2}{\sigma^2}\right) e^{\frac{-x^2}{\sigma^2}} \tag{23}$$

where σ is the standard deviation of the Gaussian distribution (the mean is 0). It is an analog of the scaling factor in wavelet analysis. For an optimal filtering of $s(\omega)$, one needs to define a σ which is not constant with frequency, it should in fact, be a ratio of the frequency ω. This is because the width of a peak in $s(\omega)$ is proportional to the central frequency of that peak. As shown in Figure 29, the width of the filtering function is close to 2σ, and if one assumes that the width of a peak in $s(\omega)$ is approximately 20% of the central frequency, one has,

$$\sigma \simeq 0.1\omega \tag{24}$$

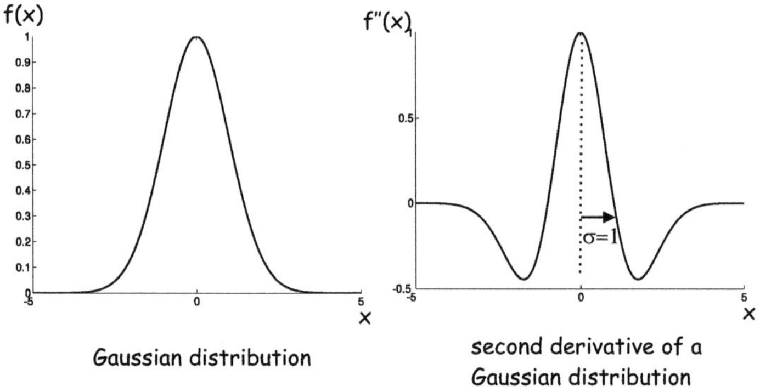

Figure 29. Gaussian distribution and its second derivative

The approach adopted to filter the signal $s(\omega)$ is as follows,

- compute $s^*(\omega) = log(s(\omega))$
- Apply a second-derivative gaussian matched filter to $s^*(\omega)$;

$$s_f^*(\omega) = \int_{-\infty}^{\infty} \left[\left(1 - \frac{(\omega - \Omega)^2}{\sigma^2}\right) e^{\frac{-(\omega-\Omega)^2}{\sigma^2}}\right] s^*(\Omega)\, d\Omega \tag{25}$$

with $\sigma = 0.1\omega$

- recover the filtered signal;

$$s_f = 10^{s_f^*} \tag{26}$$

The results of the filtering for both the undamaged and the damaged structure are shown in Figure 30. The peak indicators are computed based on the filtered signals and shown in Figure 31. The results show that peaks 6 and 7 are now well detected.

a)

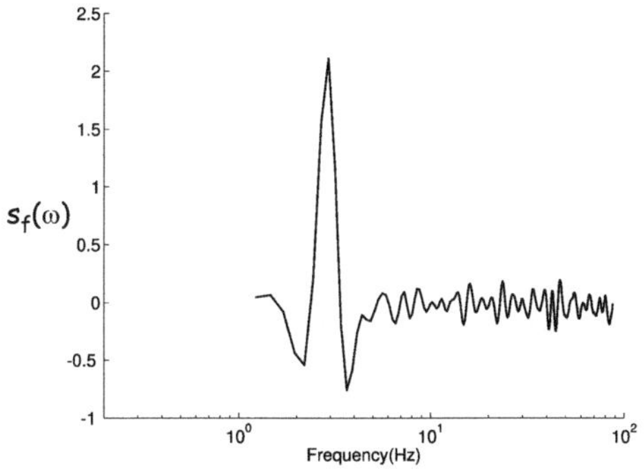

b)

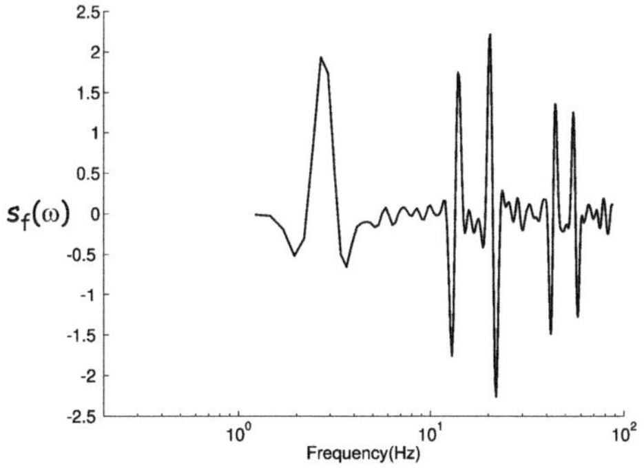

Figure 30. Filtered signal $s_f(\omega)$ computed from $s(\omega)$: a) undamaged structure; b) damaged structure

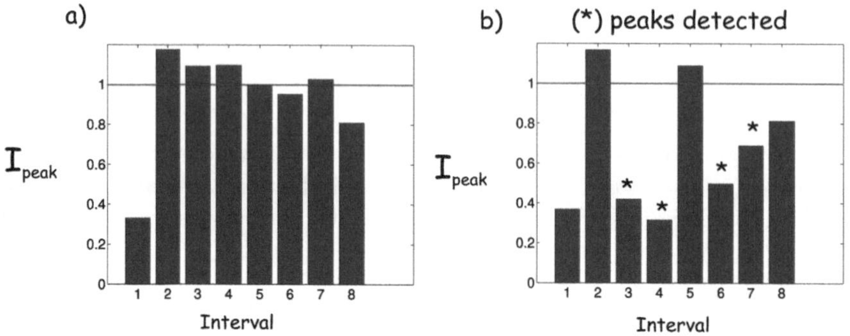

Figure 31. Value of the peak indicator in each interval for: a) the undamaged structure; b) the damaged structure, computed on the filtered signal $s_f(\omega)$

2.5 Numerical Example

The structure considered is a three-span bridge similar to the one presented in Yan et al. (2005) and Deraemaeker et al. (2008) (Figure 32). The motion is restricted to in-plane vibrations. The structure is made of two materials: steel and concrete; it is excited by a uniform pressure acting on the first span of the bridge. The excitation is a band-limited white noise in the frequency band 0-100 Hz, containing the first 10 eigenfrequencies of the structure.

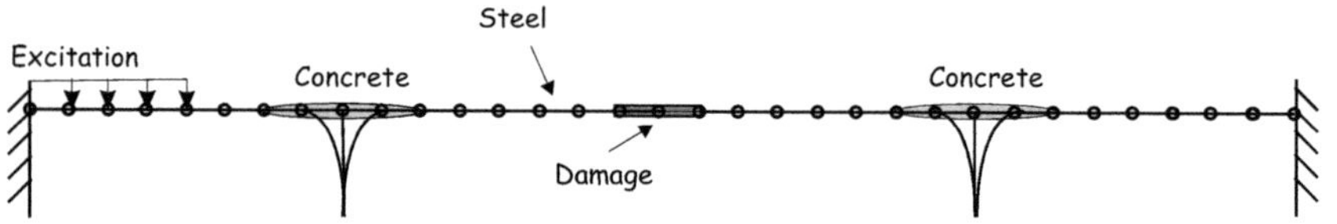

Figure 32. Three-span bridge subject to different temperature gradients and damage

Generation of Pseudo-measurements The bridge is discretised with 32 Euler-Bernoulli finite elements using the Structural Dynamics Toolbox (SDTools) under Matlab. The response is computed in the time domain using an in-house time integration scheme based on Duhamel's formula (Preumont (1982)). A total of 29 accelerometers are placed, one at each node of the finite element model (except boundary conditions). The time domain response (200 seconds of measurement at a sampling rate of 1000 Hz) is computed for each accelerometer.

In order to study the impact of damage on the peak indicators I_{peak} proposed as new features for damage detection, the response of the bridge is generated for (i) the undamaged structure (ii) the damaged structure in which the damage is gradually introduced in the central span of the bridge (Figure 32). The damage is modelled by a stiffness reduction going from 2% to 30%. In total, 25 sets of acceleration measurements are made, 10 on the undamaged structure, and 15 on the gradually damaged structure (increase of 2% at each sample).

Feature Extraction for Damage Detection The modal filter coefficients α_k (k=1..10) are computed for the 10 modeshapes in the frequency band of interest (0-100 Hz). This example focuses on the output of the modal filter tuned to mode 1. The following steps are applied in order to extract features from the acceleration measurements,

- compute the time domain output of the modal filter tuned to mode 1 (linear combination of the sensor outputs),

$$y(t) = \alpha_k y_k(t) \tag{27}$$

- compute $s(\omega)$,

$$s(\omega) = \sqrt{S_{yy}(\omega)} \tag{28}$$

  where S_{yy} is the power spectral density estimated using Welch's average periodogram method (Oppenheim and Schafer (1975)).
- filter the signal $s(\omega)$ in order to get $s_f(\omega)$, as described in section 2.4,
- compute the peak indicators for the intervals represented in Figure 27b).

Figure 33 shows the evolution of the peak indicators for intervals 3 and 7 as a function of the sample number (samples 1-10 are undamaged, then damage is gradually increased). The figures show that I_{peak} gradually decreases from sample 11, following the evolution of the amplitude of damage. The figure also shows a comparison between the peak indicator extracted from the unfiltered signal $s(\omega)$ and the filtered signal $s_f(\omega)$. In interval 7, filtering helps in detecting smaller damage levels.

Summary Modal filters can be used to efficiently combine in real time the information from very large arrays of sensors. Using the modal sensor outputs, it is shown here that it is possible to detect changes in modeshapes which are indicative of damage in structures. The proposed technique relies on signal processing techniques and allows one to program a

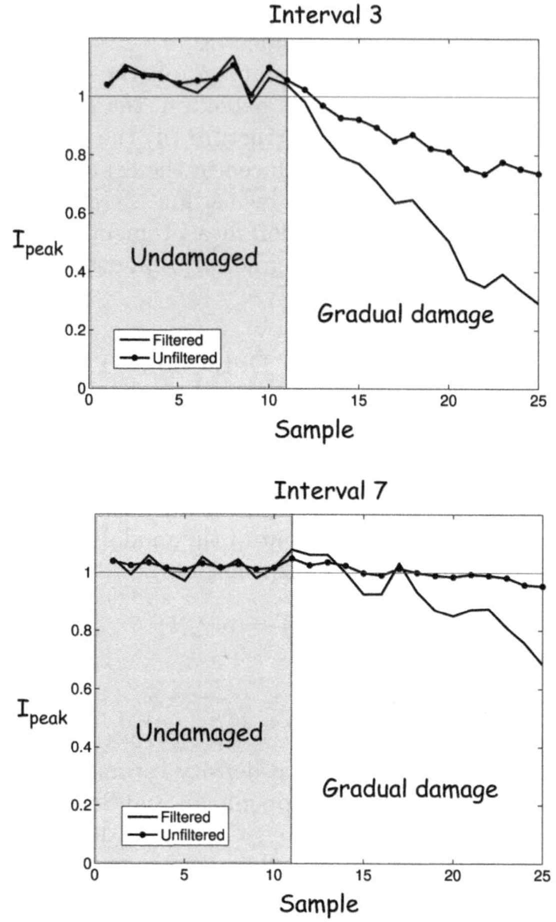

Figure 33. Evolution of I_{peak} for intervals 3 and 7 as a function of the sample number: samples 1-10: undamaged structure; samples 11-25: gradually increasing damage. I_{peak} is extracted both from the unfiltered signal $s(\omega)$ and the filtered signal $s_f(\omega)$

fully automated procedure for feature extraction based on output-only measurements on structures excited by ambient vibrations. The features extracted from modal filters using the technique described here above can then be used for damage detection, as illustrated for example in Deraemaeker et al. (2008), where a comparison with traditional features (modeshapes and eigen-frequencies) is made. The main advantages of this technique are the very low computational cost and the automation of the procedure.

Bibliography

R. Danielsson, D. Bylund, and K.E. Markides. Matched filtering with background suppression for improved quality of base peak chromatograms and mass spectra in liquid chromatography-mass spectrometry. *Analytica Chemica Acta*, 454:167–184, 2002.

I. Daubechies. *Ten Lectures on Wavelets*. SIAM, 1992.

A. Deraemaeker and A. Preumont. Vibration based damage detection using large array sensors and spatial filters. *Mechanical Systems and Signal Processing*, 20:1615–1630, 2006.

A. Deraemaeker, E. Reynders, G. De Roeck, and J. Kullaa. Vibration-based structural health monitoring using output-only measurements under changing environment. 22:34–56, 2008.

A. Deraemaeker, H. Nasser, A. Benjeddou, and A. Preumont. Mixing rules for the piezoelectric properties of Macro Fiber Composites. *Journal of Intelligent Material Systems and Structures*, 2009. to be publised.

S.W. Doebling, C.R. Farrar, and M.B. Prime. A summary review of vibration-based damage identification methods. *Shock and Vibration Digest*, 30(2):91–105, 1998.

R. Geier, G. De Roeck, and J. Petz. Cable force determination for the Danube channel bridge in Vienna. *Structural Engineering International*, 15(3):181–185, 2005.

B. Glisic and D. Inaudi. *Fibre optic methods for Structural Health Monitoring*. John Wiley and Sons, 2007.

K.H. Jarman, D.S. Daly, K.K. Anderson, and K.L. Wahl. A new approach to automated peak detection. *Chemometrics and intelligent laboratory systems*, 69:61–76, 2003.

J.P. Lynch and K.J. Loh. A summary review of wireless sensors and sensor networks for structural health monitoring. *Shock and Vibration Digest*, 38(2):91–128, 2006.

X. Meng. *Real-time Deformation Monitoring of Bridges Using GPS/Accelerometers*. PhD thesis, University of Nottingham, 2002.

A. V. Oppenheim and R.W. Schafer. *Digital Signal Processing*. Prentice-Hall, 1975.

B. Peeters and G. De Roeck. Reference-based stochastic subspace identification for output-only modal analysis. *Mechanical Systems and Signal processing*, 13(6):855–878, 1999.

A. Preumont. Frequency domain analysis of time integration operators. *Earthquake Engineering and Structural Dynamics*, 10:691–697, 1982.

A. Preumont. *Random vibrations and spectral analysis.* Kluwer, 1994.

A. Preumont, A. François, P. De Man, and V. Piefort. Spatial filters in structural control. *Journal of Sound and Vibration*, 265:61–79, 2003.

SDTools. Structural dynamics toolbox. http://www.sdtools.com/.

Ashwin Arunkumar Seshia. *Integrated Micromechanical Resonant Sensors for Inertial Measurement Systems.* PhD thesis, University of California at Berkeley, 2002.

G. J. Stein. Some recent developments in acceleration sensors. *Measurement science review*, 1(1):183–186, 2001.

G. Wang, M.L. Wang, Y. Zhao, Y. Chen, and B. Sun. Application of EM stress sensors in large steel cables. *Proceedings of the SPIE*, 5765:395–406, 2005.

H. Wenzel and D. Pichler. *Ambient vibration monitoring.* Wiley, 2005.

A.M. Yan, G. Kerschen, P. De Boe, and J.C. Golinval. Structural damaged diagnosis under varying environmental conditions - Part I : A linear analysis. *Mechanical Systems and Signal Processing*, 19(4):847–864, 2005.

F. Yu and N. Gupta. An efficient model for improving performance of vibrating-wire instruments. *Measurement*, 37:278–283, 2005.

Subspace identification for operational modal analysis

Edwin Reynders and Guido De Roeck
Katholieke Universiteit Leuven, Structural Mechanics Division
Kasteelpark Arenberg 40, B-3001 Leuven

Abstract This chapter deals with the estimation of modal parameters from measured vibration data using subspace techniques. An in-depth review of subspace identification for operational modal analysis is provided. In addition, two recent developments are emphasised: the estimation of the probability density function of the modal parameters, and the use of an exogenous force in addition to the unmeasured operational excitation.

1 Introduction

Vibration-based SHM methods very often rely on modal parameters that are estimated from measured vibration data. Classical *Experimental Modal Analysis* (EMA) techniques obtain the modal parameters from input-output measurements, i.e., measured, artificial forces are applied to the structure, and the response to these forces is recorded. The response to unmeasured, ambient forces is considered as unwanted *noise*. In general, such EMA methods are not suitable for large structures and buildings because these structures are inherently tested in operational rather than in laboratory conditions, and the contribution of the measured forces to the total structural response is usually rather low. A bridge for instance can only be excited to a limited vibration level by an artificial excitation source such as a shaker, unless it has a very heavy mass. This implies that the ever-present ambient excitation, due to for example wind or traffic, can most often not be neglected, especially at low frequencies. Output-only or *Operational Modal Analysis* (OMA) techniques have therefore been developed. They extract the modal parameters from the dynamic response to operational forces. The unmeasured, ambient forces are usually modeled as stochastic quantities with unknown parameters but with known behavior, for example, as white noise time-series with zero-mean and unknown covariances. Peeters and De Roeck (2001) provide a review of operational modal analysis techniques.

However, the OMA approach has two disadvantages when compared to EMA: the mode shapes can not be mass-normalised, and the frequency content of the excitation is usually narrow-banded. For these reasons, there has been an increasing interest in the last few years towards combined modal testing techniques, also called hybrid vibration testing or *Operational Modal Analysis with eXogenous inputs* (OMAX), where an artificial force is used in operational conditions. The main difference between OMAX and the traditional EMA approach is that the operational forces are included in the identified system model: they are not considered as noise, but as useful excitation. As a consequence, the amplitude of the artificial forces can be equal to, or even lower than the amplitude of the operational forces. This is of crucial importance for the modal testing of large structures. It allows the use of excitation devices that are small and practical when compared to the actuators that are needed for EMA testing such as electromechanical or hydraulic shakers, which are heavy and difficult to transport.

This chapter deals with the estimation of modal parameters from measured vibration data using time-domain subspace identification methods, both from the OMA and the OMAX perspective. These methods identify a discrete-time state-space model. In Section 2, it is demonstrated that this is a valid model for a vibrating structure, by deriving it from a finite element description. The deterministic state-space model is then extended with unobserved inputs and output disturbances, which are both modeled as stochastic quantities. Section 3 provides insight into the basic ideas that lie behind subspace methods; the subspace identification algorithms themselves are presented in Section 4. Two important recent developments are discussed in detail: the estimation of the uncertainty on the identified system parameters, and the use of an exogenous input in addition to the unmeasured operational excitation, for OMAX testing. In Section 5, the derivation of the modal parameters from the identified state-space model is treated, as well as the estimation of their probability density function. Finally, in Section 6, two real-life applications are discussed in detail.

2 State-space models of vibrating structures

2.1 Introduction

In this section, the use of a state-space model for operational modal analysis, with or without exogenous inputs, is discussed. Starting from a finite element description, that is commonly used for realistic physical modeling of structures in forward vibration problems, a state-space model, that is more convenient for solving inverse problems, is derived. This deterministic state-space model is then extended with unobserved inputs and output disturbances, which are both modeled as stochastic quantities.

2.2 Linear dynamic finite element model

The finite element method is the most common tool for forward modeling of vibrating structures. In the case of a linear dynamic model with general viscous damping, one has the following system of ordinary differential equations:

$$\boldsymbol{M}\frac{d^2\boldsymbol{v}(t)}{dt^2} + \boldsymbol{C^v}\frac{d\boldsymbol{v}(t)}{dt} + \boldsymbol{K}\boldsymbol{v}(t) = \boldsymbol{B_{sel}}\boldsymbol{u}(t) \tag{1}$$

where $\boldsymbol{v}(t) \in \mathbb{R}^{n_{fe}}$ is the vector with nodal displacements, $\boldsymbol{M} \in \mathbb{R}^{n_{fe}\times n_{fe}}$, $\boldsymbol{C^v} \in \mathbb{R}^{n_{fe}\times n_{fe}}$ and $\boldsymbol{K} \in \mathbb{R}^{n_{fe}\times n_{fe}}$ are the mass, viscous damping, and stiffness matrices, respectively, and $\boldsymbol{B_{sel}} \in \mathbb{R}^{n_{fe}\times n_u}$ is a selection matrix such that the vector with externally applied forces, $\boldsymbol{u}(t) \in \mathbb{R}^{n_u}$, has only elements that are not identically zero.

2.3 Continuous-time state-space model

State-space equation. By rearranging (1) and assuming that $\boldsymbol{M}$ has full rank[1], a continuous-time state space model

$$\frac{d\boldsymbol{x}(t)}{dt} = \boldsymbol{A_c}\boldsymbol{x}(t) + \boldsymbol{B_c}\boldsymbol{u}(t), \tag{2}$$

where

$$\boldsymbol{x}(t) = \begin{bmatrix} \boldsymbol{v}(t) \\ \frac{d\boldsymbol{v}(t)}{dt} \end{bmatrix}$$
$$\boldsymbol{A_c} = \begin{bmatrix} \boldsymbol{0} & \boldsymbol{I} \\ -\boldsymbol{M}^{-1}\boldsymbol{K} & -\boldsymbol{M}^{-1}\boldsymbol{C^v} \end{bmatrix}$$
$$\boldsymbol{B_c} = \begin{bmatrix} \boldsymbol{0} \\ \boldsymbol{M}^{-1}\boldsymbol{B_{sel}} \end{bmatrix},$$

is obtained. The vector $\boldsymbol{x}(t) \in \mathbb{R}^n$ is called the *state* of the structure. The number of elements of $\boldsymbol{x}(t)$, n, is called the *model order*. If the state-space model is derived from a finite element model, as in this case, one has $n = 2n_{fe}$. When the state at $t = 0$ is known, the system of ordinary differential equations (2) can be solved for $\boldsymbol{x}(t)$:

$$\boldsymbol{x}(t) = e^{\boldsymbol{A_c}t}\boldsymbol{x}(0) + \int_{[0,t]} e^{\boldsymbol{A_c}(t-\tau)}\boldsymbol{B_c}\boldsymbol{u}(\tau)d\tau. \tag{3}$$

[1] In a beam model, for example, this implies that rotational inertia is included.

Input-output equation. If the output quantities of interest are linear combinations of nodal displacements, velocities or accelerations, one has

$$\begin{aligned} \boldsymbol{y}(t) &= \boldsymbol{C}_{\ddot{\boldsymbol{v}}}\frac{d^2\boldsymbol{v}(t)}{dt^2} + \boldsymbol{C}_{\dot{\boldsymbol{v}}}\frac{d\boldsymbol{v}(t)}{dt} + \boldsymbol{C}_{\boldsymbol{v}}\boldsymbol{v}(t) \\ &= \left[\ \boldsymbol{C}_{\boldsymbol{v}} - \boldsymbol{C}_{\ddot{\boldsymbol{v}}}\boldsymbol{M}^{-1}\boldsymbol{K} \mid \boldsymbol{C}_{\dot{\boldsymbol{v}}} - \boldsymbol{C}_{\ddot{\boldsymbol{v}}}\boldsymbol{M}^{-1}\boldsymbol{C}^{\boldsymbol{v}}\ \right]\boldsymbol{x}(t) + \boldsymbol{C}_{\ddot{\boldsymbol{v}}}\boldsymbol{M}^{-1}\boldsymbol{B}_{\boldsymbol{sel}}\boldsymbol{u}(t) \\ &= \boldsymbol{C}_{\boldsymbol{c}}\boldsymbol{x}(t) + \boldsymbol{D}_{\boldsymbol{c}}\boldsymbol{u}(t) \end{aligned} \tag{4}$$

where $\boldsymbol{C}_{\ddot{\boldsymbol{v}}} \in \mathbb{R}^{n_y \times n}$, $\boldsymbol{C}_{\dot{\boldsymbol{v}}} \in \mathbb{R}^{n_y \times n}$ and $\boldsymbol{C}_{\boldsymbol{v}} \in \mathbb{R}^{n_y \times n}$ are selection matrices. Finite strains can be included in $\boldsymbol{y}(t)$, since they can be obtained by dividing the difference between two displacement DOFs by the initial distance between their nodes.

Transfer function - poles. A Laplace transform of both sides of (2) and (4) leads to a parametrisation of the transfer function:

$$\boldsymbol{y}(s) = (\boldsymbol{C}_{\boldsymbol{c}}(s\boldsymbol{I} - \boldsymbol{A}_{\boldsymbol{c}})^{-1}\boldsymbol{B}_{\boldsymbol{c}} + \boldsymbol{D}_{\boldsymbol{c}})\boldsymbol{u}(s) = \boldsymbol{H}(s)\boldsymbol{u}(s). \tag{5}$$

Following Cramer's rule, one has

$$(s\boldsymbol{I} - \boldsymbol{A}_{\boldsymbol{c}})^{-1} = \frac{\operatorname{adj}(s\boldsymbol{I} - \boldsymbol{A}_{\boldsymbol{c}})}{\det(s\boldsymbol{I} - \boldsymbol{A}_{\boldsymbol{c}})},$$

where $\det(\square)$ denotes the determinant and $\operatorname{adj}(\square)$ the adjoint matrix of a square matrix $\square$. Since $\det(s\boldsymbol{I} - \boldsymbol{A}_{\boldsymbol{c}})$ is the characteristic polynomial of $\boldsymbol{A}_{\boldsymbol{c}}$, the poles of the transfer function are the eigenvalues of $\boldsymbol{A}_{\boldsymbol{c}}$.

Change of basis - decoupling. When the state is transformed to a new basis, $\boldsymbol{x} \mapsto \boldsymbol{T}^{-1}\boldsymbol{x}$ with $\boldsymbol{T} \in \mathbb{C}^{n \times n}$ nonsingular, the input-output map provided by the state-space description is preserved when $(\boldsymbol{A}_{\boldsymbol{c}}, \boldsymbol{B}_{\boldsymbol{c}}, \boldsymbol{C}_{\boldsymbol{c}}, \boldsymbol{D}_{\boldsymbol{c}}) \mapsto (\boldsymbol{T}^{-1}\boldsymbol{A}_{\boldsymbol{c}}\boldsymbol{T}, \boldsymbol{T}^{-1}\boldsymbol{B}_{\boldsymbol{c}}, \boldsymbol{C}_{\boldsymbol{c}}\boldsymbol{T}, \boldsymbol{D}_{\boldsymbol{c}})$, as follows from (2) and (4). In particular, when $\boldsymbol{A}_{\boldsymbol{c}}$ has a similarity transform,

$$\boldsymbol{A}_{\boldsymbol{c}} = \boldsymbol{\Psi}_{\boldsymbol{c}}\boldsymbol{\Lambda}_{\boldsymbol{c}}\boldsymbol{\Psi}_{\boldsymbol{c}}^{-1}, \tag{6}$$

with $\boldsymbol{\Lambda}_{\boldsymbol{c}}$ a diagonal matrix, (2) and (4) are decoupled by putting $\boldsymbol{T} = \boldsymbol{\Psi}_{\boldsymbol{c}}$:

$$\frac{d\boldsymbol{x}_{\boldsymbol{m}}(t)}{dt} = \boldsymbol{\Lambda}_{\boldsymbol{c}}\boldsymbol{x}_{\boldsymbol{m}}(t) + \boldsymbol{L}_{\boldsymbol{c}}^T\boldsymbol{u}(t) \tag{7}$$

$$\boldsymbol{y}(t) = \boldsymbol{\Phi}_{\boldsymbol{c}}\boldsymbol{x}_{\boldsymbol{m}}(t) + \boldsymbol{D}_{\boldsymbol{c}}\boldsymbol{u}(t). \tag{8}$$

The subscript m denotes *modal*, as will be explained shortly.

2.4 Discrete-time state-space model

Definition - ZOH assumption. Since for a given input $\boldsymbol{u}(t)$, solving the continuous-time state-space description analytically is usually impossible in the time domain, it seems natural to convert this model to discrete time:

$$\boldsymbol{x_{k+1}} = \boldsymbol{A}\boldsymbol{x_k} + \boldsymbol{B}\boldsymbol{u_k} \tag{9}$$
$$\boldsymbol{y_k} = \boldsymbol{C}\boldsymbol{x_k} + \boldsymbol{D}\boldsymbol{u_k}. \tag{10}$$

For free vibration problems, where the inputs are identically zero, an exact discretisation is possible by solving the system of equations (2,4) using (3):

$$\boldsymbol{x}((k+1)T) = e^{\boldsymbol{A}_c T}\boldsymbol{x}(kT)$$
$$\boldsymbol{y}(kT) = \boldsymbol{C_c}\boldsymbol{x}(kT),$$

where T denotes the sampling period. This leads to an exact map with $(\boldsymbol{A}, \boldsymbol{C}) = (e^{\boldsymbol{A}_c T}, \boldsymbol{C_c})$. The map is very important when solving the inverse modal analysis problem by fitting (9-10) to measured sampled data. When converting the fitted discrete-time state-space model to a continuous-time equivalent, the inverse map leads to the exact continuous-time equivalents of the discrete poles and mode shapes.

For forced vibration problems, a *Zero-Order-Hold* (ZOH) assumption is often made, which means that the force is assumed constant (equal to $\boldsymbol{u}(kT)$) in $[kT, (k+1)T)$. With this assumption, the following map is obtained from (3):

$$\boldsymbol{A} = e^{\boldsymbol{A}_c T}, \quad \boldsymbol{B} = \int_{kT}^{(k+1)T} e^{\boldsymbol{A}_c((k+1)T-\tau)} d\tau \boldsymbol{B_c} = (\boldsymbol{A} - \boldsymbol{I})\boldsymbol{A_c}^{-1}\boldsymbol{B_c} \tag{11}$$
$$\boldsymbol{C} = \boldsymbol{C_c}, \qquad \boldsymbol{D} = \boldsymbol{D_c}.$$

A proof for the second equality in the expression for $\boldsymbol{B}$ can be found in (Juang, 1994, p. 20). Although alternative discretisation strategies are possible (Franklin et al., 1998, ch. 6), the ZOH discretisation has the advantage that it is exact when $\boldsymbol{u}(t) = \boldsymbol{0}$, as discussed above. However, the input matrix $\boldsymbol{B_c}$, that is calculated through the inverse map, is not a good approximation of the true $\boldsymbol{B_c}$ when the sampling frequency is not much larger than twice the largest important frequency that is present in the spectra of the input and output signals.

Solution of the state equations - impulse response. The state equations (9-10) can be solved by simple forward calculation:

$$\boldsymbol{y_k} = \boldsymbol{C}\boldsymbol{A}^k\boldsymbol{x_0} + \sum_{l=1}^{k} \boldsymbol{C}\boldsymbol{A}^{l-1}\boldsymbol{B}\boldsymbol{f_{k-l}} + \boldsymbol{D}\boldsymbol{f_k}.$$

From this result, the impulse response is readily obtained as

$$\boldsymbol{H}_0 = \boldsymbol{D}, \qquad \boldsymbol{H}_k = \boldsymbol{C}\boldsymbol{A}^{k-1}\boldsymbol{B},\ k > 1. \tag{12}$$

Transfer function - poles. By taking the z-transform of both sides of (9-10), a parametrisation of the transfer function is obtained:

$$\boldsymbol{y}(z) = (\boldsymbol{C}(z\boldsymbol{I} - \boldsymbol{A})^{-1}\boldsymbol{B} + \boldsymbol{D})\boldsymbol{u}(z) = \boldsymbol{H}(z)\boldsymbol{u}(z) \tag{13}$$

Just as for the continuous-time state-space model (see Section 2.3), it follows from Cramer's rule that the poles of the transfer function are the eigenvalues of $\boldsymbol{A}$.

Change of basis - decoupling. Following the same lines as for the continuous-time case, one has that the input-output map provided by the discrete-time state-space description is preserved when $(\boldsymbol{A},\boldsymbol{B},\boldsymbol{C},\boldsymbol{D}) \mapsto (\boldsymbol{T}^{-1}\boldsymbol{A}\boldsymbol{T},\boldsymbol{T}^{-1}\boldsymbol{B},\boldsymbol{C}\boldsymbol{T},\boldsymbol{D})$, and that, when $\boldsymbol{A}$ has a similarity transform,

$$\boldsymbol{A} = \boldsymbol{\Psi}_d\boldsymbol{\Lambda}_d\boldsymbol{\Psi}_d^{-1}, \tag{14}$$

where $\boldsymbol{\Lambda}_d$ is a diagonal matrix, (9-10) is decoupled by putting $\boldsymbol{T} = \boldsymbol{\Psi}_d$:

$$\boldsymbol{x}_{m,k+1} = \boldsymbol{\Lambda}_d\boldsymbol{x}_{m,k} + \boldsymbol{L}_d^T\boldsymbol{u}_k \tag{15}$$

$$\boldsymbol{y}_k = \boldsymbol{\Phi}_d\boldsymbol{x}_{m,k} + \boldsymbol{D}\boldsymbol{u}_k. \tag{16}$$

2.5 Modeling loads and sensor noise

In this section, a step closer to the experimental world is made. The goal is to obtain a more realistic description for the *measured* input-output behavior of real structures. Therefore, measurement noise is taken into account, and the inputs $\boldsymbol{u}(t)$ that were considered in the previous section, are split into two parts: a part that can be measured in an operational vibration test, and a part that can not be measured.

Unobserved loads. When the loads can not be measured, they have to be identified together with the system, from the measured response. The concerned discipline is called output-only or blind system identification. When identifying the input and the system at the same time, a problem of identifiability occurs: the system can not be determined unless extra assumptions are made concerning the unknown inputs.

Most ambient excitation sources, such as seismic waves (Clough and Penzien, 1995), turbulent wind or water pressure (Durbin and Petterson Reif,

2001; Durbin and Medic, 2007), or road or railway traffic (Braun and Hellenbroich, 1991; ORE, 1971), are often modeled as stochastic loads in forward calculations. An exception is excitation due to rotating machinery, such as wind turbines, which usually has a harmonic nature, i.e., its frequency content consists approximately of discrete peaks that occur at integer multiples of a fundamental frequency.

When unobserved ambient or harmonic loads are taken into account during system identification, the presented model structures need to be extended. Pintelon et al. (2008) present a model for a structure's response to non-stationary harmonic excitation, and use it for operational modal analysis. This model is generally applicable and could be combined with the presented discrete-time state-space model. Stochastic load modeling is discussed in detail in the next paragraph.

Stationarity, ergodicity, and zero mean. When a system that is driven by an unmeasured, stochastic input, needs to be identified, extra assumptions on the input, that is, other that stochasticity, are needed. Wide-sense stationarity, which means that the covariance between two time samples depends only on the time difference, not on the time instances at which the samples were taken, and quadratic mean ergodicity, i.e., ensemble averaging can be replaced by time averaging, are classical assumptions (Dougherty, 1999). They are mild in the sense that if they are not valid, they only increase the variance errors of the identified system description. The zero mean assumption holds exactly when the constant trend is removed from the outputs (hence also from the unmeasured inputs) and from the measured inputs. In this case, the covariance functions of the inputs and outputs equal their correlation functions.

Discrete-time white noise. Another classical, but more restrictive assumption, is that the stochastic unobserved input is a white noise vector. The sampled, stationary stochastic input sequence $(\boldsymbol{u}_{\boldsymbol{k}}^{s})$ is said to be a zero-mean discrete-time white noise sequence when its correlation function obeys (Dougherty, 1999, p. 154)

$$\boldsymbol{R}_{\boldsymbol{u}^s\boldsymbol{u}^s,j} = \mathcal{E}\left(\boldsymbol{u}_{\boldsymbol{k}+j}^{s}\boldsymbol{u}_{\boldsymbol{k}}^{s\,T}\right) = \begin{cases} \mathrm{Cov}(\boldsymbol{u}_{\boldsymbol{k}}^{s}), & j=0 \\ \boldsymbol{0}, & j \neq 0 \end{cases},$$

where $\mathcal{E}$ denotes the expectation operator and Cov the covariance operator. The discrete *Power Spectral Density* (PSD) of $(\boldsymbol{u}_{\boldsymbol{k}}^{s})$, which is defined as the *Discrete Fourier Transform* (DFT) of $\boldsymbol{R}_{\boldsymbol{u}^s\boldsymbol{u}^s,j}$, obeys

$$\boldsymbol{S}_{\boldsymbol{u}^s\boldsymbol{u}^s,j} = \boldsymbol{R}_{\boldsymbol{u}^s\boldsymbol{u}^s,0}.$$

This means that the discrete PSD of a discrete-time white noise sequence is real and constant.

Colored noise. Whether the white noise assumption is approximately true for the unobserved stochastic inputs entirely depends on the PSD of the inputs in the considered frequency band. Often coloured noise is a more realistic assumption: the unobserved inputs are assumed to be white noise that has passed through a linear time-invariant system, called the noise colour. If in that case, white noise inputs are assumed, the noise colour of the unmeasured excitation is part of the identified system model (see Figure 1). It can only be separated from the true system model if some prior knowledge is available. For instance, if it can be assumed that the modes of the vibrating structure are lowly damped real normal modes, then highly damped and complex modes can be assumed to represent the input noise colour.

Sensor noise. Even if all inputs could be measured and the vibrating structure would obey all assumptions, that is, no model errors are made, there would still be a discrepancy between the *measured* inputs and outputs and their true values. This is due to electric disturbances in the measurement equipment. These disturbances can be important when the amplitudes of the measured signals are low compared to the noise floor of the equipment.

Based on physical principles, it can be shown that many important electr(on)ic disturbances have a white noise nature. Nyquist (1928) derived that at room temperature, the thermal noise voltage at both ends of a resistor is approximately white below 1000GHz. However, the measurement equipment makes up a dynamical system of its own, and the disturbances that it generates are not perfectly white. When nonwhite output measurement noise is modeled as white, the noise colour becomes part of the identified system, just as with nonwhite unmeasured inputs (see Figure 1). Harmonic measurement noise is often due to harmonic AC components in the electricity grid, and could be modeled just as harmonic unobserved loads.

2.6 A combined deterministic-stochastic state-space model

Continuous combined state-space model. When the observed (deterministic) inputs $\boldsymbol{u}(t)$ and outputs $\boldsymbol{y}(t)$ are corrupted by additive sensor noise, denoted as $\boldsymbol{u}^{\boldsymbol{n}}(t) \in \mathbb{R}^{n_u}$ and $\boldsymbol{y}^{\boldsymbol{n}}(t) \in \mathbb{R}^{n_y}$, respectively, and when unobserved stochastic inputs $\boldsymbol{u}^{\boldsymbol{s}}(t) \in \mathbb{R}^{n_u^s}$ are present, the state-space model

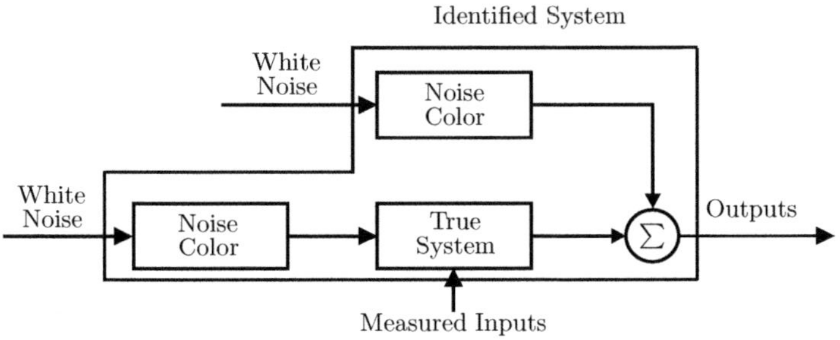

Figure 1. Block diagram of the identified system description when the unmeasured excitation is coloured noise.

(2, 4) can be extended to

$$\frac{d\boldsymbol{x}(t)}{dt} = \boldsymbol{A_c}\boldsymbol{x}(t) + \boldsymbol{B_c}\boldsymbol{u}(t) + \boldsymbol{w}(t) \qquad (17)$$

$$\boldsymbol{y}(t) = \boldsymbol{C_c}\boldsymbol{x}(t) + \boldsymbol{D_c}\boldsymbol{u}(t) + \boldsymbol{v}(t), \qquad (18)$$

where

$$\boldsymbol{w}(t) \triangleq \boldsymbol{B_c^s}\boldsymbol{u^s}(t) - \boldsymbol{B_c}\boldsymbol{u^n}(t), \qquad \boldsymbol{v}(t) \triangleq \boldsymbol{D_c^s}\boldsymbol{u^s}(t) - \boldsymbol{D_c}\boldsymbol{u^n}(t) + \boldsymbol{y^n}(t), \qquad (19)$$

$$\boldsymbol{B_c^s} = \begin{bmatrix} \boldsymbol{0} \\ \boldsymbol{M}^{-1}\boldsymbol{B_{sel2}} \end{bmatrix}, \quad \text{and} \quad \boldsymbol{D_c^s} = \boldsymbol{C_{\ddot{v}}}\boldsymbol{M}^{-1}\boldsymbol{B_{sel2}},$$

with $\boldsymbol{B_{sel2}} \in \mathbb{R}^{n_{fe} \times n_u^s}$ a selection matrix.

Discrete combined state-space model - decomposition. With a discretisation scheme such as ZOH, (17-18) can be converted to

$$\boldsymbol{x_{k+1}} = \boldsymbol{A}\boldsymbol{x_k} + \boldsymbol{B}\boldsymbol{u_k} + \boldsymbol{w_k} \qquad (20)$$

$$\boldsymbol{y_k} = \boldsymbol{C}\boldsymbol{x_k} + \boldsymbol{D}\boldsymbol{u_k} + \boldsymbol{v_k}. \qquad (21)$$

When it is assumed that the samples of $\boldsymbol{u^n}(t)$, $\boldsymbol{y^n}(t)$, and $\boldsymbol{u^s}(t)$ make up discrete-time white noise sequences, $\boldsymbol{w_k}$ and $\boldsymbol{v_k}$ are discrete-time white noise sequences as well:

$$\mathcal{E}\left(\begin{bmatrix} \boldsymbol{w_{k+l}} \\ \boldsymbol{v_{k+l}} \end{bmatrix} \begin{bmatrix} \boldsymbol{w_k} \\ \boldsymbol{v_k} \end{bmatrix}^T \right) = \begin{bmatrix} \boldsymbol{Q} & \boldsymbol{S} \\ \boldsymbol{S}^T & \boldsymbol{R} \end{bmatrix} \delta_1(l), \qquad (22)$$

where $\delta_1(\square)$ is the unit impulse function, i.e., $\delta_1(0) = 1$ and $\delta_1(\square) = 0$ if $\square \neq 0$. With the decomposition of the states and outputs in a deterministic and a stochastic part,

$$\boldsymbol{x}_k = \boldsymbol{x}_k^d + \boldsymbol{x}_k^s \qquad \text{and} \qquad \boldsymbol{y}_k = \boldsymbol{y}_k^d + \boldsymbol{y}_k^s,$$

(20-20) is decomposed into a deterministic subsystem

$$\boldsymbol{x}_{k+1}^d = \boldsymbol{A}\boldsymbol{x}_k^d + \boldsymbol{B}\boldsymbol{u}_k \tag{23}$$

$$\boldsymbol{y}_k^d = \boldsymbol{C}\boldsymbol{x}_k^d + \boldsymbol{D}\boldsymbol{u}_k, \tag{24}$$

and a stochastic subsystem

$$\boldsymbol{x}_{k+1}^s = \boldsymbol{A}\boldsymbol{x}_k^s + \boldsymbol{w}_k \tag{25}$$

$$\boldsymbol{y}_k^s = \boldsymbol{C}\boldsymbol{x}_k^s + \boldsymbol{v}_k. \tag{26}$$

Just as for the deterministic subsystem, the eigenvalue decomposition of $\boldsymbol{A}$, (14), decouples the stochastic subsystem:

$$\boldsymbol{x}_{m,k+1}^s = \boldsymbol{\Lambda}_d \boldsymbol{x}_{m,k}^s + \boldsymbol{w}_{m,k} \tag{27}$$

$$\boldsymbol{y}_k^s = \boldsymbol{\Phi}_d \boldsymbol{x}_{m,k}^s + \boldsymbol{v}_{m,k}, \tag{28}$$

where $\boldsymbol{w}_{m,k} = \boldsymbol{\Psi}_d^{-1}\boldsymbol{w}_k$.

Correlation matrices. The following definitions of correlation matrices of the stochastic subsystem (25-26) and the relationships between them are very frequently used, both in solving forward and inverse problems:

$$\begin{aligned}
\boldsymbol{\Sigma}^s &\triangleq \mathcal{E}\left(\boldsymbol{x}_{k+1}^s \boldsymbol{x}_{k+1}^{s\,T}\right) = \mathcal{E}\left(\left(\boldsymbol{A}\boldsymbol{x}_k^s + \boldsymbol{w}_k\right)\left(\boldsymbol{A}\boldsymbol{x}_k^s + \boldsymbol{w}_k\right)^T\right) \\
&= \boldsymbol{A}\boldsymbol{\Sigma}^s\boldsymbol{A}^T + \boldsymbol{Q} &&(29)\\
\boldsymbol{G} &\triangleq \mathcal{E}\left(\boldsymbol{x}_{k+1}^s \boldsymbol{y}_k^{s\,T}\right) = \mathcal{E}\left(\left(\boldsymbol{A}\boldsymbol{x}_k^s + \boldsymbol{w}_k\right)\left(\boldsymbol{C}\boldsymbol{x}_k^s + \boldsymbol{v}_k\right)^T\right) \\
&= \boldsymbol{A}\boldsymbol{\Sigma}^s\boldsymbol{C}^T + \boldsymbol{S} &&(30)\\
\boldsymbol{\Lambda}_l &\triangleq \mathcal{E}\left(\boldsymbol{y}_{k+l}^s \boldsymbol{y}_k^{s\,T}\right) \\
&= \begin{cases}
\mathcal{E}\left(\left(\boldsymbol{C}\boldsymbol{x}_k^s + \boldsymbol{v}_k\right)\left(\boldsymbol{C}\boldsymbol{x}_k^s + \boldsymbol{v}_k\right)^T\right) = \boldsymbol{C}\boldsymbol{\Sigma}^s\boldsymbol{C}^T + \boldsymbol{R}, & l = 0 \\
\mathcal{E}\left(\left(\boldsymbol{C}\boldsymbol{x}_{k+l}^s + \boldsymbol{v}_{k+l}\right)\boldsymbol{y}_k^{s\,T}\right) = \boldsymbol{C}\boldsymbol{A}^{l-1}\boldsymbol{G}, & l > 0 \\
\mathcal{E}\left(\boldsymbol{y}_k^s \boldsymbol{y}_{k+l}^{s\,T}\right)^T = \boldsymbol{G}^T\left(\boldsymbol{A}^{l-1}\right)^T\boldsymbol{C}^T, & l < 0
\end{cases} &&(31)
\end{aligned}$$

where, as before, stationarity and ergodicity of all stochastic sequences was assumed, as well as the fact that $\boldsymbol{x}_k$ is independent of $\boldsymbol{w}_k$ and $\boldsymbol{v}_k$, which follows immediately from equations (20-21).

Positive output correlation function and positive output power spectral density. When comparing (12) and (31), it follows that the impulse response function of the deterministic subsystem (23-24) and the positive output correlation function of the stochastic subsystem (25-26) have the same structure: the quadruplet $(\boldsymbol{A}, \boldsymbol{G}, \boldsymbol{C}, \boldsymbol{\Lambda_0}/2)$ of the stochastic positive correlation function plays the role of the quadruplet $(\boldsymbol{A}, \boldsymbol{B}, \boldsymbol{C}, \boldsymbol{D})$ of the deterministic impulse response function. Since the transfer function of the deterministic subsystem is the z-transform of its impulse response function and the positive power spectral density of the stochastic subsystem is the z-transform of its positive correlation function, it follows from (13) that

$$\boldsymbol{S}^{+}_{\boldsymbol{y}^s\boldsymbol{y}^s}(z) = \boldsymbol{C}(z\boldsymbol{I} - \boldsymbol{A})^{-1}\boldsymbol{G} + \frac{1}{2}\boldsymbol{\Lambda_0}. \tag{32}$$

State estimation - the reference-based Kalman filter. Because the stochastic terms $\boldsymbol{w_k}$ and $\boldsymbol{v_k}$ are unknown, the state $\boldsymbol{x_k}$ cannot be calculated exactly from (20-21). Nevertheless, a one-step ahead estimate of $\boldsymbol{x^s_{k+1}}$ can be calculated if the current output vector $\boldsymbol{y_k}$ is known. From $\boldsymbol{y_k}$, $\boldsymbol{y^s_k}$ is obtained after subtracting the deterministic part $\boldsymbol{y^d_k}$. The Kalman filter offers a technique for determining the optimal linear estimate because the estimator is unbiased and has minimum variance (Kalman, 1960).

Reynders and De Roeck (2008) worked out a reference-based Kalman filter. Reference outputs form a subset (containing n_r elements) of the complete set of n_y outputs:

$$\boldsymbol{y^{ref}_k} \triangleq \boldsymbol{S^r}\boldsymbol{y_k},$$

where $\boldsymbol{S^r} \in \mathbb{N}^{n_r \times n_y}$ is a selection matrix. In order not to lose information, it is important that any mode of interest is clearly present in at least one reference output. Good reference output candidates are usually driving point outputs, or, in case of measurements performed in different setups, the output channels common to each setup. The reference outputs can be written as

$$\boldsymbol{y^{ref}_k} = \boldsymbol{S^r}\boldsymbol{C}\boldsymbol{x_k} + \boldsymbol{S^r}\boldsymbol{D}\boldsymbol{u_k} + \boldsymbol{S^r}\boldsymbol{v_k} \triangleq \boldsymbol{C^{ref}}\boldsymbol{x_k} + \boldsymbol{D^{ref}}\boldsymbol{u_k} + \boldsymbol{v^{ref}_k}.$$

Reynders and De Roeck (2008) showed that the optimal linear one-step-ahead state estimate $\hat{\boldsymbol{x}}^s_{k+1}$ and the reference-based non-stationary Kalman filter $\boldsymbol{K_k}$ can be calculated from the following set of equations, using the current state estimate $\hat{\boldsymbol{x}}^s_k$ and the measured current reference output vector

$\boldsymbol{y}_k^{s,ref}$ and assuming $\hat{\boldsymbol{x}}_0^s = \mathcal{E}\left(\boldsymbol{x}_0^s\right)$:

$$\begin{aligned}
\hat{\boldsymbol{x}}_{k+1}^s &= (\boldsymbol{A} - \boldsymbol{K}_k\boldsymbol{C}^{ref})\hat{\boldsymbol{x}}_k^s + \boldsymbol{K}_k\boldsymbol{y}_k^{s,ref} && (33)\\
\boldsymbol{K}_k &= (\boldsymbol{A}\boldsymbol{P}_k\boldsymbol{C}^{ref^T} + \boldsymbol{S}^{ref})(\boldsymbol{R}^{ref} + \boldsymbol{C}^{ref}\boldsymbol{P}_k\boldsymbol{C}^{ref^T})^{-1} && (34)\\
\boldsymbol{P}_{k+1} &= \boldsymbol{A}\boldsymbol{P}_k\boldsymbol{A}^T + \boldsymbol{Q} - (\boldsymbol{A}\boldsymbol{P}_k\boldsymbol{C}^{ref^T} + \boldsymbol{S}^{ref})\\
&\quad\cdot(\boldsymbol{R}^{ref} + \boldsymbol{C}^{ref}\boldsymbol{P}_k\boldsymbol{C}^{ref^T})^{-1}(\boldsymbol{A}\boldsymbol{P}_k\boldsymbol{C}^{ref^T} + \boldsymbol{S}^{ref})^T && (35)
\end{aligned}$$

where $\boldsymbol{R}^{ref} = \boldsymbol{S}^r\boldsymbol{R}\boldsymbol{S}^{r^T}$, $\boldsymbol{S}^{ref} = \boldsymbol{S}\boldsymbol{S}^{r^T}$ and $\boldsymbol{P}_k$ is correlation of the state estimation error:

$$\boldsymbol{e}_k \triangleq \boldsymbol{x}_k - \hat{\boldsymbol{x}}_k = \boldsymbol{x}_k^s - \hat{\boldsymbol{x}}_k^s, \qquad \boldsymbol{P}_k \triangleq \mathcal{E}\left(\boldsymbol{e}_k\boldsymbol{e}_k^T\right). \tag{36}$$

After solving (35) for $\boldsymbol{P}_k$, $\boldsymbol{K}_k$ can be calculated with (34), and the optimal estimate of $\hat{\boldsymbol{x}}_{k+1}$ is then obtained using (33). It is usually assumed that $\hat{\boldsymbol{x}}_0^s = \boldsymbol{0}$ and $\boldsymbol{P}_0 = \boldsymbol{0}$, see Van Overschee and De Moor (1996), but other choices can be made.

Since the stochastic part of the model (20-21) is driven by stationary random processes, the Kalman filter $\boldsymbol{K}_k$ is equal to the time-invariant matrix $\boldsymbol{K}$ when no initial conditions are taken into account (as is the case when taking a double-sided z transform), which is formally shown in (Anderson and Moore, 1979, ch. 4). In this case, $\boldsymbol{P}_k$ equals the time-invariant matrix $\boldsymbol{P}$, and the subscript k in (34) and (35) disappears. Since the effect of initial conditions dies out for a stable system when $k \to \infty$, it also follows that $\boldsymbol{K}_k \to \boldsymbol{K},\ k \to \infty$.

Reference-based forward innovation model. The (reference-based) forward innovation $\boldsymbol{e}_k^{(ref),f}$ is defined as

$$\boldsymbol{e}_k^{(ref),f} \triangleq \boldsymbol{y}_k^{(ref)} - \boldsymbol{C}^{(ref)}\hat{\boldsymbol{x}}_k - \boldsymbol{D}^{(ref)}\boldsymbol{u}_k.$$

By decomposing the Kalman filter state $\hat{\boldsymbol{x}}_{k+1}$ into its deterministic and stochastic components, one has

$$\begin{aligned}
\hat{\boldsymbol{x}}_{k+1} &= \boldsymbol{x}_{k+1}^d + \hat{\boldsymbol{x}}_{k+1}^s = \boldsymbol{A}\boldsymbol{x}_k^d + \boldsymbol{B}\boldsymbol{u}_k + (\boldsymbol{A} - \boldsymbol{K}_k\boldsymbol{C}^{ref})\hat{\boldsymbol{x}}_k^s + \boldsymbol{K}_k\boldsymbol{y}_k^{ref,s}\\
&= \boldsymbol{A}\hat{\boldsymbol{x}}_k + \boldsymbol{B}\boldsymbol{u}_k + \boldsymbol{K}_k(\boldsymbol{y}_k^{ref} - \boldsymbol{C}^{ref}\hat{\boldsymbol{x}}_k - \boldsymbol{D}^{ref}\boldsymbol{u}_k)\\
&= \boldsymbol{A}\hat{\boldsymbol{x}}_k + \boldsymbol{B}\boldsymbol{u}_k + \boldsymbol{K}_k\boldsymbol{S}^r\boldsymbol{e}_k^f, && (37)
\end{aligned}$$

where the second equality follows from (33). Obviously, one has

$$\begin{aligned}
\boldsymbol{y}_k &= \boldsymbol{C}\hat{\boldsymbol{x}}_k + \boldsymbol{D}\boldsymbol{u}_k + (\boldsymbol{y}_k - \boldsymbol{C}\hat{\boldsymbol{x}}_k - \boldsymbol{D}\boldsymbol{u}_k)\\
&= \boldsymbol{C}\hat{\boldsymbol{x}}_k + \boldsymbol{D}\boldsymbol{u}_k + \boldsymbol{e}_k^f. && (38)
\end{aligned}$$

Equations (37-38) represent the reference-based forward innovation model of the structure. It follows that for large k, the Kalman filter states constitute a particular state-space basis, since when $k \to \infty$, $\boldsymbol{K_k} \to \boldsymbol{K}$ as explained above, and $(\boldsymbol{K S^r e^f_k})$ and $(\boldsymbol{e^f_k})$ are system and measurement noise sequences, respectively, that obey the discrete-time white noise assumption.

3 Subspace identification: principles and strategies

3.1 Introduction

System Identification can be defined as the field of study where models are fitted to measured data. It involves three basic entities (Ljung, 1999, p. 13):

- designing an experiment that is as informative as possible and obtaining the data;
- choosing a set of candidate models, like for instance a stochastic state space model structure;
- choosing an identification method, i.e., a strategy for determining the model in the set that explains the data 'best', and an identification algorithm, i.e., a numerical algorithm that calculates the actual estimate.

A myriad of system identification algorithms is available from the literature, but, as shown by Ljung (1999), they can be considered as particular implementations of just a few general ideas. This section introduces the main ideas that lie behind the class of subspace identification algorithms for fitting the discrete-time state-space model of a vibrating structure, presented in the previous section, to measured data.

In the remainder of this chapter, the following assumptions are made.

Assumption 3.1. The stochastic output sequence $\boldsymbol{y^s_k}$ is generated by (25-26). The white process and measurement noise sequences are not identically zero.

Assumption 3.2. When exogenous forces are measured, the force sequence $(\boldsymbol{u_k})$, $k = 0, \dots, N-1$, is observed free of noise and it is persistently exciting of order $\geq 2\imath + n$. The latter is a technical assumption which ensures that a block Hankel input matrix with $2\imath + n$ block rows has full row rank. The response sequence $(\boldsymbol{y^d_k})$ due to the input sequence $(\boldsymbol{u_k})$ is generated by the deterministic subsystem (23-24), which is controllable.

Assumption 3.3. When exogenous forces are measured, they are uncorrelated with the stochastic outputs, i.e.,

$$\forall k, l : \mathcal{E}\left(\boldsymbol{y^s_k u^T_l}\right) = \boldsymbol{0}.$$

3.2 System realisation

Introduction. The realisation problem was originally defined by Kalman (1963) as the problem of identifying a linear dynamic system from a non-parametric impulse response sequence that was generated by this system. Ho and Kalman (1966) found an efficient way to solve this problem, beginning with a finite-dimensional block Hankel matrix composed of noise-free impulse responses and ending with the system matrices of a deterministic state-space model. Later, Zeiger and McEwen (1974) and Kung (1978) proposed to perform the factorisation step of the Ho-Kalman procedure by singular value decomposition, where only the significant singular values and the corresponding singular vectors are retained. This truncation enabled one to deal with noise on the impulse responses, which is always present in experimental data. Another variant called the *Eigensystem Realisation Algorithm* (ERA), developed by Juang and Pappa (1985), introduced the idea of reference outputs into the realisation procedure. Later, Juang et al. (1988) proposed a variant of ERA that starts from correlations of impulse responses instead of the impulse response matrices themselves. This version is called the ERA with data correlations (ERA/DC). Akaike (1974) extended the realisation theory to stochastic systems and also gave a stochastic interpretation of the Ho-Kalman algorithm.

Stochastic system realisation. Suppose a nonparametric estimate of the stochastic output correlation sequence ($\boldsymbol{\Lambda}_k^{ref}$) is available. From (31), it follows that these correlation matrices can be parametrised as follows:

$$\boldsymbol{\Lambda}_k^{ref} = \boldsymbol{C}\boldsymbol{A}^{k-1}\boldsymbol{G}^{ref}, \; k \geq 1.$$

Stochastic realisation starts with gathering the correlation matrices in a block Hankel matrix:

$$\boldsymbol{L}_{1|\imath}^{ref} \triangleq \begin{bmatrix} \boldsymbol{\Lambda}_1^{ref} & \boldsymbol{\Lambda}_2^{ref} & \dots & \boldsymbol{\Lambda}_\imath^{ref} \\ \boldsymbol{\Lambda}_2^{ref} & \boldsymbol{\Lambda}_3^{ref} & \dots & \boldsymbol{\Lambda}_{\imath+1}^{ref} \\ \vdots & \vdots & \dots & \vdots \\ \boldsymbol{\Lambda}_\imath^{ref} & \boldsymbol{\Lambda}_{\imath+1}^{ref} & \dots & \boldsymbol{\Lambda}_{2\imath-1}^{ref} \end{bmatrix}, \tag{39}$$

where $\imath$ is chosen in such a way that, if n is the expected system order, $n_y \imath \geq n$, $n_r \imath \geq n$ and $\imath \geq 2$. The block Hankel matrix decomposes into the extended observability matrix $\boldsymbol{\mathcal{O}}_\imath$ and the reference-based extended stochastic controllability matrix $\mathcal{C}_i^S \in \mathbb{R}^{n_y \times i n_r}$:

$$\boldsymbol{L}_{1|i}^{ref} = \underbrace{\begin{bmatrix} \boldsymbol{C} \\ \boldsymbol{CA} \\ \dots \\ \boldsymbol{CA}^{i-1} \end{bmatrix}}_{=\mathcal{O}_i} \underbrace{\begin{bmatrix} \boldsymbol{G}^{ref} & \boldsymbol{AG}^{ref} & \dots & \boldsymbol{A}^{i-1}\boldsymbol{G}^{ref} \end{bmatrix}}_{\triangleq \mathcal{C}_i^S}. \tag{40}$$

The matrices $\mathcal{O}_i$ and $\mathcal{C}_i^S$ can be obtained from $\boldsymbol{L}_{1|i}^{ref}$, up to a similarity transformation of the $\boldsymbol{A}$ matrix, using reduced singular value decomposition:

$$\boldsymbol{L}_{1|i}^{ref} = \boldsymbol{USV}^T, \qquad \mathcal{O}_i = \boldsymbol{US}^{1/2}, \qquad \mathcal{C}_i^S = \boldsymbol{S}^{1/2}\boldsymbol{V}^T, \tag{41}$$

where $\boldsymbol{S} \in \mathbb{R}^{n \times n}$ contains only the nonzero singular values and $\boldsymbol{U} \in \mathbb{R}^{i n_y \times n}$ and $\boldsymbol{V} \in \mathbb{R}^{i n_r \times n}$ contain the corresponding singular vectors. If the stochastic output correlation matrices $\boldsymbol{\Lambda}_k^{ref}$ are exact, the number of nonzero singular values equals the system order n. If not, the system order is lower than the number of nonzero singular values. It then needs to be estimated as the number of significant singular values. In this case $\boldsymbol{S}$ contains only the significant singular values and $\boldsymbol{U}$ and $\boldsymbol{V}$ contain the corresponding singular vectors.

The $\boldsymbol{C}$ matrix can be determined as the first n_y rows of $\mathcal{O}_i$ and the $\boldsymbol{G}^{ref}$ matrix can be determined as the first n_r columns of $\mathcal{C}_i$. Different algorithms have been proposed for the determination of $\boldsymbol{A}$. The algorithm of Kung (1978) is computationally the most efficient one. It makes use of the shift structure of the matrix $\mathcal{O}_i$:

$$\boldsymbol{A} = \underline{\mathcal{O}_i}^{\dagger}\overline{\mathcal{O}_i}. \tag{42}$$

where $\underline{\mathcal{O}_i}$ is equal to $\mathcal{O}_i$ without the last n_y rows and $\overline{\mathcal{O}_i}$ is equal to $\mathcal{O}_i$ without the first n_y rows. $\square^{\dagger}$ denotes the Moore-Penrose pseudo-inverse of the matrix $\square$, see Ben-Israel and Greville (1974).

3.3 Subspace identification

Introduction. As shown in this section, subspace identification can be considered as an extension of system realisation. Instead of starting from a nonparametric impulse response or stochastic correlation function, subspace methods start directly from the measured data samples, which are put in different Hankel matrices. Projections between the data Hankel matrices then lead to a matrix from which the extended observability matrix $\mathcal{O}_i$ is derived by performing a singular value decomposition, just as in the system realisation technique discussed above. In a second step, the state-space matrices are identified using $\mathcal{O}_\imath$. Several algorithms exist for this, and some of them are discussed in the next section. The discussion hereunder is restricted to the time domain, but as shown by Cauberghe (2004), the same techniques can be employed starting from frequency-domain data.

Orthogonal and oblique projections. Subspace identification methods make extensive use of geometric projections. This paragraph provides a concise review on the topic. Consider the matrices $\boldsymbol{M_1} \in \mathbb{R}^{m_1 \times \jmath}$, $\boldsymbol{M_2} \in \mathbb{R}^{m_2 \times \jmath}$ and $\boldsymbol{M_3} \in \mathbb{R}^{m_3 \times \jmath}$. The row space spanned by the rows of $\boldsymbol{M_2}$, denoted as $\mathcal{M}_2$, is defined as the set of all possible linear combinations of the rows of $\boldsymbol{M_2}$:

$$\mathcal{M}_2 \triangleq \left\{\boldsymbol{y} \in \mathbb{R}^{\jmath} | \boldsymbol{y} = \boldsymbol{M_2}^T \boldsymbol{x},\ \forall \boldsymbol{x} \in \mathbb{R}^{m_2}\right\}.$$

The orthogonal projection of $\boldsymbol{M_1}$ onto $\mathcal{M}_2$ is defined as

$$\boldsymbol{M_1}/\mathcal{M}_2 \triangleq \boldsymbol{M_1}\boldsymbol{M_2}^T \left(\boldsymbol{M_2}\boldsymbol{M_2}^T\right)^\dagger \boldsymbol{M_2}.$$

When $\boldsymbol{M_1}/\mathcal{M}_2 = \boldsymbol{0}$, the rows of $\boldsymbol{M_1}$ and $\boldsymbol{M_2}$ are said to be orthogonal to each other. The orthogonal projection $\boldsymbol{M_1}/\mathcal{M}_2$ is graphically depicted in Figure 2 for $m_1 = m_2 = 1$ and $\jmath = 2$.

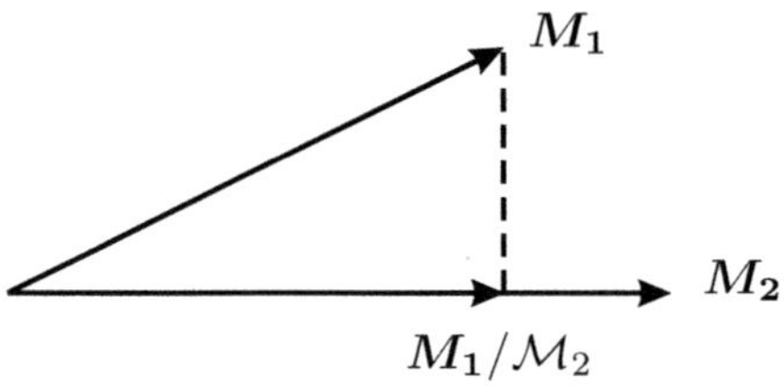

Figure 2. The orthogonal projection of $\boldsymbol{M_1}$ on the row space of $\boldsymbol{M_2}$, for the case $m_1 = m_2 = 1$ and $\jmath = 2$.

The oblique projection of $\boldsymbol{M_1}$ onto $\mathcal{M}_2$ along the row space of $\boldsymbol{M_3}$, denoted as $\mathcal{M}_3$, is defined as

$$\boldsymbol{M_1}/_{\mathcal{M}_3}\mathcal{M}_2 \triangleq \boldsymbol{M_1} \begin{bmatrix} \boldsymbol{M_2} \\ \boldsymbol{M_3} \end{bmatrix}^T \left(\begin{bmatrix} \boldsymbol{M_2} \\ \boldsymbol{M_3} \end{bmatrix} \begin{bmatrix} \boldsymbol{M_2} \\ \boldsymbol{M_3} \end{bmatrix}^T \right)^\dagger \begin{bmatrix} \boldsymbol{M_2} \\ \boldsymbol{0} \end{bmatrix}.$$

From the definition, it follows immediately that $\boldsymbol{M_3}/_{\mathcal{M}_3}\mathcal{M}_2 = \boldsymbol{0}$. The oblique projection $\boldsymbol{M_1}/_{\mathcal{M}_3}\mathcal{M}_2$ is graphically depicted in Figure 3 for $m_1 = m_2 = m_3 = 1$ and $\jmath = 2$.

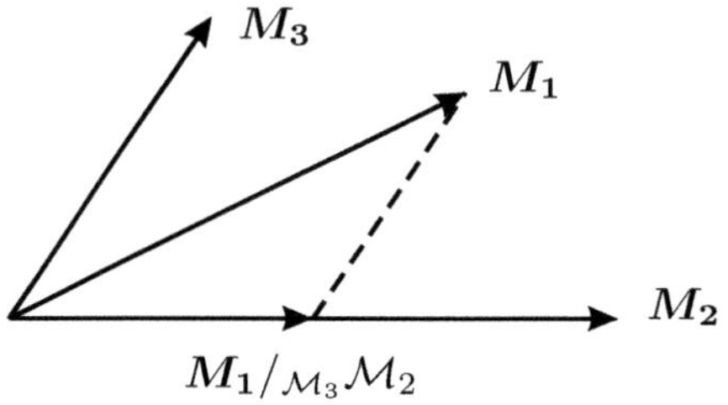

Figure 3. The oblique projection of M_1 on the row space of M_2 along M_3, for the case $m_1 = m_2 = m_3 = 1$ and $j = 2$.

Matrix input-output equations. From (23-24), one has the following relationship for the deterministic subsystem of a combined deterministic-stochastic state-space model:

$$\begin{bmatrix} y_0^d \\ y_1^d \\ \vdots \\ y_{i-1}^d \end{bmatrix} = \underbrace{\begin{bmatrix} C \\ CA \\ \vdots \\ CA^{i-1} \end{bmatrix}}_{=\mathcal{O}_i} x_0^d + \underbrace{\begin{bmatrix} D & 0 & \dots & 0 \\ CB & D & \dots & 0 \\ \vdots & \vdots & \dots & 0 \\ CA^{i-2}B & CA^{i-3}B & \dots & D \end{bmatrix}}_{\triangleq \mathcal{F}_i^d} \begin{bmatrix} u_0 \\ u_1 \\ \vdots \\ u_i \end{bmatrix}. \quad (43)$$

Define a block Hankel matrix of outputs as

$$Y_{k_1|k_2} \triangleq \begin{bmatrix} y_{k_1} & y_{k_1+1} & \cdots & y_{k_1+j-1} \\ y_{k_1+1} & y_{k_1+2} & \cdots & y_{k_1+j} \\ \vdots & \vdots & \dots & \vdots \\ y_{k_2} & y_{k_2+1} & \cdots & y_{k_2+j-1} \end{bmatrix},$$

a block Hankel matrix of inputs as

$$U_{k_1|k_2} \triangleq \begin{bmatrix} u_{k_1} & u_{k_1+1} & \cdots & u_{k_1+j-1} \\ u_{k_1+1} & u_{k_1+2} & \cdots & u_{k_1+j} \\ \vdots & \vdots & \dots & \vdots \\ u_{k_2} & u_{k_2+1} & \cdots & u_{k_2+j-1} \end{bmatrix},$$

and a block row vector of states as

$$X_{k_1|k_1} \triangleq \begin{bmatrix} x_{k_1} & x_{k_1+1} & \dots & x_{k_1+j-1} \end{bmatrix}.$$

With these definitions, (43) generalises to

$$\boldsymbol{Y}_{k_1|k_2} = \boldsymbol{\mathcal{O}}_{k_2-k_1+1}\boldsymbol{X}^d_{k_1|k_1} + \boldsymbol{\mathcal{F}}^d_\imath \boldsymbol{U}_{k_1|k_2} + \boldsymbol{Y}^s_{k_1|k_2}, \tag{44}$$

where $\boldsymbol{Y}^s_{k_1|k_2}$ denotes the stochastic part of $\boldsymbol{Y}_{k_1|k_2}$ and $\boldsymbol{X}^d_{k_1|k_1}$ the deterministic part of $\boldsymbol{X}_{k_1|k_1}$. As a special case, one has

$$\boldsymbol{Y}_{\imath|2\imath-1} = \boldsymbol{\mathcal{O}}_\imath \boldsymbol{X}^d_{\imath|\imath} + \boldsymbol{\mathcal{F}}^d_\imath \boldsymbol{U}_{\imath|2\imath-1} + \boldsymbol{Y}^s_{\imath|2\imath-1}. \tag{45}$$

As noted by Goethals (2005), the primary objective of subspace identification is to extract the extended observability matrix $\boldsymbol{\mathcal{O}}_\imath$ from this equation. The system matrices are obtained in a second step.

Obtaining the observability matrix Starting from the forward innovation description (37-38), one has

$$\boldsymbol{Y}_{\imath|2\imath-1} = \boldsymbol{\mathcal{O}}_\imath \hat{\boldsymbol{X}}_{\imath|\imath} + \boldsymbol{\mathcal{F}}^d_\imath \boldsymbol{U}_{\imath|2\imath-1} + \boldsymbol{Y}^f_{\imath|2\imath-1}, \tag{46}$$

where $\boldsymbol{Y}^f_{\imath|2\imath-1}$ contains the contributions of the forward innovations:

$$\boldsymbol{Y}^f_{k_1|k_2} \triangleq \underbrace{\begin{bmatrix} \boldsymbol{I}_{n_y} & \boldsymbol{0} & \dots & \boldsymbol{0} \\ \boldsymbol{C}_s\boldsymbol{K}^s_{k_1}\boldsymbol{S}^r & \boldsymbol{I} & \dots & \boldsymbol{0} \\ \vdots & \vdots & \dots & \vdots \\ \boldsymbol{C}_s\boldsymbol{A}_s^{k_2-k_1-1}\boldsymbol{K}^s_{k_2-1}\boldsymbol{S}^r & \boldsymbol{C}_s\boldsymbol{A}_s^{k_2-k_1-2}\boldsymbol{K}^s_{k_2-2}\boldsymbol{S}^r & \dots & \boldsymbol{I}_{n_y} \end{bmatrix}}_{\triangleq \boldsymbol{\mathcal{F}}^s_{k_2-k_1+1}} \boldsymbol{E}^f_{k_1|k_2} \tag{47}$$

and a block Hankel matrix of forward innovations is defined as

$$\boldsymbol{E}^f_{k_1|k_2} \triangleq \begin{bmatrix} \boldsymbol{e}^f_{k_1} & \boldsymbol{e}^f_{k_1+1} & \dots & \boldsymbol{e}^f_{k_1+\jmath-1} \\ \boldsymbol{e}^f_{k_1+1} & \boldsymbol{e}^f_{k_1+2} & \dots & \boldsymbol{e}^f_{k_1+\jmath} \\ \vdots & \vdots & \dots & \vdots \\ \boldsymbol{e}^f_{k_2} & \boldsymbol{e}^f_{k_2+1} & \dots & \boldsymbol{e}^f_{k_2+\jmath-1} \end{bmatrix}.$$

$\hat{\boldsymbol{X}}_{\imath|\imath}$ contains in each column a non-stationary reference-based Kalman filter state, see Section 2.6, that is estimated from the corresponding columns of $\boldsymbol{U}_{0|\imath-1}$ and $\boldsymbol{Y}^{ref}_{0|\imath-1}$. Since the Kalman filter is linear, it follows that $\hat{\boldsymbol{X}}_{\imath|\imath}$ lies in the row space of $\boldsymbol{U}_{0|\imath-1}$, denoted as $\mathscr{U}_{0|\imath-1}$, and the row space of $\boldsymbol{Y}^{ref}_{0|\imath-1}$, denoted as $\mathscr{Y}^{ref}_{0|\imath-1}$. As the row space of $\boldsymbol{U}_{0|\imath-1}$ is a subset of the row space of $\boldsymbol{U}_{0|2\imath-1}$, which is denoted as $\mathscr{U}_{0|2\imath-1}$, one has

$$\text{row space}\left(\hat{\boldsymbol{X}}_{\imath|\imath}\right) \subseteq \left(\mathscr{U}_{0|\imath-1} \vee \mathscr{Y}^{ref}_{0|\imath-1}\right) \subseteq \left(\mathscr{U}_{0|2\imath-1} \vee \mathscr{Y}^{ref}_{0|\imath-1}\right),$$

where $\vee$ denotes joint row space. Given that the forward innovations are uncorrelated with the measured inputs and the outputs that were used for the estimation of the Kalman filter states, projection of both sides of (46) onto the row space spanned by the inputs and the 'past' outputs yields

$$\underset{j\to\infty}{\text{a.s.lim}} \boldsymbol{Y}_{i|2i-1} / \left(\mathscr{U}_{0|2i-1} \vee \mathscr{Y}^{ref}_{0|i-1} \right) = \lim_{j\to\infty} \boldsymbol{\mathcal{O}}_i \hat{\boldsymbol{X}}_{i|i} + \boldsymbol{\mathcal{F}}^d_i \boldsymbol{U}_{i|2i-1}. \tag{48}$$

The *almost-sure limit*, denoted as a.s.lim, is a stochastic limit. A stochastic variable is said to converge almost surely to a certain value, when this convergence holds for almost all realisations of the stochastic variable, except for those with probability zero (Dougherty, 1999). When the orthogonal projection is replaced by an oblique projection along $\mathscr{U}_{i|2i-1}$, the second term of the right hand side drops:

$$\underset{j\to\infty}{\text{a.s.lim}} \boldsymbol{Y}_{i|2i-1} /_{\mathscr{U}_{i|2i-1}} \left(\mathscr{U}_{0|2i-1} \vee \mathscr{Y}^{ref}_{0|i-1} \right) = \lim_{j\to\infty} \boldsymbol{\mathcal{O}}_i \hat{\boldsymbol{X}}_{i|i}.$$

This is the rationale behind the so-called N4SID (*Numerical algorithms for Subspace State Space System IDentification*) approach of Van Overschee and De Moor (1994a). The oblique projection is depicted in Figure 4.

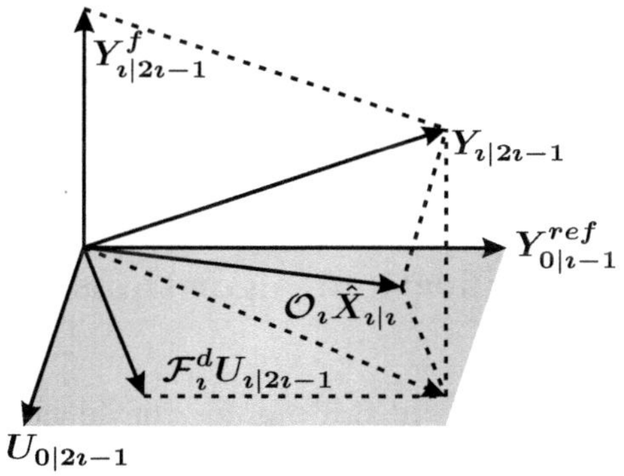

Figure 4. Subspace identification, N4SID approach: graphical representation of the estimation of $\boldsymbol{\mathcal{O}}_i \hat{\boldsymbol{X}}_{i|i}$ from $\boldsymbol{Y}_{i|2i-1}$ by oblique projection.

By introducing left and right weighting matrices $\boldsymbol{W_1}$ and $\boldsymbol{W_2}$,

$$\underset{j\to\infty}{\text{a.s.lim}} \boldsymbol{W_1} \boldsymbol{Y}_{i|2i-1} /_{\mathscr{U}_{i|2i-1}} \left(\mathscr{U}_{0|2i-1} \vee \mathscr{Y}^{ref}_{0|i-1} \right) \boldsymbol{W_2} = \boldsymbol{W_1} \lim_{j\to\infty} \boldsymbol{\mathcal{O}}_i \hat{\boldsymbol{X}}_{i|i} \boldsymbol{W_2}, \tag{49}$$

it was shown by Van Overschee and De Moor (1994b) that other subspace methods, such as the *Past-Outputs Multi-variable Output-Error State sPace* (PO-MOESP) method of Verhaegen (1994), and the *Canonical Variate Analysis* (CVA) approach of Larimore (1990), fall into this framework. From a singular value decomposition of the left hand side of this equation, which can be calculated from measured data only, $\boldsymbol{\mathcal{O}}_\imath$ can be determined up to a nonsingular right factor, after left multiplication with $\boldsymbol{W}_1^{-1}$.

Output-only case. When no inputs are measured, (49) simplifies to the following orthogonal projection:

$$\text{a.s.}\lim_{\jmath\to\infty} \boldsymbol{W}_1 \boldsymbol{Y}_{\imath|2\imath-1} / \mathscr{Y}^{ref}_{0|\imath-1} \boldsymbol{W}_2 = \boldsymbol{W}_1 \lim_{\jmath\to\infty} \boldsymbol{\mathcal{O}}_\imath \hat{\boldsymbol{X}}_{\imath|\imath} \boldsymbol{W}_2. \tag{50}$$

The projection is graphically depicted in Figure 5. The choice of the weighting matrices is further discussed in Section 4.3.

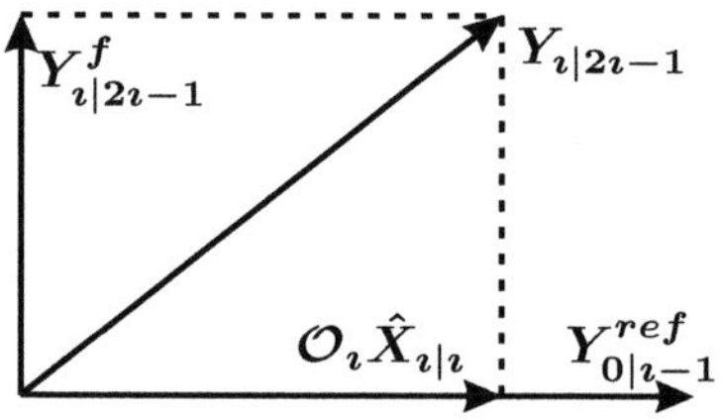

Figure 5. Output-only subspace identification: graphical representation of the estimation of $\boldsymbol{\mathcal{O}}_\imath \hat{\boldsymbol{X}}_{\imath|\imath}$ from $\boldsymbol{Y}_{\imath|2\imath-1}$ by orthogonal projection.

4 Subspace identification: algorithms

4.1 Introduction

Three powerful subspace algorithms for the identification of a state-space description from measured data are discussed. In Section 4.2, the *REFerence-based COVariance-driven Stochastic Subspace Identification* (SSI-cov/ref) algorithm is treated. It combines the nonparametric estimation of output correlations with the stochastic realisation algorithm presented in Section 3.2. Advantages of this output-only algorithm are its conceptual simplicity, and the ability to compute the *Probability Density Function* (PDF) of the identified system parameters.

The *REFerence-based DATA-driven Stochastic Subspace Identification* (SSI-data/ref) algorithm is presented in Section 4.3. It is also an output-only system identification algorithm, whose main advantage is an optimal statistical performance when the weighting matrices are properly chosen.

The *REFerence-based DATA-driven Combined deterministic-stochastic Subspace Identification* (CSI-data/ref) algorithm is discussed in Section 4.4. It allows to use exogenous forces in addition to the unmeasured operational excitation. Since these measured forces are exactly known, they reduce the uncertainty of the identified system description compared to the output-only case.

4.2 Covariance-driven stochastic subspace identification

Nonparametric estimation of output correlations. Denote $\boldsymbol{\mathcal{L}}^{ref}_{1|2\imath-1}$ as the matrix containing stacked stochastic output correlation matrices, i.e.,

$$\boldsymbol{\mathcal{L}}^{ref}_{1|2\imath-1} \triangleq \left[\boldsymbol{\Lambda}_1^{ref^T} \quad \boldsymbol{\Lambda}_2^{ref^T} \quad \dots \quad \boldsymbol{\Lambda}_{2\imath-1}^{ref\ T} \right]^T .$$

It can be easily estimated from the measured data:

$$\hat{\boldsymbol{\mathcal{L}}}^{ref}_{1|2\imath-1} \triangleq \frac{1}{\jmath} \hat{\boldsymbol{Y}}_{1|2\imath-1} \hat{\boldsymbol{Y}}^{ref^T}_{0|0} .$$

In order to derive the distribution of the correlation estimates, $\hat{\boldsymbol{Y}}_{1|2\imath-1}$ and $\hat{\boldsymbol{Y}}^{ref}_{0|0}$ are divided into n_B blocks that contain each $\jmath_B = \text{floor}(\jmath/n_B)$ columns, where floor($\square$) takes the nearest lower integer of the real number $\square$. The blocks are indicated as $\hat{\boldsymbol{Y}}_{1|2\imath-1,k}$ and $\hat{\boldsymbol{Y}}^{ref}_{0|0,k}$, respectively. If $\jmath = \jmath_B n_B$, one has[2]

$$\hat{\boldsymbol{\mathcal{L}}}^{ref}_{1|2\imath-1} = \frac{1}{\jmath} \hat{\boldsymbol{Y}}^{s}_{1|2\imath-1} \hat{\boldsymbol{Y}}^{s,ref^T}_{0|0} = \frac{1}{n_B} \sum_{k=1}^{n_B} \underbrace{\frac{1}{\jmath_B} \hat{\boldsymbol{Y}}_{1|2\imath-1,k} \hat{\boldsymbol{Y}}^{ref\ T}_{0|0,k}}_{\triangleq \hat{\boldsymbol{\mathcal{L}}}^{ref}_{1|2\imath-1,k}} . \tag{51}$$

When $\jmath_B$ is large enough in order for the $\hat{\boldsymbol{\mathcal{L}}}^{ref}_{1|2\imath-1,k}$ to be considered as independent samples, the central limit theorem ensures that $\hat{\boldsymbol{\mathcal{L}}}^{ref}_{1|2\imath-1}$ converges in law to a normal distribution when $n_B \rightarrow \infty$. The covariance of $\text{vec}\left(\hat{\boldsymbol{\mathcal{L}}}^{ref}_{1|2\imath-1}\right)$ can be calculated as the covariance of the sample mean:

$$\boldsymbol{\Sigma}_{\hat{\boldsymbol{\mathcal{L}}}^{ref}_{1|2\imath-1}} = \frac{1}{n_B(n_B-1)} \sum_{k=1}^{n_B} \text{vec}\left(\Delta\hat{\boldsymbol{\mathcal{L}}}^{ref}_{1|2\imath-1,k}\right) \text{vec}\left(\Delta\hat{\boldsymbol{\mathcal{L}}}^{ref}_{1|2\imath-1,k}\right)^T . \tag{52}$$

The operator vec($\square$) stacks the columns of the matrix $\square$ on top of each other.

[2] If $\jmath \neq \jmath_B n_B$, the last block can contain the additional $\jmath - \jmath_B n_B$ columns. The derivation still holds when this block is scaled with $\jmath_B + (\jmath - \jmath_B n_B)$ instead of $\jmath_B$.

The SSI-cov/ref algorithm. When the estimate $\hat{\mathcal{L}}^{ref}_{1|2i-1}$, is used to build the block Hankel matrix $\boldsymbol{L}^{ref}_{1|i}$ of the stochastic realisation algorithm presented in Section 3.2, the result makes up an output-only system identification algorithm, called the *(REFerence-based) COVariance-driven Stochastic Subspace Identification* (SSI-cov(/ref)) algorithm[3].

Asymptotic distribution of the estimates. Due to the noise on the measurement data and the fact that only a finite number of samples are available, only approximative estimates $\hat{\boldsymbol{\Lambda}}^{ref}_{0}$ and $\hat{\boldsymbol{L}}^{ref}_{1|i}$ for the matrices $\boldsymbol{\Lambda}^{ref}_{0}$ and $\boldsymbol{L}^{ref}_{1|i}$ are available. In addition, the number of nonzero singular values of $\hat{\boldsymbol{L}}^{ref}_{1|i}$ is larger than the system order n, as explained in Section 3.2. Therefore, the realisation algorithm does not yield the exact system matrices $(\boldsymbol{A}, \boldsymbol{G}^{ref}, \boldsymbol{C}, \boldsymbol{\Lambda}^{ref}_{0}/2)$, but only estimates $(\hat{\boldsymbol{A}}, \hat{\boldsymbol{G}}^{ref}, \hat{\boldsymbol{C}}, \hat{\boldsymbol{\Lambda}}^{ref}_{0}/2)$. If $\hat{\boldsymbol{L}}^{ref}_{1|i}$ is accurate, its estimation error, defined as

$$\Delta \boldsymbol{L}^{ref}_{1|i} \triangleq \hat{\boldsymbol{L}}^{ref}_{1|i} - \boldsymbol{L}^{ref}_{1|i},$$

is small compared to $\boldsymbol{L}^{ref}_{1|i}$ and $\hat{\boldsymbol{L}}^{ref}_{1|i}$, and the influence of this error on the system realisation $(\hat{\boldsymbol{A}}, \hat{\boldsymbol{G}}^{ref}, \hat{\boldsymbol{C}}, \hat{\boldsymbol{\Lambda}}_{0}/2)$ can be investigated using a first-order sensitivity analysis, as shown below.

From (39), one has

$$\mathrm{vec}(\boldsymbol{L}_{1|i}) = \boldsymbol{S}_1 \mathrm{vec}(\boldsymbol{\mathcal{L}}^{ref}_{0|2i-1}), \qquad \mathrm{vec}(\boldsymbol{\Lambda}_0) = \boldsymbol{S}_2 \mathrm{vec}(\boldsymbol{\mathcal{L}}^{ref}_{0|2i-1}), \tag{53}$$

where $\boldsymbol{S}_1$ and $\boldsymbol{S}_2$ are selection matrices,

$$\begin{aligned}
\boldsymbol{S}_1 &\triangleq \begin{bmatrix} \boldsymbol{S}^T_{1,1} & \boldsymbol{S}^T_{1,2} & \cdots & \boldsymbol{S}^T_{1,i} \end{bmatrix}^T \\
\boldsymbol{S}_2 &\triangleq \boldsymbol{I}_{n_u} \otimes \begin{bmatrix} \boldsymbol{I}_{n_y} & \boldsymbol{0}_{n_y \times (2i-1)n_y} \end{bmatrix} \\
\boldsymbol{S}_{1,k} &\triangleq \boldsymbol{I}_{n_u} \otimes \begin{bmatrix} \boldsymbol{0}_{in_y \times n_y k} & \boldsymbol{I}_{in_y} & \boldsymbol{0}_{in_y \times (i-k)n_y} \end{bmatrix},
\end{aligned}$$

and $\otimes$ denotes the Kronecker product.

From (41), a first-order perturbation of $\boldsymbol{\mathcal{O}}_i$ and $\boldsymbol{\mathcal{C}}^S_i$ can be written as

$$\Delta \boldsymbol{\mathcal{O}}_i \approx \boldsymbol{U} \frac{1}{2} \boldsymbol{S}^{-1/2} \Delta \boldsymbol{S} + \Delta \boldsymbol{U} \boldsymbol{S}^{1/2} \tag{54}$$

$$\Delta \boldsymbol{\mathcal{C}}^S_i \approx \frac{1}{2} \boldsymbol{S}^{-1/2} \Delta \boldsymbol{S} \boldsymbol{V}^T + \boldsymbol{S}^{1/2} \Delta \boldsymbol{V}^T. \tag{55}$$

[3]Since the stochastic outputs are assumed to have a zero mean value, their correlation matrices equal their covariance matrices, hence the name SSI-*cov*.

The first-order perturbation of $\boldsymbol{S}$ equals (Pintelon et al., 2006):

$$\operatorname{diag}(\Delta \boldsymbol{S}) \approx \begin{bmatrix} (\boldsymbol{v_1} \otimes \boldsymbol{u_1})^T \\ \vdots \\ (\boldsymbol{v_n} \otimes \boldsymbol{u_n})^T \end{bmatrix} \operatorname{vec}(\Delta \boldsymbol{L_{1|i}}), \tag{56}$$

where diag(□) takes the diagonal of a square matrix □, $\boldsymbol{u_j}$ is the jth column of $\boldsymbol{U}$, and $\boldsymbol{v_j}$ is the jth column of $\boldsymbol{V}$. Pintelon et al. (2006) derived a formula for the first-order sensitivity of the singular vectors of a matrix, which enables one to link $\Delta\boldsymbol{U}$ and $\Delta\boldsymbol{V}$ to $\Delta\boldsymbol{L_{1|i}}$:

$$\Delta \boldsymbol{Z} \approx \begin{bmatrix} \boldsymbol{B_1^\dagger C_1} \\ \vdots \\ \boldsymbol{B_n^\dagger C_n} \end{bmatrix} \operatorname{vec}(\Delta \boldsymbol{L_{1|i}}), \tag{57}$$

where

$$\boldsymbol{Z} \triangleq \operatorname{vec}\left(\begin{bmatrix} \boldsymbol{U} \\ \boldsymbol{V} \end{bmatrix}\right), \qquad \boldsymbol{C_j} \triangleq \frac{1}{\sigma_j}\begin{bmatrix} \boldsymbol{v_j^T} \otimes (\boldsymbol{I_{in_y}} - \boldsymbol{u_j u_j^T}) \\ \left(\boldsymbol{u_j^T} \otimes (\boldsymbol{I_{in_u}} - \boldsymbol{v_j v_j^T})\right)\boldsymbol{P_1} \end{bmatrix},$$

$$\boldsymbol{B_j} \triangleq \begin{bmatrix} \boldsymbol{I_{in_y}} & -\frac{\boldsymbol{L_{1|i}}}{\sigma_j} \\ -\frac{\boldsymbol{L_{1|i}^T}}{\sigma_j} & \boldsymbol{I_{in_u}} \end{bmatrix}, \qquad \boldsymbol{P_1} \triangleq \sum_{k_1=1}^{in_y}\sum_{k_2=1}^{in_u} \boldsymbol{E_{k_1k_2}^{in_y\times in_u}} \otimes \boldsymbol{E_{k_2k_1}^{im\times in_y}}.$$

Using Kronecker algebra (Brewer, 1978) and substituting equations (56-57), (54) yields

$$\begin{aligned} \operatorname{vec}(\Delta\boldsymbol{\mathcal{O}_i}) &\approx \left(\boldsymbol{I_n} \otimes \boldsymbol{U}\frac{1}{2}\boldsymbol{S}^{-1/2}\right)\operatorname{vec}(\Delta\boldsymbol{S}) + \left(\boldsymbol{S}^{1/2^T} \otimes \boldsymbol{I_{in_y}}\right)\operatorname{vec}(\Delta\boldsymbol{U}) \\ &\approx (\boldsymbol{\mathcal{A}_1} + \boldsymbol{\mathcal{A}_2})\operatorname{vec}(\Delta\boldsymbol{L_{1|i}}), \end{aligned} \tag{58}$$

where

$$\boldsymbol{\mathcal{A}_1} \triangleq \left(\boldsymbol{I_n} \otimes \left(\frac{1}{2}\boldsymbol{U}\boldsymbol{S}^{-1/2}\right)\right)\boldsymbol{S_3}\begin{bmatrix} (\boldsymbol{v_1} \otimes \boldsymbol{u_1})^T \\ \vdots \\ (\boldsymbol{v_n} \otimes \boldsymbol{u_n})^T \end{bmatrix},$$

$$\boldsymbol{\mathcal{A}_2} \triangleq \underbrace{\left(\boldsymbol{S}^{1/2^T} \otimes \boldsymbol{I_{in_y}}\right)\left(\boldsymbol{I_n} \otimes \begin{bmatrix}\boldsymbol{I_{in_y}} & \boldsymbol{0_{in_y\times in_u}}\end{bmatrix}\right)}_{=\left(\boldsymbol{S}^{1/2^T} \otimes \begin{bmatrix}\boldsymbol{I_{in_y}} & \boldsymbol{0_{in_y\times in_u}}\end{bmatrix}\right)}\begin{bmatrix} \boldsymbol{B_1^\dagger C_1} \\ \vdots \\ \boldsymbol{B_n^\dagger C_n} \end{bmatrix},$$

$$\boldsymbol{S_3} \triangleq \sum_{k=1}^{n} \boldsymbol{E_{(k-1)n+k,k}^{nn\times n}}.$$

With Kronecker algebra and (56-57), (55) can be converted to

$$\begin{aligned}\mathrm{vec}(\Delta\boldsymbol{\mathcal{C}}_{\imath}^{S}) &\approx \left(\boldsymbol{V}\otimes\frac{1}{2}\boldsymbol{S}^{-1/2}\right)\Delta\boldsymbol{S}+\left(\boldsymbol{I}_{\imath n_u}\otimes\boldsymbol{S}^{1/2}\right)\boldsymbol{P}_2\mathrm{vec}\Delta\boldsymbol{V}\\ &\approx \boldsymbol{\mathcal{A}}_3\mathrm{vec}(\Delta\boldsymbol{L}_{1|\imath})+\boldsymbol{\mathcal{A}}_4\mathrm{vec}(\Delta\boldsymbol{L}_{1|\imath}),\end{aligned} \tag{59}$$

where

$$\begin{aligned}\boldsymbol{P}_2 &\triangleq \sum_{k_1=1}^{n}\sum_{k_2=1}^{\imath n_u}\boldsymbol{E}_{k_1+(k_2-1)n,(k_1-1)\imath n_u+k_2}^{\imath n_u n\times\imath n_u n}\\ \boldsymbol{\mathcal{A}}_3 &\triangleq \left(\boldsymbol{V}\otimes\frac{1}{2}\boldsymbol{S}^{-\frac{1}{2}}\right)\boldsymbol{S}_3\begin{bmatrix}(\boldsymbol{v}_1\otimes\boldsymbol{u}_1)^T\\ \vdots\\ (\boldsymbol{v}_n\otimes\boldsymbol{u}_n)^T\end{bmatrix}\\ \boldsymbol{\mathcal{A}}_4 &\triangleq \left(\boldsymbol{I}_{\imath n_u}\otimes\boldsymbol{S}^{1/2}\right)\boldsymbol{P}_2\left(\boldsymbol{I}_n\otimes\begin{bmatrix}\boldsymbol{0}_{\imath n_u\times\imath n_y} & \boldsymbol{I}_{\imath n_u}\end{bmatrix}\right)\begin{bmatrix}\boldsymbol{B}_1^{\dagger}\boldsymbol{C}_1\\ \vdots\\ \boldsymbol{B}_n^{\dagger}\boldsymbol{C}_n\end{bmatrix}.\end{aligned}$$

The first-order perturbation of $\boldsymbol{A}$, calculated via (42), can be expressed as:

$$\mathrm{vec}(\Delta\boldsymbol{A}) \approx \boldsymbol{\mathcal{A}}_5\mathrm{vec}(\Delta\boldsymbol{\mathcal{O}}_{\imath}),$$

where

$$\begin{aligned}\boldsymbol{\mathcal{A}}_5 &\triangleq \boldsymbol{I}_n\otimes(\underline{\boldsymbol{\mathcal{O}}_{\imath}}^{\dagger}\boldsymbol{S}_4)-\boldsymbol{A}^T\otimes(\underline{\boldsymbol{\mathcal{O}}_{\imath}}^{\dagger}\boldsymbol{S}_5)\\ &\quad+\left((\overline{\boldsymbol{\mathcal{O}}_{\imath}}^T\boldsymbol{S}_5-\boldsymbol{A}^T\underline{\boldsymbol{\mathcal{O}}_{\imath}}^T\boldsymbol{S}_5)\otimes(\underline{\boldsymbol{\mathcal{O}}_{\imath}}^T\underline{\boldsymbol{\mathcal{O}}_{\imath}})^{-1}\right)\boldsymbol{P}_3\\ \boldsymbol{S}_4 &\triangleq \begin{bmatrix}\boldsymbol{0}_{(\imath-1)n_y\times l} & \boldsymbol{I}_{(\imath-1)n_y}\end{bmatrix}\\ \boldsymbol{S}_5 &\triangleq \begin{bmatrix}\boldsymbol{I}_{(\imath-1)n_y} & \boldsymbol{0}_{(\imath-1)n_y\times n_y}\end{bmatrix}\\ \boldsymbol{P}_3 &\triangleq \sum_{k_1=1}^{n}\sum_{k_2=1}^{\imath n_y}\boldsymbol{E}_{k_1+(k_2-1)n,(k_1-1)\imath n_y+k_2}^{\imath n_y n\times\imath n_y n}.\end{aligned}$$

Substitution of (58) and (53) into this result yields

$$\mathrm{vec}(\Delta\boldsymbol{A})\approx\boldsymbol{\mathcal{A}}_5\left(\boldsymbol{\mathcal{A}}_1+\boldsymbol{\mathcal{A}}_2\right)\boldsymbol{S}_1\mathrm{vec}(\Delta\boldsymbol{\mathcal{L}}_{0|2\imath-1}^{ref})=\boldsymbol{\mathcal{A}}_6\mathrm{vec}(\Delta\boldsymbol{\mathcal{L}}_{0|2\imath-1}^{ref}). \tag{60}$$

As $\boldsymbol{G}^{ref}$ corresponds to the first n_r columns of $\boldsymbol{\mathcal{C}}_{\imath}^{S}$, application of (59) and (53) yields

$$\mathrm{vec}(\Delta\boldsymbol{B})\approx\boldsymbol{\mathcal{A}}_7\left(\boldsymbol{\mathcal{A}}_3+\boldsymbol{\mathcal{A}}_4\right)\boldsymbol{S}_1\mathrm{vec}(\Delta\boldsymbol{\mathcal{L}}_{0|2\imath-1}^{ref})=\boldsymbol{\mathcal{A}}_8\mathrm{vec}(\Delta\boldsymbol{\mathcal{L}}_{0|2\imath-1}^{ref}), \tag{61}$$

where $\mathcal{A}_7 \triangleq \begin{bmatrix} \boldsymbol{I}_{nn_u} & \boldsymbol{0}_{n(\imath-1)n_u} \end{bmatrix}$. As $\boldsymbol{C}$ is determined as the first l rows of $\mathcal{O}_\imath$, application of (58) and (53) leads to

$$\text{vec}(\Delta \boldsymbol{C}) \approx \mathcal{A}_9 \left(\mathcal{A}_1 + \mathcal{A}_2\right) \boldsymbol{S}_1 \text{vec}(\Delta \mathcal{L}^{ref}_{0|2\imath-1}) = \mathcal{A}_{10} \text{vec}(\Delta \mathcal{L}^{ref}_{0|2\imath-1}), \quad (62)$$

where $\mathcal{A}_9 = \boldsymbol{I}_n \otimes \begin{bmatrix} \boldsymbol{I}_{n_y} & \boldsymbol{0}_{n_y,(\imath-1)n_y} \end{bmatrix}$. Finally, the application of (53) yields

$$\text{vec}(\Delta \boldsymbol{\Lambda}_0) = \boldsymbol{S}_2 \text{vec}(\Delta \mathcal{L}^{ref}_{0|2\imath-1}). \quad (63)$$

With the definition

$$\mathcal{A} \triangleq \begin{bmatrix} \mathcal{A}_6^T & \mathcal{A}_8^T & \mathcal{A}_{10}^T & \boldsymbol{S}_2^T \end{bmatrix}^T,$$

the following expression for the covariances of the identified system matrices is obtained (Reynders et al., 2008):

$$\boldsymbol{\Sigma}_{\hat{\boldsymbol{A}},\hat{\boldsymbol{G}}^{ref},\hat{\boldsymbol{C}},\hat{\boldsymbol{\Lambda}}_0^{ref}/2} = \mathcal{A} \boldsymbol{\Sigma}_{\hat{\mathcal{L}}^{ref}_{0|2\imath-1}} \mathcal{A}^T, \qquad \Delta \mathcal{L}^{ref}_{0|2\imath-1} \to \boldsymbol{0}, \quad (64)$$

where

$$\boldsymbol{\Sigma}_{\hat{\boldsymbol{A}},\hat{\boldsymbol{G}}^{ref},\hat{\boldsymbol{C}},\hat{\boldsymbol{\Lambda}}_0^{ref}/2} \triangleq \text{Cov}\left(\begin{bmatrix} \text{vec}(\hat{\boldsymbol{A}})^T & \text{vec}(\hat{\boldsymbol{G}}^{ref})^T & \text{vec}(\hat{\boldsymbol{C}})^T & \text{vec}(\hat{\boldsymbol{\Lambda}}_0^{ref})^T \end{bmatrix}^T \right).$$

Since the output correlations are asymptotically normally distributed, it follows from the first-order sensitivity analysis that the same holds for the estimated system matrices, when $\Delta \mathcal{L}^{ref}_{0|2\imath-1} \to \boldsymbol{0}$.

A note on the choice of $\imath$. In theory, any value $\imath$ that is larger than $\text{ceil}(n/n_y) + 1$ with $\text{ceil}(\square)$ a function that rounds real argument $\square$ to the nearest integer towards $+\infty$, can be chosen for the identification. However, the quality of the identified system model depends on this choice. If the lowest eigenfrequency f_1 of the structure is low compared to the sampling frequency, and if the value of $\imath$ is low, it is possible that each column of $\boldsymbol{H}_{1|\imath}$ contains only a small part of the corresponding eigenperiod and as a consequence the eigenfrequency is not well identified. A solution is to choose $\imath$ as large as possible, but then the calculation time and memory usage might become excessive. Therefore, Reynders and De Roeck (2008) proposed the following rule of thumb for choosing $\imath$:

$$\imath \geq \frac{1}{2 f_0 T}, \quad (65)$$

with f_0 the lowest frequency of interest.

Summary. Figure 6 lists the different steps of the SSI-cov/ref algorithm for the identification of the state-space matrices $\boldsymbol{A}$, $\boldsymbol{G}^{ref}$, $\boldsymbol{C}$, and $\boldsymbol{\Lambda}_0$ and the estimation of their joint probability density function.

SSI-cov/ref algorithm.

1. Compute $\hat{\mathcal{L}}^{ref}_{1|2\imath-1}$ and $\boldsymbol{\Sigma}_{\hat{\mathcal{L}}^{ref}_{1|2\imath-1}}$ using (51) and (52), respectively. Build $\boldsymbol{L}^{ref}_{1|\imath}$ using (39).
2. Choose the system order n and compute the extended observability matrix $\mathcal{O}_\imath$ and the reference-based extended stochastic controllability matrix $\mathcal{C}^S_\imath$ from the singular value decomposition of $\boldsymbol{L}^{ref}_{1|\imath}$ using (41).
3. Estimate the state-space matrices. $\hat{\boldsymbol{C}}$ equals the first n_y rows of $\mathcal{O}_\imath$, $\hat{\boldsymbol{G}}^{ref}$ equals the first n_r columns of $\mathcal{C}_\imath$, and $\hat{\boldsymbol{A}}$ is determined from (42).
4. Compute $\boldsymbol{\Sigma}_{\hat{\boldsymbol{A}},\hat{\boldsymbol{G}}^{ref},\hat{\boldsymbol{C}},\hat{\boldsymbol{\Lambda}}^{ref}_0/2}$ using (64), where an expression for $\mathcal{A}$ is derived in the above linear sensitivity analysis.

Figure 6. The SSI-cov/ref algorithm with estimation of the probability density function of the estimated state-space matrices.

4.3 Data-driven stochastic subspace identification

A strongly consistent subspace algorithm. In Section 3.3, it was derived that, under assumption 3.1,

$$\underset{\jmath\to\infty}{\text{a.s.lim}} \boldsymbol{W}_1 \boldsymbol{Y}^s_{\imath|2\imath-1}/\mathcal{Y}^{s,ref}_{0|\imath-1}\boldsymbol{W}_2 = \boldsymbol{W}_1 \lim_{\jmath\to\infty} \mathcal{O}_\imath \hat{\boldsymbol{X}}_{\imath|\imath}\boldsymbol{W}_2. \tag{50}$$

This is the basic relationship for the *DATA-driven Stochastic Subspace Identification* (SSI-data) class of subspace algorithms, that only differ up to specific choices for the weighting matrices $\boldsymbol{W}_1$ and $\boldsymbol{W}_2$. From the reduced singular value decomposition

$$\boldsymbol{W}_1 \boldsymbol{Y}^s_{\imath|2\imath-1}/\mathcal{Y}^{s,ref}_{0|\imath-1}\boldsymbol{W}_2 \approx \boldsymbol{U}\boldsymbol{S}\boldsymbol{V}^T,$$

where the diagonal matrix $\boldsymbol{S} \in \mathbb{R}^{n\times n}$ contains only the n highest singular values and $\boldsymbol{U} \in \mathbb{R}^{\imath n_y\times n}$ and $\boldsymbol{V} \in \mathbb{R}^{\jmath\times n}$ contain the corresponding singular vectors, it follows from (50) that the following estimates are strongly consistent[4] under assumption 3.1:

$$\hat{\mathcal{O}}_i \triangleq \boldsymbol{W}_1^{-1}\boldsymbol{U}\boldsymbol{S}^{\frac{1}{2}}, \qquad \hat{\boldsymbol{X}}^s_{i|i} \triangleq \hat{\mathcal{O}}_i^{\dagger}\boldsymbol{Y}^s_{i|2i-1}/\mathscr{Y}^{s,ref}_{0|i-1}, \qquad \text{and}$$
$$\hat{\boldsymbol{X}}^s_{i+1|i+1} \triangleq \underline{\hat{\mathcal{O}}_i}^{\dagger}\boldsymbol{Y}^s_{i+1|2i-1}/\mathscr{Y}^{s,ref}_{0|i}. \tag{66}$$

From the reference-based forward innovation model (37-38), one has

$$\begin{bmatrix}\hat{\boldsymbol{X}}^s_{i+1|i+1}\\ \boldsymbol{Y}_{i|i}\end{bmatrix} = \begin{bmatrix}\boldsymbol{A}\\ \boldsymbol{C}\end{bmatrix}\hat{\boldsymbol{X}}^s_{i|i} + \begin{bmatrix}\boldsymbol{K}_i\boldsymbol{E}^{f,ref}_{i|i}\\ \boldsymbol{E}^f_{i|i}\end{bmatrix}. \tag{67}$$

Since it can be shown that $\mathcal{E}\left(\boldsymbol{e}_k\hat{\boldsymbol{x}}_k^{sT}\right) = \boldsymbol{0}$, one has

$$\underset{j\to\infty}{\text{a.s.lim}}\begin{bmatrix}\hat{\boldsymbol{X}}^s_{i+1|i+1}\\ \boldsymbol{Y}_{i|i}\end{bmatrix}\hat{\boldsymbol{X}}^{s\,\dagger}_{i|i} = \begin{bmatrix}\boldsymbol{A}\\ \boldsymbol{C}\end{bmatrix},$$

so an estimate of $\boldsymbol{A}$ and $\boldsymbol{C}$ that is strongly consistent under assumption 3.1 is obtained from

$$\begin{bmatrix}\hat{\boldsymbol{A}}\\ \hat{\boldsymbol{C}}\end{bmatrix} \triangleq \begin{bmatrix}\hat{\hat{\boldsymbol{X}}}^s_{i+1|i+1}\\ \boldsymbol{Y}_{i|i}\end{bmatrix}\hat{\hat{\boldsymbol{X}}}^{s\,\dagger}_{i|i}. \tag{68}$$

As explained in Section 2.6, $\boldsymbol{K}_i \to \boldsymbol{K}$ when $i \to \infty$. Therefore, one has

$$\underset{i,j\to\infty}{\text{a.s.lim}}\frac{1}{j}\begin{bmatrix}\boldsymbol{K}_i\boldsymbol{E}^{f,ref}_{i|i}\\ \boldsymbol{E}^f_{i|i}\end{bmatrix}\begin{bmatrix}\boldsymbol{K}_i\boldsymbol{E}^{f,ref}_{i|i}\\ \boldsymbol{E}^f_{i|i}\end{bmatrix}^T = \begin{bmatrix}\boldsymbol{Q}^{ref} & \boldsymbol{S}^{ref}\\ \boldsymbol{S}^{ref\,T} & \boldsymbol{R}\end{bmatrix}.$$

For finite i and j, the left-hand side of this equation can be estimated from (67), after estimating $\boldsymbol{A}$ and $\boldsymbol{C}$. From the equations (29), (30), (31), (34), and (35), it is then possible to calculate estimates for $\boldsymbol{\Sigma}^s$, $\boldsymbol{G}$, $\boldsymbol{\Lambda}_0$, $\boldsymbol{K}$, and $\boldsymbol{P}$, respectively.

Asymptotic statistical efficiency and the choice of weighting matrices. The general formulation of the SSI-data/ref algorithm presented above leaves some freedom in the choice of the weighting matrices $\boldsymbol{W}_1$ and $\boldsymbol{W}_2$. In order not to lose the consistency property, it is obvious that $\boldsymbol{W}_1$ should be of full rank, and that $\boldsymbol{W}_2$ should preserve the rank of the matrices with which it is multiplied. Van Overschee and De Moor (1996) showed that by specific choices of the weighting matrices, the SSI-cov/ref algorithm,

[4] An estimate is strongly consistent when its almost-sure limit equals the exact value.

presented in Section 4.2, the *Unweighted Principal Component* (UPC) algorithm of Arun and Kung (1990), and the CVA approach (Akaike, 1974, 1975; Arun and Kung, 1990) fit into this framework.

For a correctly specified system order n, the choice of the weighting matrix $\boldsymbol{W}_2$ doesn't have an influence on the asymptotic distribution of the estimated system, see Bauer et al. (2000), but as shown by Bauer and Ljung (2002), the choice of $\boldsymbol{W}_1$ is important. Recently, Bauer (2005) proved that the output-only CVA algorithm, for which

$$\boldsymbol{W}_1 = \left(\frac{1}{j}\boldsymbol{Y}_{i|2i-1}\boldsymbol{Y}_{i|2i-1}^T\right)^{-1/2} \qquad \text{and} \qquad \boldsymbol{W}_2 = \boldsymbol{I}_j,$$

is asymptotically statistically efficient for $j \rightarrow \infty$. This means that the covariance matrix of the estimates equals asymptotically the Cramér-Rao lower bound, i.e., no estimator with lower covariance can be found. Since the choice of $\boldsymbol{W}_2$ is unimportant, this result obviously holds for a class of algorithms having the same $\boldsymbol{W}_1$ as CVA, called the *Larimore type* of algorithms. This type of algorithms leads to a theoretically optimal choice of weighting matrices. Also for SSI-data/ref, the rule of thumb (65) for the choice of i is advised.

Implementation. As shown by Peeters and De Roeck (1999), the SSI-data/ref algorithm can be efficiently implemented by making use of the LQ-factorisation technique, where the explicit computation of the Q factor can be avoided. Figure 7 contains a step-by-step overview of this implementation in case of CVA weighting.

4.4 Data-driven combined deterministic-stochastic subspace identification

Introduction. The *REFerence-based DATA-driven Combined Subspace Identification* (CSI-data/ref) algorithm of Reynders and De Roeck (2008) identifies the combined deterministic-stochastic state-space model that was presented in Section 2.6. It is the reference-based generalisation of the robust combined subspace algorithm of Van Overschee and De Moor (1996).

A strongly consistent subspace algorithm. In Section 3.3, it was derived that, under assumptions 3.1, 3.2, and 3.3,

$$\underset{j\rightarrow\infty}{\text{a.s.lim}} \boldsymbol{W}_1 \boldsymbol{Y}_{i|2i-1} / _{\mathscr{U}_{i|2i-1}} \left(\mathscr{U}_{0|2i-1} \vee \mathscr{Y}^{ref}_{0|i-1}\right) \boldsymbol{W}_2 = \boldsymbol{W}_1 lim_{j\rightarrow\infty} \boldsymbol{O}_i \hat{\boldsymbol{X}}_{i|i} \boldsymbol{W}_2. \quad (49)$$

This is the basic relationship for the CSI-data/ref class of algorithms, that only differ up to the weighting matrices $\boldsymbol{W}_1$ and $\boldsymbol{W}_2$. The initial-

SSI-data/ref algorithm (CVA weighting).

1. Construct the block Hankel matrices $\boldsymbol{Y}^{s,ref}_{0|\imath}$, $\boldsymbol{Y}^{s}_{\imath+1|2\imath-1}$, and $\boldsymbol{Y}^{s,\,ref}_{\imath|\imath}$, where $\boldsymbol{Y}^{s,\tilde{}ref}_{\imath|\imath}$ contains the non-reference rows of $\boldsymbol{Y}^{s}_{\imath|\imath}$. Perform the following LQ decomposition, with $\boldsymbol{L}$ lower triangular and $\boldsymbol{Q}$ orthonormal:

$$\begin{bmatrix} \boldsymbol{Y}^{s,ref}_{0|\imath-1} \\ \boldsymbol{Y}^{s,ref}_{\imath|\imath} \\ \boldsymbol{Y}^{s,\tilde{}ref}_{\imath|\imath} \\ \boldsymbol{Y}^{s}_{\imath+1|2\imath-1} \end{bmatrix} = \frac{1}{\sqrt{\jmath}} \underbrace{\begin{bmatrix} \boldsymbol{L}_{11} & \mathbf{0} & \mathbf{0} & \mathbf{0} \\ \boldsymbol{L}_{21} & \boldsymbol{L}_{22} & \mathbf{0} & \mathbf{0} \\ \boldsymbol{L}_{31} & \boldsymbol{L}_{32} & \boldsymbol{L}_{33} & \mathbf{0} \\ \boldsymbol{L}_{41} & \boldsymbol{L}_{42} & \boldsymbol{L}_{43} & \boldsymbol{L}_{44} \end{bmatrix}}_{\boldsymbol{L}} \underbrace{\begin{bmatrix} \boldsymbol{Q}_1 \\ \boldsymbol{Q}_2 \\ \boldsymbol{Q}_3 \\ \boldsymbol{Q}_4 \end{bmatrix}}_{\boldsymbol{Q}},$$

   where $\boldsymbol{L}_{11} \in \mathbb{R}^{\imath n_r \times \imath n_r}$, $\boldsymbol{L}_{22} \in \mathbb{R}^{n_r \times n_r}$, and $\boldsymbol{L}_{33} \in \mathbb{R}^{(n_y - n_r)\times(n_y - n_r)}$. $\boldsymbol{Q}$ does not need to be determined.
2. Compute $\boldsymbol{W}_1$ as

$$\boldsymbol{W}_1 = \begin{bmatrix} \boldsymbol{L}_{21} & \boldsymbol{L}_{22} & \mathbf{0} & \mathbf{0} \\ \boldsymbol{L}_{31} & \boldsymbol{L}_{32} & \boldsymbol{L}_{33} & \mathbf{0} \\ \boldsymbol{L}_{41} & \boldsymbol{L}_{42} & \boldsymbol{L}_{43} & \boldsymbol{L}_{44} \end{bmatrix} \begin{bmatrix} \boldsymbol{L}_{21} & \boldsymbol{L}_{22} & \mathbf{0} & \mathbf{0} \\ \boldsymbol{L}_{31} & \boldsymbol{L}_{32} & \boldsymbol{L}_{33} & \mathbf{0} \\ \boldsymbol{L}_{41} & \boldsymbol{L}_{42} & \boldsymbol{L}_{43} & \boldsymbol{L}_{44} \end{bmatrix}^T.$$

3. Compute the singular value decomposition $\boldsymbol{USV}^T = \boldsymbol{W}_1 \begin{bmatrix} \boldsymbol{L}_{21} \\ \boldsymbol{L}_{31} \\ \boldsymbol{L}_{41} \end{bmatrix}$. Choose the system order n and retain only the n highest singular values and the corresponding singular vectors.
4. Compute the observability matrix as $\hat{\boldsymbol{\mathcal{O}}}_\imath = \boldsymbol{W}_1^{-1}\boldsymbol{US}^{\frac{1}{2}}$. Determine $\underline{\boldsymbol{\mathcal{O}}_\imath}$ by deleting the last n_y rows of $\hat{\boldsymbol{\mathcal{O}}}_\imath$.
5. Finally, compute the system estimates:

$$\begin{bmatrix} \hat{\boldsymbol{A}} \\ \hat{\boldsymbol{C}} \end{bmatrix} = \begin{bmatrix} \underline{\boldsymbol{\mathcal{O}}_\imath}^\dagger \boldsymbol{L}_{41} \\ \boldsymbol{L}_{21} \\ \boldsymbol{L}_{31} \end{bmatrix} \left(\hat{\boldsymbol{\mathcal{O}}}_\imath^\dagger \begin{bmatrix} \boldsymbol{L}_{21} \\ \boldsymbol{L}_{31} \\ \boldsymbol{L}_{41} \end{bmatrix} \right)^\dagger.$$

Figure 7. Implementation of the SSI-data/ref CVA algorithm.

state matrix for the sequence of reference-based Kalman filter states $\hat{\boldsymbol{X}}_{i|i}$ is $\hat{\boldsymbol{X}}_{0|0} = \hat{\boldsymbol{X}}^{d}_{0|0}/_{\mathscr{U}_{i|2i-1}}\mathscr{U}_{0|i-1}$. Using the reduced singular value decomposition

$$\boldsymbol{W_1 Y_{i|2i-1}}/_{\mathscr{U}_{i|2i-1}}\left(\mathscr{U}_{0|2i-1} \vee \mathscr{Y}^{ref}_{0|i-1}\right)\boldsymbol{W_2} \approx \boldsymbol{USV}^T,$$

where the diagonal matrix $\boldsymbol{S} \in \mathbb{R}^{n\times n}$ contains the highest n singular values and $\boldsymbol{U} \in \mathbb{R}^{i(n_y+n_u)\times n}$ and $\boldsymbol{V} \in \mathbb{R}^{j\times n}$ contain the corresponding singular vectors, one has from (49) that the following estimate for the observability matrix is strongly consistent under assumptions 3.1, 3.2, and 3.3:

$$\hat{\boldsymbol{\mathcal{O}}}_i \triangleq \boldsymbol{W}_1^{-1}\boldsymbol{US}^{\frac{1}{2}}. \tag{69}$$

For the estimation of $\hat{\boldsymbol{X}}_{i|i}$ and $\hat{\boldsymbol{X}}_{i+1|i+1}$, (49) can unfortunately not be used since the initial state matrix of $\hat{\boldsymbol{X}}_{i+1|i+1}$ would be $\hat{\boldsymbol{X}}^{d}_{0|0}/_{\mathscr{U}_{i+1|2i}}\mathscr{U}_{0|i}$, which is different from the initial state matrix of $\hat{\boldsymbol{X}}_{i|i}$. As noted in Van Overschee and De Moor (1994a), this would lead to an inconsistent algorithm for finite values of i. Therefore, the orthogonal projection (48), which was derived in Section 3.3, is considered instead:

$$\underset{j\to\infty}{\text{a.s.lim}}\ \underbrace{\boldsymbol{Y}_{i|2i-1}/\left(\mathscr{U}_{0|2i-1} \vee \mathscr{Y}^{ref}_{0|i-1}\right)}_{\triangleq \boldsymbol{\mathcal{Z}}_i} = \lim_{j\to\infty} \boldsymbol{\mathcal{O}}_i\hat{\boldsymbol{X}}_{i|i} + \boldsymbol{\mathcal{F}}^d_i\boldsymbol{U}_{i|2i-1}. \tag{48}$$

The initial state matrix for the sequence of reference-based Kalman filter states $\hat{\boldsymbol{X}}_{i|i}$ is now $\hat{\boldsymbol{X}}_{0|0} = \hat{\boldsymbol{X}}^{d}_{0|0}/\mathscr{U}_{0|2i-1}$. Following the same lines, one has

$$\begin{aligned}\underset{j\to\infty}{\text{a.s.lim}}\ \underbrace{\boldsymbol{Y}_{i+1|2i-1}/\left(\mathscr{U}_{0|2i-1} \vee \mathscr{Y}^{ref}_{0|i}\right)}_{\triangleq \boldsymbol{\mathcal{Z}}_{i+1}} \\ = \lim_{j\to\infty} \boldsymbol{\mathcal{O}}_{i-1}\hat{\boldsymbol{X}}_{i+1|i+1} + \boldsymbol{\mathcal{F}}^d_{i-1}\boldsymbol{U}_{i+1|2i-1},\end{aligned} \tag{70}$$

where $\hat{\boldsymbol{X}}_{i+1|i+1}$ has the same initial state matrix as in the previous expression. From the reference-based forward innovation model (37-38), one has

$$\begin{bmatrix}\hat{\boldsymbol{X}}_{i+1|i+1}\\ \boldsymbol{Y}_{i|i}\end{bmatrix} = \begin{bmatrix}\boldsymbol{A}\\ \boldsymbol{C}\end{bmatrix}\hat{\boldsymbol{X}}_{i|i} + \begin{bmatrix}\boldsymbol{B}\\ \boldsymbol{D}\end{bmatrix}\boldsymbol{U}_{i|i} + \begin{bmatrix}\boldsymbol{K}_i\boldsymbol{E}^{f,ref}_{i|i}\\ \boldsymbol{E}^{f}_{i|i}\end{bmatrix}. \tag{71}$$

Substitution of (69-70) yields

$$\underset{j\to\infty}{\text{a.s.lim}}\left[\hat{\boldsymbol{\mathcal{O}}}^{\dagger}_{i-1}\boldsymbol{\mathcal{Z}}_{i+1}\right] = \lim_{j\to\infty}\begin{bmatrix}\boldsymbol{A}\\ \boldsymbol{C}\end{bmatrix}\hat{\boldsymbol{\mathcal{O}}}^{\dagger}_i\boldsymbol{\mathcal{Z}}_i + \boldsymbol{K}_i\boldsymbol{U}_{i|2i-1} + \begin{bmatrix}\boldsymbol{K}_i\boldsymbol{E}^{f,ref}_{i|i}\\ \boldsymbol{E}^{f}_{i|i}\end{bmatrix}, \tag{72}$$

where

$$\boldsymbol{K}_\imath = \begin{bmatrix} -\boldsymbol{A}\mathcal{O}_\imath^\dagger \mathcal{F}_\imath^d + \begin{bmatrix} \boldsymbol{B} & \mathcal{O}_{\imath-1}^\dagger \mathcal{F}_{\imath-1}^d \end{bmatrix} \\ -\boldsymbol{C}\mathcal{O}_\imath^\dagger \mathcal{F}_\imath^d + \begin{bmatrix} \boldsymbol{D} & \boldsymbol{0} \end{bmatrix} \end{bmatrix}. \tag{73}$$

Since the state estimation error $\boldsymbol{e_k}$ is orthogonal to the state estimate $\hat{\boldsymbol{x}}_{\boldsymbol{k}}$ (Juang, 1994), it follows that a strongly consistent estimate of $\boldsymbol{A}$, $\boldsymbol{C}$, and $\boldsymbol{K}_\imath$ is obtained from

$$\begin{bmatrix} \begin{bmatrix} \hat{\boldsymbol{A}} \\ \hat{\boldsymbol{C}} \end{bmatrix} & \hat{\boldsymbol{K}}_\imath \end{bmatrix} \triangleq$$

$$\begin{bmatrix} \hat{\mathcal{O}}_{\imath-1}^\dagger \boldsymbol{Y}_{\imath+1|2\imath} / \left(\mathscr{U}_{0|2\imath-1} \vee \mathscr{Y}_{0|\imath}^{ref} \right) \\ \boldsymbol{Y}_{\imath|\imath} \end{bmatrix} \begin{bmatrix} \hat{\mathcal{O}}_\imath^\dagger \boldsymbol{Y}_{\imath|2\imath-1} / \left(\mathscr{U}_{0|2\imath-1} \vee \mathscr{Y}_{0|\imath-1}^{ref} \right) \\ \boldsymbol{U}_{\imath|\imath} \end{bmatrix}^\dagger.$$

$\boldsymbol{K}_\imath$ is linear in $\boldsymbol{B}$ and $\boldsymbol{D}$ (see (73)), so if $\boldsymbol{K}_{\imath,\boldsymbol{k}} \in \mathbb{R}^{n \times n_u}$ is defined as the kth block column of $\boldsymbol{K}_\imath$, $\boldsymbol{\mathcal{M}_k} \in \mathbb{R}^{n \times n_y}$ as the kth block column of $\mathcal{O}_{\imath-1}^\dagger$, $\boldsymbol{\mathcal{L}_{1,k}} \in \mathbb{R}^{n \times n_y}$ as the kth block column of $\boldsymbol{A}\mathcal{O}_\imath^\dagger$ and $\boldsymbol{\mathcal{L}_{2,k}} \in \mathbb{R}^{n_y \times n_y}$ as the kth block column of $\boldsymbol{C}\mathcal{O}_\imath^\dagger$, one has:

$$\begin{aligned} \boldsymbol{K}_{\imath,\boldsymbol{k}} &= \boldsymbol{\mathcal{N}_k} \begin{bmatrix} \boldsymbol{D}^T & \boldsymbol{B}^T \end{bmatrix}^T \\ \boldsymbol{\mathcal{N}}_1 &= \begin{bmatrix} -\boldsymbol{\mathcal{L}}_{11} & \boldsymbol{\mathcal{M}}_1 - \boldsymbol{\mathcal{L}}_{12} & \dots & \boldsymbol{\mathcal{M}}_{\imath-1} - \boldsymbol{\mathcal{L}}_{1\imath} \\ \boldsymbol{I}_{n_y} - \boldsymbol{\mathcal{L}}_{21} & -\boldsymbol{\mathcal{L}}_{22} & \dots & -\boldsymbol{\mathcal{L}}_{2\imath} \end{bmatrix} \begin{bmatrix} \boldsymbol{I}_{n_y} \boldsymbol{0} & \\ \boldsymbol{0} & \mathcal{O}_i \end{bmatrix} \\ \boldsymbol{\mathcal{N}}_{k \neq 1} &= \begin{bmatrix} \boldsymbol{\mathcal{M}}_{k-1} - \boldsymbol{\mathcal{L}}_{1k} & \dots & \boldsymbol{\mathcal{M}}_{\imath-1} - \boldsymbol{\mathcal{L}}_{1\imath} & \boldsymbol{0}_{n \times l(k-1)} \\ -\boldsymbol{\mathcal{L}}_{2k} & \dots & -\boldsymbol{\mathcal{L}}_{2\imath} & \boldsymbol{0}_{l \times l(k-1)} \end{bmatrix} \begin{bmatrix} \boldsymbol{I}_{n_y} \boldsymbol{0} & \\ \boldsymbol{0} & \mathcal{O}_i \end{bmatrix}. \end{aligned}$$

With this factorisation, it follows from (72) that strongly consistent estimates of $\boldsymbol{B}$ and $\boldsymbol{D}$ can be obtained from

$$\begin{aligned} \text{vec}\left(\begin{bmatrix} \hat{\boldsymbol{D}} \\ \hat{\boldsymbol{B}} \end{bmatrix} \right) &= \left(\sum_{k=1}^{\imath+1} \boldsymbol{U}_{\imath+k-1|\imath+k-1}^T \otimes \hat{\boldsymbol{\mathcal{N}}}_{\boldsymbol{k}} \right)^\dagger \text{vec}(\hat{\boldsymbol{\mathcal{P}}}) \\ \hat{\boldsymbol{\mathcal{P}}} &\triangleq \begin{bmatrix} \hat{\mathcal{O}}_{\imath-1}^\dagger \boldsymbol{\mathcal{Z}}_{\imath+1} \\ \boldsymbol{Y}_\imath \end{bmatrix} - \begin{bmatrix} \hat{\boldsymbol{A}} \\ \hat{\boldsymbol{C}} \end{bmatrix} \mathcal{O}_\imath^\dagger \boldsymbol{\mathcal{Z}}_\imath, \end{aligned}$$

where $\boldsymbol{U}_{\imath+k-1|\imath+k-1}$ is the $(\imath+k-1)^{\text{th}}$ block row of $\boldsymbol{U}_{0|2\imath-1}$. Finally, the noise covariance matrices can be estimated from the residuals of (71):

$$\underset{\imath,\jmath\to\infty}{\text{a.s.lim}} \frac{1}{\jmath} \begin{bmatrix} \boldsymbol{K}_\imath \boldsymbol{E}_{\imath|\imath}^{f,ref} \\ \boldsymbol{E}_{\imath|\imath}^f \end{bmatrix} \begin{bmatrix} \boldsymbol{K}_\imath \boldsymbol{E}_{\imath|\imath}^{f,ref} \\ \boldsymbol{E}_{\imath|\imath}^f \end{bmatrix}^T = \begin{bmatrix} \boldsymbol{Q} & \boldsymbol{S} \\ \boldsymbol{S}^T & \boldsymbol{R} \end{bmatrix} \tag{74}$$

The fact that $\imath$ needs to go to infinity is explained by the non-stationarity of the Kalman filter in (71) for finite $\imath$ values, see Van Overschee and De Moor (1996).

Choice of weighting matrices. The general formulation of the CSI-data algorithm presented above leaves some freedom in the choice of the weighting matrices $\boldsymbol{W_1}$ and $\boldsymbol{W_2}$. In order not to lose the consistency property, it is obvious that $\boldsymbol{W_1}$ should be of full rank, and that $\boldsymbol{W_2}$ should preserve the rank of the matrices with which it is multiplied. It was shown by Van Overschee and De Moor (1994b) that by specific choices of the weighting matrices, the N4SID algorithm of Van Overschee and De Moor (1994a), the PO-MOESP algorithm of Verhaegen (1994), and the CVA algorithm of Larimore (1990) can be obtained. However, it is well known that an oblique projection, which is an essential step in the algorithm (49), is a numerically ill-conditioned problem. Goethals (2005) indicated that the ill-conditioning can be removed by choosing $\boldsymbol{W_1}$ and $\boldsymbol{W_2}$ in such a way that $\boldsymbol{W_1}\boldsymbol{\mathcal{O}_i}\boldsymbol{W_2}$ consists of an orthogonal projection. A possible choice is $\boldsymbol{W_1} = \boldsymbol{I}$ and $\boldsymbol{W_2} = \boldsymbol{P}^{\perp}_{\mathcal{U}_{\imath|2\imath-1}}$, with $\boldsymbol{P}^{\perp}_{\mathcal{U}_{\imath|2\imath-1}}$ the orthogonal projector onto the orthogonal complement of $\mathcal{U}_{\imath|2\imath-1}$. This corresponds to the PO-MOESP algorithm. Also for CSI-data/ref, the rule of thumb (65) for the choice of $\imath$ is advised.

Implementation. The CSI-data/ref algorithm can, like the SSI-data/ref algorithm, be efficiently implemented by making use of the LQ-factorisation technique, where the explicit computation of the Q factor can be avoided. The interested reader is referred to (Reynders and De Roeck, 2008) for the implementation details.

5 Estimation of the modal parameters

5.1 Introduction

When a state-space model is identified from measured input-output or output-only data, a free vibration analysis and a modal decomposition of the identified model results in eigenfrequencies, damping ratios, mode shapes, and modal participation vectors of the structure. If a driving point measurement is made, absolute mode scaling is possible as well.

5.2 Estimation of the modal parameters

When a discrete-time deterministic $(\boldsymbol{A}, \boldsymbol{B}, \boldsymbol{C}, \boldsymbol{D})$ or stochastic state-space model $(\boldsymbol{A}, \boldsymbol{G}^{(ref)}, \boldsymbol{C}, \boldsymbol{\Lambda_0}/2)$ is identified, the modal parameters can be estimated as follows. The eigenvalue decomposition (14) leads to the system description in modal form (15-16), from which the unscaled mode shapes $\boldsymbol{\phi_j}$ and the discrete-time modal participation vectors $\boldsymbol{l_{dj}}$ are immediately obtained. From the ZOH map (11), it follows that

$$\lambda_{cj} = \frac{\ln(\lambda_{dj})}{T}. \tag{75}$$

The undamped frequency and damping ratio are then obtained as

$$f_{uj} = \frac{|\lambda_{cj}|}{2\pi} \quad \text{and} \quad \xi_j = \frac{-\lambda_{cjR}}{|\lambda_{cj}|}, \tag{76}$$

respectively. Reynders and De Roeck (2008) showed that, when at least one driving point measurement is made, and the response DOFs are displacements, velocities, or accelerations, the modal scaling factors can be obtained from the following expression:

$$q_j = \text{vec}\left(\frac{i\omega - \lambda_{cj}}{e^{i\omega T} - \lambda_{dj}} \begin{bmatrix} \boldsymbol{\phi}_{jv} \\ i\omega\boldsymbol{\phi}_{j\dot{v}} \\ \omega^2\boldsymbol{\phi}_{j\ddot{v}} \end{bmatrix} \boldsymbol{\phi}_{je}^T\right)^{\dagger} \text{vec}\left(\begin{bmatrix} \boldsymbol{\phi}_{jv} \\ \boldsymbol{\phi}_{j\dot{v}} \\ \frac{1-e^{i\omega T}}{\lambda_{dj}-1}\boldsymbol{\phi}_{j\ddot{v}} \end{bmatrix} \boldsymbol{l}_{dj}^T\right), \tag{77}$$

where the subscripts $\square_v$, $\square_{\dot{v}}$, and $\square_{\ddot{v}}$ select the displacement, velocity, and acceleration response DOFs, respectively. Since in an identified model, it is most probable that the contribution of a mode is modeled best near its resonance frequency, it is suggested to choose $\omega = \omega_j$ for the estimation. Alternatively, a series of frequencies, containing for example also the half power points, could be used. For the scaling of the mode shapes, two schemes are quite popular.

- In the *unit modal displacement* weighting scheme, the mode shape $\boldsymbol{\phi}_j$ is scaled in such a way that one of the elements, usually the one with the largest amplitude, of the scaled mode shape $\boldsymbol{\varphi}_j$ is unity:

$$\boldsymbol{\varphi}_j = \frac{\boldsymbol{\phi}_j}{\phi_{j,k}}. \tag{78}$$

- When a driving point FRF can be calculated, the *unity modal mass* weighting scheme leads to (Heylen et al., 1997)

$$\boldsymbol{\varphi}_j = \sqrt{2\lambda_{cj}q_j}\boldsymbol{\phi}_j. \tag{79}$$

5.3 Distribution of the estimates

In Section 4.2, the distributions of the parameters estimated with SSI-cov/ref were found to be asymptotically normally distributed because of the central limit theorem and the fact that a linear combination of normally distributed random variables is again normally distributed. Since the coefficient of variation of the estimated parameters is usually low, a linear sensitivity analysis of the modal parameters as a function of the model parameters is sufficiently accurate to determine the covariances of the modal parameters. Their distribution is then asymptotically normal.

Discrete-time and continuous-time poles. As shown by Reynders et al. (2008), one has that, for the discrete-time poles λ_{dj},

$$\Delta\lambda_{dj,re} \approx \underbrace{\left(\frac{(\boldsymbol{\psi_{dj}} \otimes \overline{\boldsymbol{\chi_{dj}}})^T}{\boldsymbol{\chi}_{dj}^* \boldsymbol{\psi_{dj}}}\right)_{re}}_{\triangleq \boldsymbol{J}_{\lambda_{dj}}} \mathrm{vec}(\Delta \boldsymbol{A}),$$

where $\boldsymbol{\chi_{dj}}$ and $\boldsymbol{\psi_{dj}}$ denote the left and right eigenvectors corresponding to the eigenvalue λ_{dj} of $\boldsymbol{A}$, respectively.

The relationship between the discrete-time system poles and the continuous-time system poles is given by (75). It follows that the real and imaginary parts of λ_{cj} are given as

$$\lambda_{cjR} = \frac{\ln|\lambda_{dj}|^2}{2T} \qquad \text{and} \qquad \lambda_{cjI} = \frac{1}{T}\tan^{-1}\left(\frac{\lambda_{djI}}{\lambda_{djR}}\right). \tag{80}$$

A linear sensitivity analysis of these expressions leads to (Pintelon et al., 2007)

$$\Delta\boldsymbol{\lambda_{cj,re}} = \boldsymbol{J}_{\boldsymbol{\lambda}_{cj}} \Delta\boldsymbol{\lambda_{dj,re}}, \qquad \text{where} \qquad \boldsymbol{J}_{\boldsymbol{\lambda}_{cj}} \triangleq \frac{1}{T|\lambda_{dj}|^2}\begin{bmatrix} \lambda_{djR} & \lambda_{djI} \\ -\lambda_{djI} & \lambda_{djR} \end{bmatrix}.$$

Eigenfrequencies and damping ratios. The relationship between the eigenfrequencies, damping ratios and the continuous-time system poles is given by (76). A linear sensitivity analysis of these expressions yields (Pintelon et al., 2007)

$$\begin{bmatrix} \Delta f_j \\ \Delta \xi_j \end{bmatrix} = \boldsymbol{J}_{\boldsymbol{f_j \xi_j}} \Delta\boldsymbol{\lambda_{cj,re}}, \quad \text{where} \quad \boldsymbol{J}_{\boldsymbol{f_j \xi_j}} \triangleq \frac{1}{|\lambda_{cj}|}\left[\begin{array}{c|c} \frac{\lambda_{cjR}}{2\pi} & \frac{\lambda_{cjI}}{2\pi} \\ \hline -\frac{\lambda_{cjI}^2}{|\lambda_{cj}|^2} & \frac{\lambda_{cjR}\lambda_{cjI}}{|\lambda_{cj}|^2} \end{array}\right].$$

Mode shapes. In order to derive the first-order sensitivity of the mode shapes, the first-order sensitivity of the eigenvectors $\boldsymbol{\psi_{dj}}$ of $\boldsymbol{A}$ is needed. For the generic case, where $\boldsymbol{A}$ has n different eigenvalues, a first-order perturbation of $\boldsymbol{\psi_{dj}}$ with respect to the elements of $\boldsymbol{A}$ was derived by Reynders et al. (2008):

$$\Delta\boldsymbol{\psi}_{dj} \approx \underbrace{\left(\boldsymbol{I} - \frac{\boldsymbol{A}}{\lambda_{dj}}\right)^{\dagger} \frac{1}{\lambda_{dj}} \left(\boldsymbol{\psi}_{dj}^T \otimes \left(\boldsymbol{I} - \frac{\boldsymbol{\psi}_{dj}\boldsymbol{\chi}_{dj}^*}{\boldsymbol{\chi}_{dj}^*\boldsymbol{\psi}_{dj}}\right)\right)}_{\triangleq \boldsymbol{J}_{\boldsymbol{\psi}_{dj}}} \mathrm{vec}(\boldsymbol{\Delta A}). \tag{81}$$

Using Kronecker algebra, the first-order perturbation of the corresponding mode shape $\boldsymbol{\phi}_j$ can be calculated:

$$\Delta\boldsymbol{\phi}_{j,re} \approx \left(\boldsymbol{C}\Delta\boldsymbol{\psi}_{dj} + \Delta\boldsymbol{C}\boldsymbol{\psi}_{dj}\right)_{re} = \left[\; \boldsymbol{J}_{\boldsymbol{\psi}_{dj}} \mid \boldsymbol{\psi}_{dj}^T \otimes \boldsymbol{I}_{n_y} \;\right]_{re} \begin{bmatrix} \mathrm{vec}(\Delta\boldsymbol{A}) \\ \mathrm{vec}(\Delta\boldsymbol{C}) \end{bmatrix}.$$

When the modes are scaled to a unit modal displacement in one of the degrees of freedom, a first-order sensitivity analysis of (78) leads to

$$\Delta\boldsymbol{\varphi}_{j,re} \approx \underbrace{\left(\frac{1}{\phi_{j,k}}\left(\boldsymbol{I} - \boldsymbol{\varphi}_j \boldsymbol{S}_{\boldsymbol{\phi}_{j,k}}\right)\right)_{Re}}_{\boldsymbol{J}_{\boldsymbol{\varphi}_j}} \Delta\boldsymbol{\phi}_{j,re},$$

where $\boldsymbol{S}_{\boldsymbol{\phi}_{j,k}} \triangleq \begin{bmatrix}\boldsymbol{0}_{1,k-1} & 1 & \boldsymbol{0}_{1,n_y-k}\end{bmatrix}$.

Covariance matrix of the modal parameters. Combining the sensitivity formulae for the modal parameters derived above, their complete covariance matrix is obtained:

$$\mathrm{Cov}\left(\begin{bmatrix} f_j \\ \xi_j \\ \boldsymbol{\varphi}_{j,re} \end{bmatrix}, \begin{bmatrix} f_l \\ \xi_l \\ \boldsymbol{\varphi}_{l,re} \end{bmatrix}\right) = \boldsymbol{J}_{f\xi\varphi,j} \mathrm{cov}\left(\begin{bmatrix} \mathrm{vec}(\hat{\boldsymbol{A}}) \\ \mathrm{vec}(\hat{\boldsymbol{C}}) \end{bmatrix}\right) \boldsymbol{J}_{f\xi\varphi,l}^T, \tag{82}$$

where

$$\boldsymbol{J}_{f\xi\varphi,j} \triangleq \begin{bmatrix} \boldsymbol{J}_{f_j\xi_j}\boldsymbol{J}_{\boldsymbol{\lambda}_{cj}}\boldsymbol{J}_{\boldsymbol{\lambda}_{dj}} & \boldsymbol{0}_{2\times n_y n} \\ \boldsymbol{J}_{\boldsymbol{\psi}_{dj}}\boldsymbol{J}_{\boldsymbol{\varphi}_j} & \left(\boldsymbol{\psi}_{dj}^T \otimes \boldsymbol{I}_{n_y}\right)\boldsymbol{J}_{\boldsymbol{\varphi}_j} \end{bmatrix}.$$

Distribution of damping ratios for lightly damped modes. When a mode is lightly damped, it follows from (80) that the real part of its continuous-time pole, λ_{cjR}, is not only close to zero, but it also has a potentially large coefficient of variation, such that a linear sensitivity analysis, as presented above, is not sufficiently accurate for determining its asymptotic *Probability Density Function* (PDF). This affects significantly the PDF of the damping ratio, see (76), but not of the other modal parameters. Since the damping ratio ξ_j is a nonlinear function of the discrete-time system pole λ_{dj} only, and since the real and imaginary parts of λ_{dj} are to a good approximation asymptotically jointly normally distributed, the PDF of ξ_j can be calculated using a Monte Carlo simulation, where the joint PDF of $\lambda_{dj,re}$, estimated using the linear sensitivity approach, is sampled. Because each Monte Carlo simulation step is computationally very cheap, the total computational cost of this approach is sufficiently low for modal testing.

6 Applications

6.1 Introduction.

In this section, the application of operational modal analysis techniques, with and without exogenous forces, is investigated for two different types of structures.

In Section 6.2, the feasibility of using an exogenous force in operational modal testing of a prestressed concrete bridge is investigated. The performance of the CSI-data/ref algorithm on the shaker data of the Z24 bridge, that have been proposed as a benchmark for testing modal parameter estimation algorithms, is investigated and compared with previously reported results.

The second application deals with operational modal analysis of a steel transmitter mast under wind loading, and is presented in Section 6.3. The goal is to demonstrate the feasibility of estimating confidence intervals on modal parameters obtained from a *single* operational modal test.

6.2 Z24 bridge

Introduction. The Z24 bridge was part of the road connection between the villages of Koppigen and Utzenstorf, Switzerland, over-passing the A1 highway between Bern and Zürich. It was a classical post-tensioned concrete two-cell box-girder bridge with a main span of 30m and two side spans of 14m, see Figure 8. The bridge was built as a free standing frame with the approaches backfilled later. Both abutments consisted of triple concrete columns connected with concrete hinges to the girder. Both intermediate supports were concrete piers clamped into the girder. An extension of the bridge girder at the approaches provided a sliding slab. All supports were rotated with respect to the longitudinal axis which yielded a skew bridge. The bridge, that dated from 1963, was demolished at the end of 1998, because a new railway adjacent to the highway required a new bridge with a larger side span.

Before complete demolition, the bridge was subjected to a long-term continuous monitoring test and a short-term progressive damage test in the

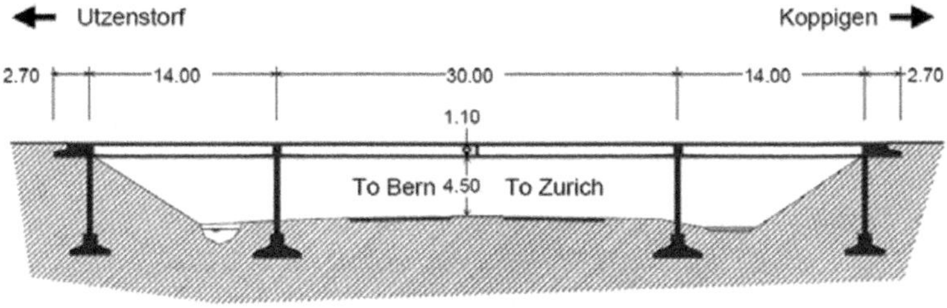

Figure 8. Side view of the Z24 bridge. Distances are in m.

framework of the Brite-EuRam project CT96 0277 SIMCES (De Roeck, 2003).

- A *long-term continuous monitoring test* took place during the year before demolition. The aim was to quantify the environmental variability of the bridge dynamics.
- A *Progressive Damage Test* (PDT) took place in a one-month time period, shortly before complete demolition. The aim was to prove experimentally that realistic damage has a measurable influence on bridge dynamics. Each PDT step alternated with short-term modal tests, while the continuous monitoring system was still running during these tests.

In order for the applied damage scenarios to be significant and realistic, it was made sure that (1) they were relevant for the safety of the bridge and (2) the simulated damage occurred frequently, a condition which was checked in the literature and by questioning Swiss bridge owners. Since the A1 highway was never closed to traffic, some damage scenarios that meet these criteria could not be applied without reducing the safety of the traffic which was considered of paramount importance. The traffic on the Z24 bridge was diverted to the A36 highway. Table 1 gives a complete overview of all progressive damage tests that were performed.

Before and after each applied damage scenario, the bridge was subjected to a forced and an ambient vibration test. Since the ambient forces such as wind excitation or traffic under the bridge could not be excluded during the vibration measurements, all modal tests can be considered as operational tests, with or without the use of artificial (exogenous) forces. With a measurement grid consisting of a regular 3×45 grid on top of the bridge deck and a 2×8 grid on each of the two pillars, 291 degrees of freedom were measured: all three displacement components on the pillars, and mainly vertical and lateral displacements on the bridge deck. Because of the limited num-

No.	Date (1998)	Scenario	Description / Simulation of real damage cause
1	04.08	1st reference measurement	Healthy structure
2	09.08	2nd reference measurement	After installation of lowering system
3	10.08	Lowering of pier, 20 mm	Settlement of subsoil, erosion
4	12.08	Lowering of pier, 40 mm	
5	17.08	Lowering of pier, 80 mm	
6	18.08	Lowering of pier, 95 mm	
7	19.08	Tilt of foundation	Settlement of subsoil, erosion
8	20.08	3rd reference measurement	After lifting of bridge to initial position
9	25.08	Spalling of concrete, 12 m^2	Vehicle impact, carbonisation and
10	26.08	Spalling of concrete, 24 m^2	subsequent corrosion of reinforcement
11	27.08	Landslide at abutment	Heavy rainfall, erosion
12	31.08	Failure of concrete hinge	Chloride attack, corrosion
13	02.09	Failure of anchor heads I	Corrosion, overstress
14	03.09	Failure of anchor heads II	
15	07.09	Rupture of tendons I	Erroneous or forgotten injection of
16	08.09	Rupture of tendons II	tendon tubes, chloride influence
17	09.09	Rupture of tendons III	

Table 1. Z24 progressive damage test: overview of applied damage scenarios.

ber of accelerometers and acquisition channels, the data were collected in 9 setups using 5 reference channels, see Figure 9. The forced excitation was applied by two vertical shakers of EMPA, Switzerland, placed on the bridge deck. A 1kN shaker was placed on the middle span and a 0.5kN shaker was placed at the Koppigen side span, see figure 8. The shaker input signals were generated with an inverse FFT algorithm, resulting in a fairly flat force spectrum between 3 and 30Hz. After scenario 8, a drop weight test was also performed. The applied shaker and drop weight forces were periodic with 8 periods. In each modal test, a total of 65536 samples were collected at a sampling rate of 100Hz, using an anti-aliasing filter with a 30Hz cut-off frequency. Krämer et al. (1999) provide a detailed description of all vibration tests on the Z24 bridge.

The measurement data have been used for two benchmarks.

- The shaker, ambient and drop weight vibration data from the third reference measurement on the Z24 bridge (scenario 8, Table 1) were presented as a benchmark study for system identification methods for operational modal analysis at the IMAC XIX conference in 2001.
- The data from the long-term continuous monitoring test and the progressive damage test were presented as benchmark data for al-

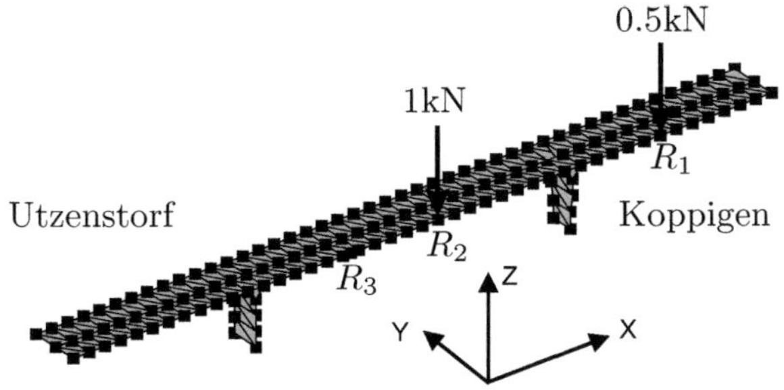

Figure 9. Z24 bridge: Measurement grid, reference positions and shaker positions.

gorithms for structural health monitoring and damage identification in the framework of the European Cost Action F3.

Reynders and De Roeck (2009) provide a literature review of benchmark results. Here, the benchmark shaker data from PDT scenario 8 are used to test the performance of the CSI-data/ref algorithm with respect to the results presented in the literature.

Previous results. Peeters and Ventura (2003) compared the modal parameter estimates obtained by 7 different research teams in the framework of the system identification benchmark with the data from scenario 8. In addition, new modal parameter estimation techniques have been validated on the benchmark data, such as a parametric and nonparametric setup assembly approach followed by maximum likelihood estimation, proposed by Parloo et al. (2003), and an iterative SDOF technique, proposed by Allen and Ginsberg (2006). The best reported result was obtained by applying a subspace identification algorithm (Peeters and Ventura, 2003) and a maximum likelihood algorithm (Parloo et al., 2003) to the shaker data. In this way, 10 modes could be determined.

Results obtained with CSI-data/ref. When analyzing the shaker data of damage scenario 8 for the system identification benchmark, both the classical CSI-data algorithm and the reference-based version CSI-data/ref were used. For CSI-data, $\imath = 30$ was chosen, and a stabilisation diagram of good quality was constructed up to a model order of 160. For CSI-data/ref, the 5 acceleration channels common to every setup were chosen to be the references for the identification, $\imath = 50$ was chosen, and a stabilisation diagram of good quality was constructed up to a model order of 160. Note that, in order to obtain similar computation time and memory usage, the values of $\imath$ differ for CSI-data and CSI-data/ref. A choice of $\imath = 30$ for both CSI-data and CSI-data/ref would have resulted in a faster calculation and a smaller memory usage for CSI-data/ref. According to the rule of thumb (65), the choice of $\imath$ corresponds to $f_0 = 1.67$Hz for CSI and $f_0 = 1$Hz for CSI/ref.

From the stabilisation diagram constructed with the CSI-data and CSI-data/ref methods, 13 and 14 modes could be identified, respectively. Table 2 shows the sample mean values of the eigenfrequencies and damping ratios, obtained with each method, as well as the sample standard deviations, calculated from the 9 different setups. The CSI-data/ref method clearly yields more accurate estimates of both eigenfrequencies and damping ratios for modes 3, 6 and 8, while the opposite is true for mode 7. With the CSI-data/ref method, one more mode could be obtained than with the CSI-data method. Table 2 also shows the MAC values between the corresponding mode shapes determined using CSI-data and CSI-data/ref. The mode shapes are all very well correlated. An almost perfect correlation is observed for modes 1, 2, 5, 6, 7, 8 and 10. Mode 9 has the lowest MAC value, which indicates that the mode shape estimated with one or both methods is of a lower quality than the other modes. This corresponds to a relatively high uncertainty on the eigenfrequency and damping ratio of mode 9 for both CSI-data and CSI-data/ref.

Figure 10 shows the identified bending modes. Mode 13 could only be identified with CSI-data/ref. The identified lateral modes are plotted in Figure 11. The identification of these modes proves experimentally that the combined deterministic-stochastic subspace identification method enables one to identify modes excited by both forced or ambient loading or a combination of both. Indeed, these lateral modes are almost exclusively excited by ambient forces. Modes 3 and 4 were not detected in previous studies. The quality of the mode shape of mode 4 is lower because the only horizontal reference DOF is located near the center of the middle span, in the zone with almost zero modal displacement. This results in a 'stepped' mode shape.

Due to the skewness of the bridge supports, some modes occur where bending and torsion are combined. Two of these modes are shown in Figure 12. Their eigenfrequencies are closely spaced. Although they look very similar, they are truly different, as is confirmed by their experimental MAC value (Heylen et al., 1997) of 0.18 for CSI-data/ref.

Figure 13 shows the higher torsion modes that were identified. The mode

mode	CSI-data				CSI-data/ref				MAC
	f(Hz)	σ_f(Hz)	ξ(%)	σ_d(%)	f(Hz)	σ_f(Hz)	ξ(%)	σ_d(%)	
1	3.871	0.001	0.89	0.05	3.871	0.002	0.88	0.04	1.00
2	4.823	0.008	1.63	0.06	4.818	0.011	1.66	0.04	1.00
3	6.697	0.127	4.23	1.45	6.722	0.028	3.82	0.62	0.98
4	8.355	0.059	8.91	1.77	8.346	0.104	9.37	1.33	0.96
5	9.769	0.005	1.54	0.03	9.772	0.005	1.57	0.02	1.00
6	10.51	0.011	1.45	0.06	10.50	0.007	1.43	0.04	1.00
7	12.42	0.020	3.11	0.03	12.42	0.025	3.15	0.12	1.00
8	13.21	0.033	4.76	0.29	13.21	0.018	4.72	0.17	1.00
9	17.45	0.212	4.34	0.38	17.52	0.169	3.64	1.39	0.92
10	19.27	0.019	2.43	0.10	19.28	0.022	2.46	0.06	1.00
11	19.68	0.080	5.58	0.31	19.65	0.113	5.51	0.29	0.98
12	26.64	0.054	3.20	0.11	26.62	0.055	3.12	0.11	0.95
13	/	/	/	/	33.18	0.202	4.33	1.78	/
14	37.25	0.198	3.69	0.48	37.20	0.106	3.94	0.61	0.95

Table 2. Z24 bridge, scenario 8: eigenfrequencies f_j and damping ratios ξ_j determined with CSI-data and CSI-data/ref, and MAC values between corresponding mode shapes.

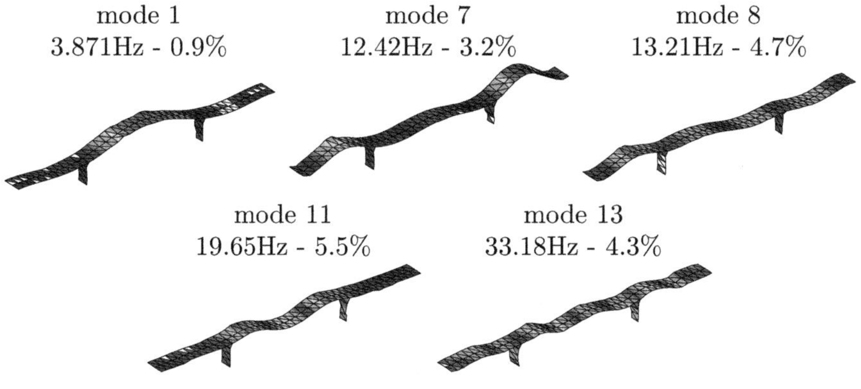

Figure 10. Z24 bridge, scenario 8: bending modes identified with CSI-data(/ref).

shape of mode 9 is less smooth than the mode shapes of the other modes; this corresponds to the higher uncertainty on the corresponding eigenfrequency and damping ratio (Table 2). In mode 12, the Koppigen pier (Figure 9) has the highest participation, which could be due to the fact that a cut

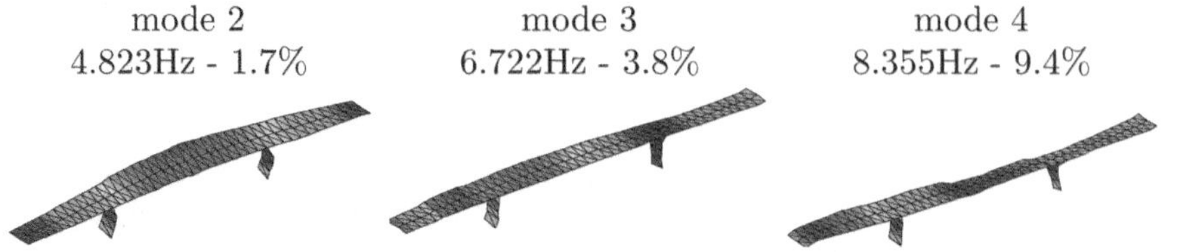

Figure 11. Z24 bridge, scenario 8: lateral modes identified with CSI-data(/ref).

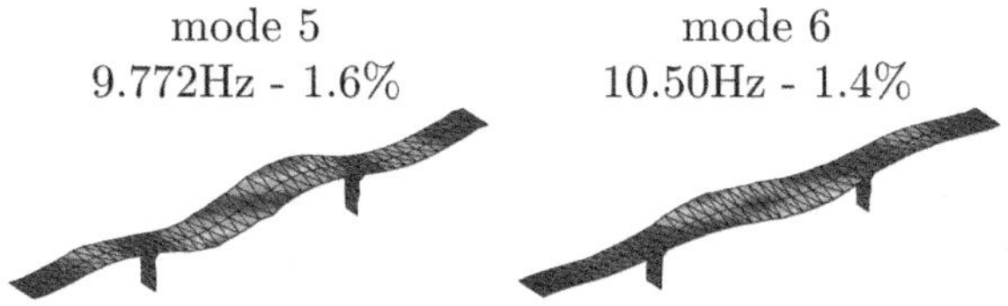

Figure 12. Z24 bridge, scenario 8: two closely spaced mixed torsion/bending modes identified with CSI-data(/ref).

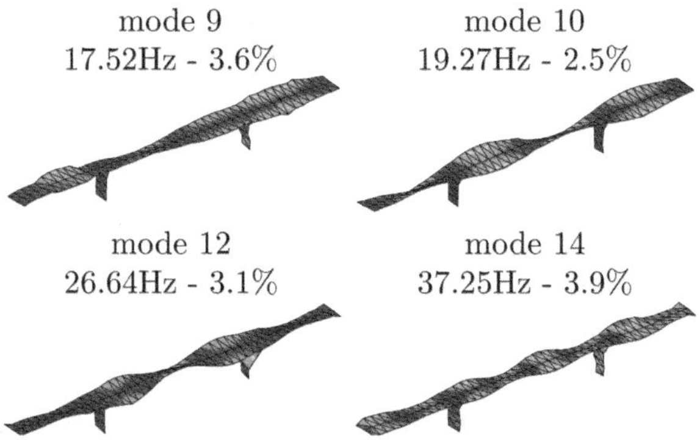

Figure 13. Z24 bridge, scenario 8: higher torsion modes identified with CSI-data(/ref)

was made through this pier for the simulation of a settlement: although the

mode	DOF	CSI		CSI/ref	
		u(mm)	σ_u(mm)	u(mm)	σ_u(mm)
1	R_2z	1.28 + 0.06i	0.05 + 0.04i	1.30 + 0.07i	0.03 + 0.03i
2	R_2y	-1.38 + 0.38i	0.18 + 0.21i	1.38 + 0.15i	0.13 + 0.06i
3	R_2y	-0.22 - 0.22i	0.10 + 0.19i	0.15 + 0.13i	0.09 + 0.30i
4	R_3z	0.27 + 0.05i	0.18 + 0.13i	0.46 + 0.13i	0.20 + 0.16i
5	R_3z	-3.69 - 0.42i	0.42 + 0.14i	3.75 + 0.55i	0.36 + 0.22i
6	R_2z	1.16 + 0.29i	0.05 + 0.07i	-1.11 - 0.25i	0.05 + 0.05i
7	R_1z	-4.74 - 0.83i	0.23 + 0.09i	-4.65 - 0.75i	0.27 + 0.19i
8	R_1z	-3.01 - 0.63i	0.41 + 0.19i	2.96 + 0.70i	0.16 + 0.13i
9	R_1z	-3.11 - 1.35i	0.55 + 0.59i	-2.34 - 1.04i	1.29 + 0.71i
10	R_3z	3.08 + 1.05i	0.33 + 0.19i	3.12 + 1.16i	0.31 + 0.07i
11	R_2z	-0.92 - 0.33i	0.38 + 0.18i	-0.94 - 0.19i	0.23 + 0.18i
12	R_2z	-1.36 - 0.42i	0.08 + 0.11i	1.31 + 0.32i	0.11 + 0.12i
13	R_2z	/	/	-0.39 - 0.07i	0.19 + 0.54i
14	R_1z	1.53 + 1.55i	0.58 + 0.34i	-1.73 - 1.63i	0.47 + 0.39i

Table 3. Z24 bridge, scenario 8: modal displacements, scaled to unity modal mass. The standard deviations consist of the standard deviation of the real part and the standard deviation of the imaginary part of the corresponding displacement. They are the sample standard deviations for the total of nine setups.

settlement was removed, the original state was only approximately reached. The identification of mode 14 at 37.25Hz, despite the fact that the cut-off frequency of the analog anti-aliasing filter was set to 30Hz, indicates that the method is able to identify modes that are only very weakly present in the data.

Modes 1, 2, 5, 6, 7, 8, 9, 10, 11 and 12 have been detected in previous benchmark analyses; the other modes were only detected in the present study. Because CSI-data and CSI-data/ref identify a combined deterministic-stochastic state-space model, they enable a mass-normalisation of the modes that are excited by the deterministic forces. Table 3 shows, for each scaled mode, the largest of the reference displacement DOFs (Figure 9). From this table, bearing in mind that none of the identified modes is truly complex, the following can be noticed.

- For modes 1, 5, 6, 7, 8, 10, 11 and 12, all in the frequency band of forced excitation (3-30Hz), the largest reference displacements are in the direction of the forced excitation and the real part of the scaled mode shapes is much larger than the imaginary part. This indicates that they are properly scaled.

- Although mode 2 is mainly a lateral mode and the forced excitation is purely vertical, the real part of the scaled mode shape is much larger than the imaginary part. Together with the small sample variance of the modal displacement, this is an indication that the mode is properly scaled.
- As indicated by the large imaginary part and the large sample variance, the lateral modes 3 and 4 are not properly scaled because they are not well excited by the artificial forces. However, using an OMAX approach, it is possible to identify these modes because they are excited by ambient forces.
- The rather large uncertainty on the scaled displacements of mode 9 corresponds to the large uncertainty on the frequency and damping ratio (Table 2) and the quality of the mode shape (Figure 13).
- Modes 13 and 14 can not be properly scaled, as their eigenfrequencies lie outside the frequency band of the forced excitation and above the cut-off frequency of the anti-aliasing filter.

6.3 Steel transmitter mast

Introduction. In this section, the SSI-cov/ref algorithm is applied to the operational modal analysis of a truss structure, situated in the harbor of Antwerp, Belgium. It is a mast of 30m height which contains sectorial antennae for a cellular phone network at the top (Figure 14). The antennae comprise about 10% of the total weight of the structure.

Vibration measurements and signal preprocessing. On March 26, 1998, an ambient vibration test was performed on the structure. Peeters and De Roeck (1999) presented the results of this test. Here, their uncertainty is investigated.

The aim of the test was to investigate the structure's modal parameters, particularly the damping ratios, in the frequency range of $0 - 5$Hz. 17 degrees of freedom (DOFs), all horizontal accelerations, have been measured in 3 setups using 3 reference DOFs that were common to each setup. Three horizontal accelerations were measured at heights of 6.17m, 12.17m, 18.17m, 24.17m and 29.90m. If it is assumed that the cross-section of the mast remains undeformed during the vibration test, which is a reasonable assumption in the frequency band considered, three DOFs are sufficient to describe the complete horizontal movement of the cross-section. Two orthogonal accelerations at the top of the mast (at a height of 33.00m) were measured as well.

Figure 14. Steel mast with sectorial antennae at the top.

The data were sampled at a rate of 100Hz. The cut-off frequency of the anti-aliasing filter was set to 20Hz. The number of samples was set to $N = 30720$, which resulted in a measurement time of approximately 5 minutes. Afterwards, the data were digitally filtered with a low-pass filter with a cut-off frequency of 5Hz and re-sampled at 12.5Hz, which reduced the number of samples to $N = 3840$.

Modal parameters. With SSI-cov/rev, the eigenfrequencies, damping ratios, and mode shapes, as well as covariances of these modal parameters, are identified for each individual setup. The identification parameters are $\imath = 10$, $n = 20$ and $n_b = 100$. According to the rule of thumb (65), the choice of $\imath$ corresponds to $f_0 = 0.63$Hz. The parts of the mode shapes that result from the different setups are combined with linear least squares. They are scaled to 1 in one of the reference DOFs. The variances of the different modal displacements are adopted accordingly.

The identified eigenfrequencies and their estimated 95% confidence intervals are shown in Table 4. The eigenfrequency estimates obtained in each setup match very well. The estimated confidence intervals correspond very well with the variation of the actual values from setup to setup.

The identified damping ratios and their estimated 95% confidence intervals are shown in Table 5. The damping ratio estimates differ quite a lot from setup to setup, which is also reflected in the high values of the

mode	setup 1	setup 2	setup 3
1	1.172 ± 0.003	1.168 ± 0.006	1.166 ± 0.004
2	1.178 ± 0.005	1.178 ± 0.005	1.180 ± 0.005
3	1.948 ± 0.009	1.956 ± 0.008	1.951 ± 0.004
4	2.603 ± 0.006	2.599 ± 0.008	2.601 ± 0.006
5	2.711 ± 0.002	2.712 ± 0.007	2.711 ± 0.004
6	3.686 ± 0.008	3.689 ± 0.005	3.685 ± 0.006
7	4.632 ± 0.010	4.633 ± 0.011	4.631 ± 0.008

Table 4. Steel mast: identified eigenfrequencies (in Hz) with their estimated 95% confidence interval.

estimated confidence intervals. Because of their high relative uncertainty, the confidence intervals of the damping ratios are calculated using a Monte Carlo sampling of the discrete-time poles, which have a low relative uncertainty. The values obtained from a linear sensitivity approach are given between brackets. Despite the large relative uncertainty on the damping ratios, both estimates for the confidence intervals agree very well.

The identified mode shapes and their estimated 95% confidence interval are shown in Figure 15. Only the third mode is a torsion mode, the other modes are bending modes. It can be seen that the mode shapes of modes 3 through 6 are very accurate, mode 7 is fairly accurate and modes 1 and 2 are inaccurate. This difference in mode shape quality is also apparent when the real and imaginary parts of the identified modes are plotted in the complex plane (see Figure 16): while modes 3 through 7 are fairly real, modes 1 and 2 have important imaginary components.

This can be physically understood from the fact that the structure is almost triply symmetric, since perfect multiple symmetric structures have pairs of eigenmodes with identical eigenfrequencies, reflecting that the mode shapes are indefinite. Dooms et al. (2006) illustrate this for a nearly axisymmetric structure. Here, the symmetry is somewhat disturbed by the diagonals and a ladder at one of the masts. Only the first two modes are very close, with estimated eigenfrequencies of 1.172 and 1.178Hz in the first setup. The next pair of bending modes are already better separated, thanks to the symmetry disruptions, with eigenfrequencies at 2.602 and 2.711Hz.

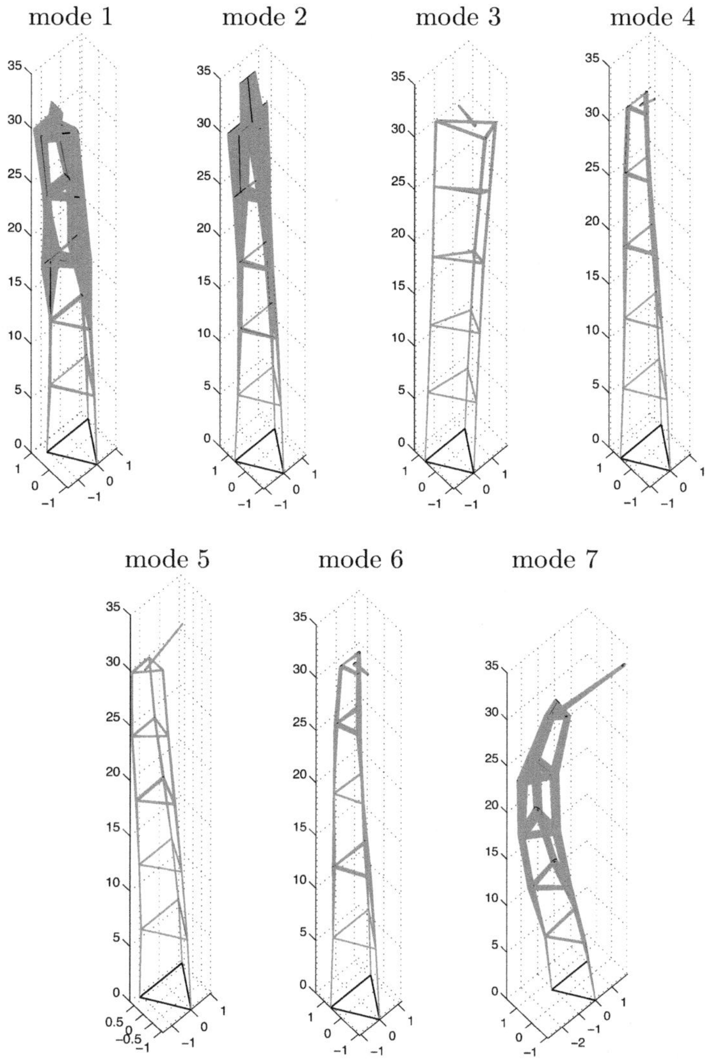

Figure 15. Steel mast: identified mode shapes (black) with their estimated 95% confidence interval (grey).

mode	setup 1	setup 2	setup 3
1	0.14 ± 0.26 (0.26)	0.55 ± 0.42 (0.42)	0.60 ± 0.35 (0.37)
2	0.67 ± 0.33 (0.37)	0.57 ± 0.42 (0.43)	0.95 ± 0.45 (0.45)
3	0.64 ± 0.32 (0.31)	0.80 ± 0.27 (0.26)	0.47 ± 0.22 (0.22)
4	0.36 ± 0.26 (0.28)	0.57 ± 0.39 (0.39)	0.35 ± 0.21 (0.21)
5	0.08 ± 0.09 (0.10)	0.27 ± 0.19 (0.22)	0.19 ± 0.11 (0.12)
6	0.28 ± 0.32 (0.33)	0.17 ± 0.18 (0.19)	0.27 ± 0.19 (0.20)
7	0.27 ± 0.23 (0.23)	0.16 ± 0.46 (0.51)	0.18 ± 0.15 (0.17)

Table 5. Steel mast: identified damping ratios (in %) with their estimated 95% confidence interval, estimated from a Monte Carlo sampling of the discrete-times poles and using a linear sensitivity analysis (between brackets).

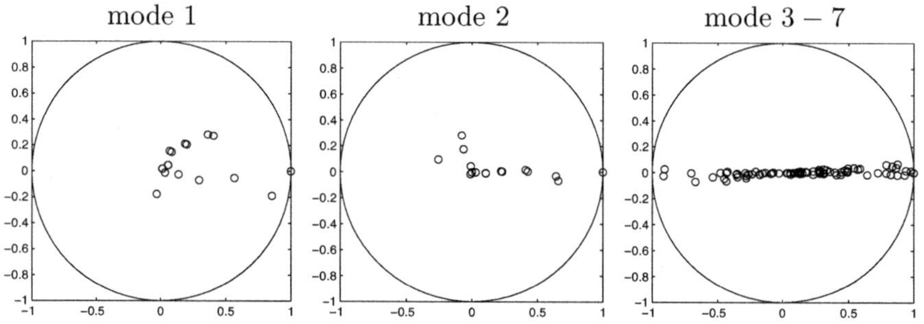

Figure 16. Steel mast: real vs. imaginary part of the identified modal displacements, scaled to 1 at the largest displacement value.

7 Conclusions

This chapter addressed the extraction of eigenfrequencies, damping ratios, mode shapes and, whenever possible, modal scaling factors, from measured operational data.

A conversion between finite element models, that are frequently used in forward modeling of vibrating structures, and discrete-time state-space models, that are identified from measured data, was derived. This confirms that the latter models are valid also from a forward point of view. Furthermore, it was shown that modeling the operational excitation due to turbulent fluid pressure, micro-tremors, or roadway or footfall excitation, as stochastic quantities, has a clear physical basis. This justifies the use of stochastic system identification algorithms for OMA, and the use of combined deterministic-stochastic system identification algorithms for OMAX.

The use of subspace identification methods and their predecessors, system realisation methods, for the identification of stochastic and combined deterministic-stochastic state-space models from operational data, was discussed. The general ideas that lie behind this class of system identification methods were separated from the algorithms themselves. Two subspace algorithms that are commonly used for OMA were discussed in detail: SSI-cov/ref and SSI-data/ref. For SSI-cov/ref, the probability density function of the identified state-space matrices was derived. Recently, it was shown that the SSI-data/ref is statistically asymptotically efficient, i.e., no estimator with lower asymptotic covariance can be found. This confirms previous experience with subspace identification algorithms, that have shown to outperform alternative system identification algorithms in several comparative studies, see, e.g., Peeters and De Roeck (2001). Since OMAX testing can enhance the current ambient vibration testing practice in the sense that it allows the mode shapes to be mass-normalised and the frequency content of the excitation to be extended, a recently developed subspace algorithm that can be used for OMAX, was presented: CSI-data/ref.

In a first real-life application, this CSI-data/ref algorithm was seen to yield the most complete set of modal parameters for the Z24 bridge benchmark data reported so far. In a second real-life application, the estimation of confidence intervals of modal parameters identified with SSI-cov/ref was illustrated for a steel transmitter mast.

Bibliography

H. Akaike. Stochastic theory of minimal realization. *IEEE Transactions on Automatic Control*, 19(6):667–674, 1974.

H. Akaike. Markovian representation of stochastic processes by canonical variables. *SIAM journal on control*, 13(1):162–173, 1975.

M.S. Allen and J.H. Ginsberg. A global, single-input-multiple-output (SIMO) implementation of the algorithm of mode isolation and application to analytical and experimental data. *Mechanical Systems and Signal Processing*, 20(5):1090–1111, 2006.

B.D.O. Anderson and J.B. Moore. *Optimal filtering*. Prentice Hall, Englewood Cliffs, NJ, 1979.

K.S. Arun and S.Y. Kung. Balanced approximation of stochastic systems. *SIAM journal on matrix analysis and applications*, 11:42–68, 1990.

D. Bauer. Comparing the CCA subspace method to pseudo maximum likelihood methods in the case of no exogenous inputs. *Journal of Time Series Analysis*, 26(5):631–668, 2005.

D. Bauer and L. Ljung. Some facts about the choice of the weighting matrices in Larimore type of subspace algorithms. *Automatica*, 38(5):763–773, 2002.

D. Bauer, M. Deistler, and W. Scherrer. On the impact of weighting matrices in subspace algorithms. In *Proceedings of the 12th IFAC symposium on system identification SYSID 2000*, Santa Barbara, CA, June 2000.

A. Ben-Israel and T. Greville. *Generalized inverses.* John Wiley, New York, NY, 1974.

H. Braun and T. Hellenbroich. Messergebnisse von Strassenunebenheiten. *VDI Berichte*, 877:47–80, 1991.

J.W. Brewer. Kronecker products and matrix calculus in system theory. *IEEE Transactions on Circuits and Systems*, 25(9):772–781, 1978.

B. Cauberghe. *Applied frequency-domain system identification in the field of experimental and operational modal analysis.* PhD thesis, Vrije Universiteit Brussel, 2004.

A. Chiuso and G. Picci. The asymptotic variance of subspace estimates. *Journal of Econometrics*, 118(1–2):257–291, 2004.

R.W. Clough and J. Penzien. *Dynamics of structures.* McGraw-Hill, New York, third edition, 1995.

G. De Roeck. The state-of-the-art of damage detection by vibration monitoring: the SIMCES experience. *Journal of Structural Control*, 10:127–143, 2003.

D. Dooms, G. Degrande, G. De Roeck, and E. Reynders. Finite element modelling of a silo based on experimental modal analysis. *Engineering Structures*, 28(4):532–542, 2006.

E.R. Dougherty. *Random processes for image and signal processing.* SPIE Press, Bellingham, WA, 1999.

P.A. Durbin and G. Medic. *Fluid dynamics with a computational perspective.* Cambridge University Press, New York, NY, 2007.

P.A. Durbin and B.A. Petterson Reif. *Statistical theory and modeling for turbulent flows.* John Wiley & Sons, New York, NY, 2001.

G.F. Franklin, J.D. Powell, and M. Workman. *Digital control of dynamic systems.* Addison-Wesley, Menlo Park, CA, 1998.

I. Goethals. *Subspace identification for linear, Hammerstein and Hammerstein-Wiener systems.* PhD thesis, Katholieke Universiteit Leuven, 2005.

W. Heylen, S. Lammens, and P. Sas. *Modal analysis theory and testing.* Department of Mechanical Engineering, Katholieke Universiteit Leuven, Leuven, Belgium, 1997.

B. Ho and R. Kalman. Effective recontruction of linear state-variable models from input/output functions. *Regelungstechnik*, 14(12):545–548, 1966.

J.-N. Juang. *Applied system identification.* Prentice-Hall, Englewood Cliffs, NJ, 1994.

J.-N. Juang and R.S. Pappa. An eigensystem realization algorithm for modal parameter identification and model reduction. *Journal of Guidance, Control and Dynamics*, 8(5):620–627, 1985.

J.-N. Juang, J.E. Cooper, and J.R. Wright. An eigensystem realization algorithm using data correlations (ERA/DC) for modal parameter identification. *Control-theory and advanced technology*, 4(1):5–14, 1988.

R. Kalman. A new approach to linear filtering and prediction problems. *Journal of Basic Engineering, Transactions of the ASME*, 82D:35–45, 1960.

R. Kalman. Mathematical description of linear dynamical systems. *SIAM journal on control*, 1(2):152–192, 1963.

C. Krämer, C.A.M. de Smet, and G. De Roeck. Z24 bridge damage detection tests. In *Proceedings of IMAC 17, the International Modal Analysis Conference*, pages 1023–1029, Kissimmee, FL, USA, February 1999.

S.Y. Kung. A new identification and model reduction algorithm via singular value decomposition. In *Proceedings of the 12th Asilomar conference on circuits, systems and computers*, pages 705–714, Pacific Grove, CA, 1978.

W.E. Larimore. Canonical variate analysis in identification, filtering and adaptive control. In *Proceedings of the 29th conference on decision and control*, pages 596–604, Honolulu, HI, December 1990.

L. Ljung. *System identification: theory for the user.* Prentice Hall, Upper Saddle River, NJ, second edition, 1999.

H. Nyquist. Thermal agitation of electric charge in conductors. *Physical Review*, 32:110–113, 1928.

ORE. Question C116: Wechselwirkung zwischen Fahrzeugen und gleis, Bericht Nr. 1: Spektrale Dichte der Unregelmässigkeiten in der Gleislage. Technical report, Office for Research and Experiments of the International Union of Railways, Utrecht, NL, 1971.

E. Parloo, P. Guillaume, and B. Cauberghe. Maximum likelihood identification of non-stationary operational data. *Journal of Sound and Vibration*, 268(5):971–991, 2003.

B. Peeters and G. De Roeck. Stochastic system identification for operational modal analysis: A review. *ASME Journal of Dynamic Systems, Measurement, and Control*, 123(4):659–667, 2001.

B. Peeters and G. De Roeck. Reference-based stochastic subspace identification for output-only modal analysis. *Mechanical Systems and Signal Processing*, 13(6):855–878, 1999.

B. Peeters and C. Ventura. Comparative study of modal analysis techniques for bridge dynamic characteristics. *Mechanical Systems and Signal Processing*, 17(5):965–988, 2003.

R. Pintelon, J. Schoukens, and P. Guillaume. Box-Jenkins identification revisited - part III: Multivariable systems. *Automatica*, 43(5):868–875, 2006.

R. Pintelon, P. Guillaume, and J. Schoukens. Uncertainty calculation in (operational) modal analysis. *Mechanical Systems and Signal Processing*, 21(6):2359–2373, 2007.

R. Pintelon, B. Peeters, and P. Guillaume. Continuous-time operational modal analysis in the presence of harmonic disturbances. *Mechanical Systems and Signal Processing*, 22(5):1017–1035, 2008.

E. Reynders and G. De Roeck. Reference-based combined deterministic-stochastic subspace identification for experimental and operational modal analysis. *Mechanical Systems and Signal Processing*, 22(3):617–637, 2008.

E. Reynders and G. De Roeck. Continuous vibration monitoring and progressive damage testing on the Z24 bridge. In C. Boller, F.K. Chang, and Y. Fujino, editors, *Encyclopedia of Structural Health Monitoring*, pages 2149–2158. John Wiley & Sons, 2009.

E. Reynders, R. Pintelon, and G. De Roeck. Uncertainty bounds on modal parameters obtained from Stochastic Subspace Identification. *Mechanical Systems and Signal Processing*, 22(4):948–969, 2008.

P. Van Overschee and B. De Moor. *Subspace identification for linear systems.* Kluwer Academic Publishers, Dordrecht, The Netherlands, 1996.

P. Van Overschee and B. De Moor. N4SID: Subspace algorithms for the identification of combined deterministic-stochastic systems. *Automatica*, 30(1):75–93, 1994a.

P. Van Overschee and B. De Moor. A unifying theorem for three subspace system identification algorithms. In *American Control Conference*, pages 1645–1649, Baltimore, MD, June–July 1994b.

M. Verhaegen. Identification of the deterministic part of MIMO state space models given in innovations from input-output data. *Automatica*, 30(1): 61–74, 1994.

H.P. Zeiger and A.J. McEwen. Approximate linear realizations of given dimension via Ho's algorithm. *IEEE Transactions on Automatic Control*, 19:153, 1974.

Vibration-Based Structural Health Monitoring Under Variable Environmental or Operational Conditions

Jyrki Kullaa *‡

* Department of Applied Mechanics, Helsinki University of Technology, Espoo, Finland

‡ Department of Mechanical and Production Engineering, Helsinki Metropolia University of Applied Sciences, Helsinki, Finland

Abstract The main postulate in vibration-based structural health monitoring (SHM) is that structural damage can be detected from changes in the damage-sensitive features extracted from vibration measurements. In order to detect damage with a high sensitivity and reliability, several functions are needed. Control charts are applied to detect statistically significant changes in the features, sensor faults are identified using the minimum mean square error (MMSE) estimation, and the undesired effects of environmental or operational variations are removed using the linear factor analysis or the nonlinear mixture of linear factor analysers model. Different applications and data sets are analysed, including a wooden bridge and a vehicle crane.

1 Introduction

Structural health monitoring (SHM) of aerospace, civil, and mechanical engineering systems is becoming increasingly important from both the economic and life-safety viewpoints. In vibration-based SHM, it is assumed that the vibration characteristics of the structure change due to damage, and by identifying the new characteristics and comparing them to those of the healthy structure, the existence of damage can be detected. Vibration-based damage detection is considered an attractive technique, because (1) it is a non-destructive testing method, and (2) it is global, i.e. the existence of structural damage can be detected remote from the sensor.

This chapter introduces the different functions needed in a structural health monitoring system. Some functions are studied in more detail, while the others are merely introduced. The underlying idea in SHM is that structural damage can be detected from changes in the monitored damage-sensitive features. These features are often extracted from vibration measurements.

Section 2 gives an overview of the functions needed in a structural health monitoring system. The characteristics of the system are its modularity and automation. The following aspects are discussed: instrumentation, excitation, data acquisition, signal processing, sensor fault identification, feature extraction, feature processing, damage detection, and alarms and reports. An example of a vibration-based structural health monitoring system is presented. Many of the details are postponed to later sections.

Section 3 introduces a technique for sensor validation using the minimum mean square error estimation. Experimental multichannel acceleration measurements with different sensor faults are used to verify the proposed method to detect, identify, and correct the faulty sensor.

Section 4 studies damage detection with an outlier analysis using control charts - one of the primary techniques of statistical process control. An advantage of control charts is that they can be automated for on-line structural health monitoring. A comparison of different control charts is performed using numerical modal data. The following control charts are studied: univariate Shewhart, x, CUSUM, and EWMA charts, and multivariate Shewhart T, Hotelling T, MCUSUM, and MEWMA control charts. The reliability and sensitivity of the control charts are investigated. Differences between off-line and on-line results are discussed. Dimensionality reduction, an essential function in damage detection, is implemented using principal component analysis.

In practice, damage detection may be difficult due to environmental or operational variations influencing the damage-sensitive features and causing false indications of damage. In Section 5, factor analysis is applied to remove these undesired effects. Factor analysis is a latent variable model, which has several advantages. The underlying variables need not be measured and the affecting physical quantities need not be known. The estimation of the model parameters and the elimination of the environmental or operational effects can be automated. Also, several environmental or operational variables can act simultaneously. The proposed method is applied to numerical data and to a wooden bridge in the laboratory.

While Section 5 studies a linear model to take into account environmental or operational variability, Section 6 introduces a mixture of linear factor analysers model, which is used to compensate nonlinear effects. The learning algorithm combines factor analysis and clustering. The model is first applied to numerical data and next to experimental data of the Z24 Bridge, in which the natural frequencies vary due to the temperature.

Section 7 studies SHM in a mechanical engineering application. Vibration-based structural health monitoring is applied to a vehicle crane. The performance of two different features to detect damage is investigated after eliminating the normal operational variations using factor analysis. Damage is introduced by adding additional masses at different locations of the structure.

Finally, Section 8 gives concluding remarks on different functions in SHM and some suggestions for an SHM system design.

2 Functions of a Structural Health Monitoring System

SHM is not just a single technique, but it comprises several functions, each of which must be carefully designed. These functions include (1) instrumentation, (2) excitation, (3) data acquisition, (4) signal processing, (5) sensor fault identification, (6) feature extraction, (7) feature processing, (8) damage detection, and (9) alarms and reports (Figure 1). Each function is briefly discussed in the following. An example of a complete vibration-based structural health monitoring system is presented. The system is modular, which means that the implementation of any function is easy without any need to design the whole system starting from scratch. It is also relatively easy to design an SHM system for a new structure by utilising the modularity.

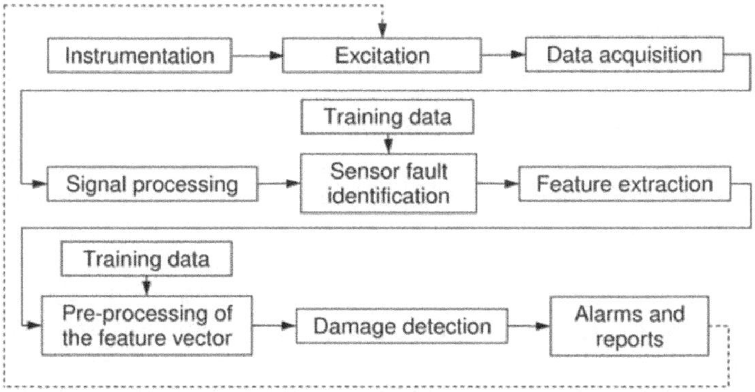

Figure 1. Functions and data flow in a structural health monitoring system.

2.1 SHM Functions

Instrumentation. The question of instrumentation includes the decision of which quantities should be measured and the selection of the transducers and their locations. In SHM with repeated measurements, permanent instrumentation is usually implemented, while temporary solutions are also possible. Traditionally, wired sensors are used, but wireless sensors and sensor networks are undergoing active research. The advantage of wireless sensors is that no wires have to be assembled. This is especially true with a monitoring system consisting of numerous sensors. The wires and connectors are also sensitive to wear. The main issues of wireless sensors are the power consumption, data transmission, and synchronisation. Therefore, special features and algorithms are studied for such systems.

Because vibration measurements can also be used to detect damage remote from the sensor, the number of sensors can vary from one to several hundreds. The sensor type depends on the application. Typical sensors are accelerometers, strain gauges, fiber optics, and lasers (as discussed in the chapter here by Deraemaeker). The measured motion is typically very small and at a low frequency; therefore, the sensitivity of the sensor has to be high. The number and positions of sensors can be determined using a finite element model of the structure and the anticipated damage scenarios and by utilising optimisation or heuristic methods.

Excitation. Excitation is either ambient or artificial. In a typical health monitoring application for civil engineering structures, the excitation cannot be measured. Vibration is caused by wind, traffic, waves, ice, or microseismic tremors. Artificial excitation must be used if ambient excitation is not available. The advantage of artificial excitation is that it can be controlled and measured. This is especially true, because it is essential that the whole frequency range of interest is excited using a broad-band excitation. This can be arranged using a shaker, an impact device, or an already available actuator in the machine. Artificial excitation can be used with smaller structures and in the laboratory. However, with large civil engineering structures with low natural frequencies, artificial excitation is difficult and expensive with a need for large hydraulic shakers and a lot of power. Also the normal operation must be interrupted in order to acquire reliable data. However, there are always disturbances due to wind, ground motion, or other sources making the controlled measurement difficult. The advantages of using ambient excitation are that loading is always available, it is inexpensive, and it easily excites the low frequencies. The main disadvantages are that the force cannot be measured and the frequency content can be an issue.

Data acquisition. A sensor converts a measured quantity to an analogue electrical signal. The objective of the data acquisition system is to store the time series in a computer for further analysis. There are several issues to consider. If the extracted features depend on more than a single sensor, the sensors have to be sampled simultaneously. The sampling rate and the record length must be adjusted to the extraction of the selected features. Also, automatic data acquisition can be based on a clock or a trigger.

Signal digitisation also needs attention. If the sampling rate is f_s, the maximum frequency $f_{\max}$ that can be observed in a sampled signal is,

$$f_{\max} = \frac{f_s}{2} \tag{1}$$

If the sampling rate is lower, frequencies above $f_s/2$ are incorrectly interpreted as lower frequencies. This phenomenon is called *aliasing*. After the A/D converter, the aliasing cannot be corrected. An analogue anti-aliasing filter is therefore always required before the A/D converter.

The analogue signal is converted (*quantised*) to the closest discrete value of the A/D converter. Typically the A/D converters have $n = 10 - 24$ bit resolution. The measurement range is divided into 2^n uniform intervals. For the best resolution, the whole dynamic range should be used. However, the signal must not exceed the limits, which would lead to an overload and consequently clip the signal. The data acquisition system should therefore be able to detect an overload event and reject the measurement. The dynamic range can be found in preliminary tests of the monitored structure.

Signal processing. Signal processing can be used to extract features directly or to act as a pre-processing step in the feature extraction. Typically, signal processing extracts some useful information from the time-histories using their stochastic properties or other assumptions. Filtering is often used to limit the frequency range of the signal; however, it should be noted that the aliasing effect, if present, cannot be removed. In vibration measurements the mean or trend in the signal is usually removed as they normally contain no useful information. The Fast Fourier Transform (FFT) is an important tool to convert signals between the time domain and the frequency domain. The method is fast and no information is lost in the transformation. However, because the Fourier transform is defined for time series of infinite length, the FFT results in a *leakage* effect. The FFT assumes a periodic time series, and if the signal is not periodic, some energy leaks to the adjacent spectral lines. Leakage can be decreased but not completely eliminated using *windowing*. The time series is multiplied with a window function forcing the signal to be (almost) periodic.

Several functions can be extracted from the time series. Some of the most often used functions are the correlation function, power spectral density function, impulse response function, and the frequency response function (FRF). They can be used directly as features or as intermediate variables to extract other features.

Sensor fault identification. Vibration-based structural health monitoring relies heavily on the measurement data recorded during the structure's life. It is therefore most important to detect possible sensor faults to maintain the reliability of the system. Monitoring systems typically include several sensors measuring the global motion, or the lowest modes, of the structure. The sensor system is therefore redundant. Cross-correlation of the sensors can be utilised to detect sensor malfunction or failure, to identify the faulty sensor, and to correct the sensor signal. Section 3 is devoted to sensor validation.

Feature extraction. Time histories include a lot of data, which has to be compressed, resulting in some characteristic properties of the structure. These *features* are considered as "fingerprints" of the system and they are extracted from time-history measurements. The objective is to find features that are sensitive to damage but insensitive to natural variability of the environment (loading, temperature, etc.). A change in the features is then an indication of damage. Because monitoring systems perform repeated measurements during the structure's life, the feature extraction must be automatic. Some of the features are easily extracted automatically, while others may need a lot of supervision and rules to automate. The feature extraction (in some cases equivalent to system identification) can be classified into parametric or non-parametric methods. Another classification is into input/output or output-only methods, depending on the measurability of the excitation. In many civil engineering applications, only output-only identification is possible. The features should then be independent of the amplitude of excitation, which makes some features less attractive for SHM.

Typical features of a structure are the natural frequencies, modeshapes, and modal dampings. Other features include spectral functions, e.g. power spectra, FRF, AR coefficients, transmissibilities, wavelets, and modal filters.

Feature processing. The dimensionality of the feature vector is considerably lower than that of the time series. Each single feature could be used for damage detection, or a multivariate statistic can be utilised. The first step is to select the training data acquired from the undamaged structure which represents the reference structural condition. The training data must also consist of a full range of normal environmental or operational conditions, because they also affect the features and can result in false indications of damage if not properly taken into account.

There can be different types of features, the ranges of which depend on the selected units and can vary considerably. For example natural frequencies may vary from 0.2 to 200 Hz, while the damping ratios are typically between 0.001 and 0.1. When considering damage detection, features with large absolute values or variation tend to dominate. Therefore each feature is usually standardised to have a zero mean and a unit variance within the training data to make the features equally important.

The number of features can still be too high for a reliable statistical analysis (curse of dimensionality). Even ten features may be too many in practice, with a finite amount of training data. Those features can be selected which are most sensitive to damage. However, the required information is usually not available. Another possibility is to let the data decide which features are the most sensitive by selecting all features and utilising all available information. Only features with frequent outliers in the training data and constant features containing no information should be removed. The resulting dimensionality problem can be solved by applying dimensionality reduction techniques, for example principal component analysis (PCA). Later sections discuss different techniques for feature processing.

Damage detection. The objective of damage detection is to determine if damage is present in the structure. The decision must be made using statistical methods, because all features vary between the measurements. A statistically significant change is an indication of damage. In most cases data from the undamaged structure are only available in the training phase and therefore unsupervised learning has to be used. A control chart (Montgomery, 1997) is one of the primary techniques of statistical process control and can also be used for structural health monitoring (Sohn et al., 2000a). It plots a quality characteristic as a function of the sample number. The chart has lower and upper control limits, which are computed from those samples when the process is assumed to be in control. When unusual sources of variability are present, sample statistics will plot outside the control limits and an alarm signal is produced. Different control charts are available; they can be univariate or multivariate. Damage detection using control charts is discussed in Section 4 and utilised in all the examples in this chapter.

Alarms and reports. The monitoring system has to be connected to the network to transfer information about the condition of the structure and the monitoring system itself. At least three functions are necessary. First, the monitoring system must be able to send alarms if it has detected a possible damage. Second, the engineer should have an access to the remote computer to transfer data to the office for an off-line analysis or to change the parameters of the monitoring system. It is also essential that the monitoring system periodically informs the engineer about the state of the system. An automatic alarm and warning system can be designed depending on the application. A good rule of thumb is that at least two consecutive damage detection events are needed for an alarm. Warning messages and reports can be sent on less evident occasions.

2.2 Example

The structure under investigation in the laboratory was the wooden bridge shown in Figure 2. It was supported by a flexible rubber bearing at each corner. The total mass of the structure was 36 kg. The healthy structure was monitored for one day to acquire training data. Two artificial damage scenarios were then introduced by adding small point masses of different size to the structure. During the measurements labelled 95–103, two small weights were inserted on the top flanges. The second damage scenario during measurements 104–112 included weights of different sizes at different locations of the structure. The last measurements 113–119 were again from a healthy structure. The added masses were small compared to the total weight of the structure.

Instrumentation. The bridge was permanently instrumented with fifteen Kistler 8712A5M1 piezo-electric accelerometers (Figure 2). The locations of the accelerometers were determined from an eigenvalue analysis using the finite element model of the bridge. After studying the lowest modeshape vectors it was decided to measure accelerations at three different positions in the longitudinal direction. Each of the three longitudinal positions had five accelerometers: vertical and transverse accelerations were measured at each top flange but transverse acceleration only at one of the bottom flanges.

Excitation. Natural excitation, typical for civil engineering structures, was not available in the laboratory. Therefore a random excitation signal was generated with an electrodynamic shaker to excite the vertical, transverse, and torsional modes (Figure 2). It should be noted that the excitation

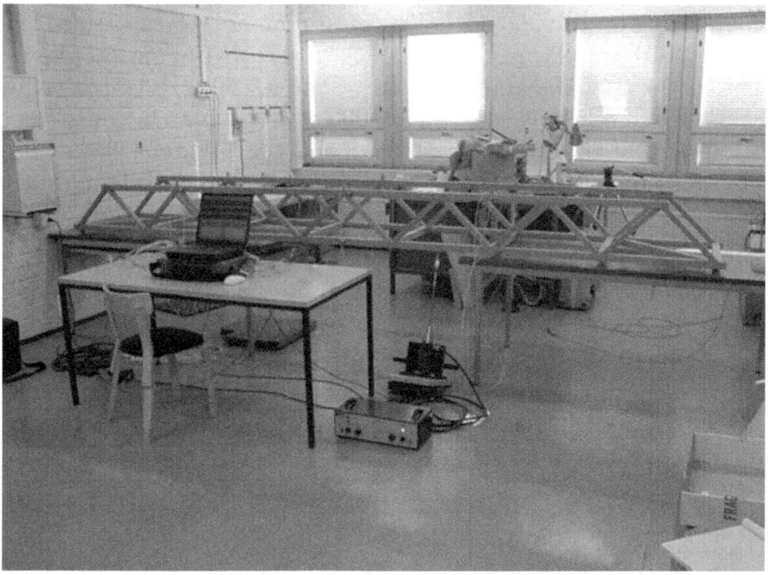

Figure 2. Wooden bridge with a monitoring system.

signal was not used in the analysis, representing a realistic monitoring of civil structures.

Data acquisition and signal processing. Data acquisition was performed using an OROS OR25 PC-Pack II 16-channel dynamic signal analyser with simultaneous sampling and an integrated anti-aliasing filter. The process was automated using a pre-defined time interval between measurements. After digitising, the signals were stored on a hard disk for further processing. The record length of the signals was 32 s with a sampling rate of 256 Hz. The total number of samples in a single measurement with 15 sensors was then 122880. Before feature extraction, the data were digitally low-pass filtered below 128 Hz and re-sampled. The number of samples decreased to half of the original number. An example of the time-histories from the first eight sensors is shown in Figure 3.

Sensor fault identification. Sensor validation was done using the minimum mean square error (MMSE) estimation (Kullaa, 2006). First, the training data and the test data were selected. The training data were three subsequent measurements from the undamaged structure (12226 samples per channel). The test data were a measurement with an added mass of size 197 g (4076 samples per channel). Also the environmental effects were different to those of the training data. Samples 5001–10000 were used to build the model, whereas samples 1–5000 were used as in-control samples to design the control charts. The control chart in Figure 4 shows no sensor fault, because the statistic stays below the upper control limit (horizontal line).

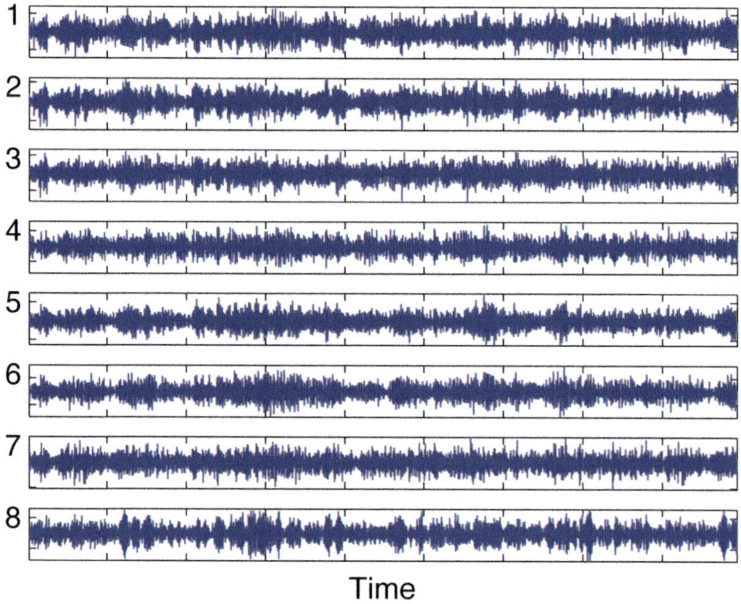

Figure 3. Time histories of the first eight channels.

Feature extraction. The objective of the study was to develop a fully automatic health monitoring system; therefore the structural features had to be extracted automatically from the response data only. The features selected for damage detection were the modal parameters of the structure: natural frequencies, modeshapes, and damping ratios. The identification of the modal parameters was performed using the stochastic subspace identification technique (Overschee and Moor, 1996; Peeters, 2000); it was made automatic by using a stabilisation diagram to find the structural modes.

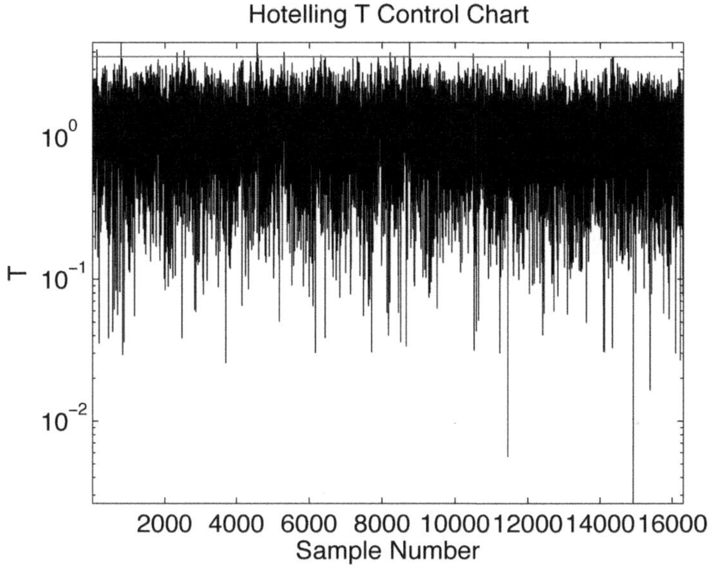

Figure 4. Control chart for sensor fault detection. Training data are the first 12226 samples.

After the identification, mode pairing had to be performed to find similar modes from different measurements. An example of the mode pairing is shown in Figure 5. The identified stabilised frequencies are shown at the top of Figure 5. The result of the classification (mode pairing) is shown at the bottom of Figure 5. The dots in the bottom diagram represent modes not identified from the corresponding measurement.

Feature processing. The features used in damage detection were the natural frequencies of modes 7 and 9–16 resulting in a feature vector with a dimensionality of $p = 9$. The lowest five modes were discarded as they represented the rigid body modes. Modes 6 and 8 were not properly identified and were therefore discarded. Every other sample between 1 and 70 was used for training data. After eliminating the environmental effects using the MMSE estimation (Kullaa, 2005), a further dimensionality reduction was made using principal component analysis (PCA) (Sharma, 1996). These terms will be defined in later sections.

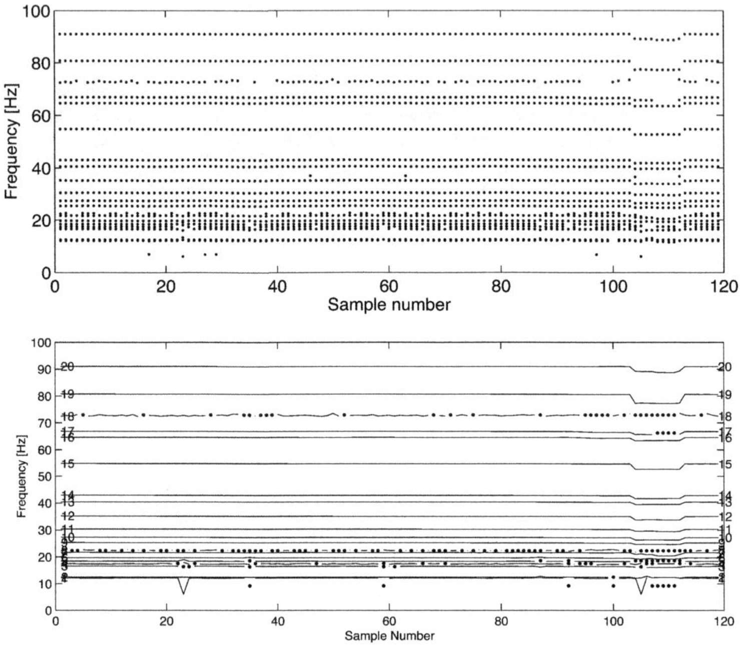

Figure 5. Mode pairing. Top: identified frequencies; bottom: paired modes. The dots in the bottom diagram represent missing modes.

Damage detection. After MMSE estimation, the largest principal value of the multivariate feature vector was used for damage detection using the Hotelling T control chart (Montgomery, 1997). Every other sample between 2 and 70 was used as in-control samples to design the control charts. The control chart after MMSE estimation and the subsequent principal component analysis is shown in Figure 6 showing an excellent detection performance. All damage levels were clearly detected and the damage statistic T increased with an increasing damage level. No false indications of damage were present. After removing the point masses, the damage statistic returned to the original level.

Alarms and reports. The remote measurement computer was connected to the GSM network with a mobile phone (Figure 7). If the system found a possible damage, a short text message was sent to the operator's mobile phone. The system also informed periodically about its running, also using

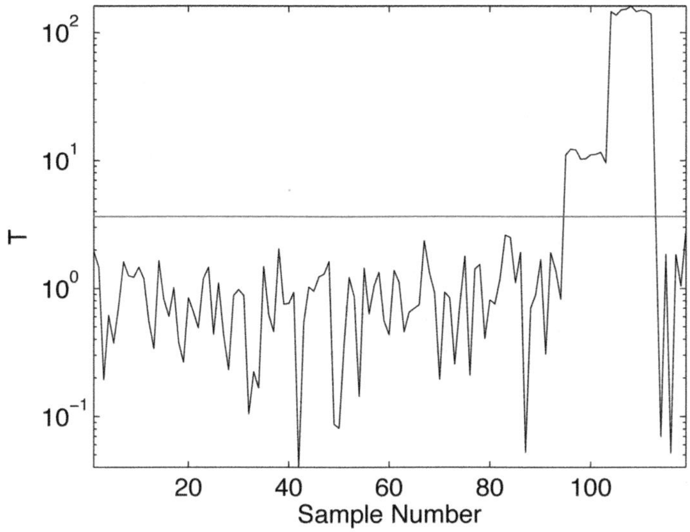

Figure 6. Control chart for damage detection.

SMS.

2.3 Summary

Functions for a vibration-based structural health monitoring system were introduced. The system was made modular in order to facilitate the system design and the introduction of new ideas without a need to re-design the whole system. A complete vibration-based SHM system was built in the laboratory. The basic elements of the system were described. The system was designed to work autonomously sending an alarm signal if the structure is damaged. The selection of the features for the damage detection is a difficult problem, because no information is available on the damaged structure. The features should be sensitive to possible changes due to damage, which suggests using a large number of features. On the other hand, a high-dimensional feature vector results in a low accuracy in the damage detection. The lowest natural frequencies can be identified relatively accurately and their number is usually low, whereas the dimensionality of the complex modeshape vectors may be high depending on the number of transducers in the system. The dimensionality can be reduced using principal component analysis. (Damping ratios were not found to be good indicators

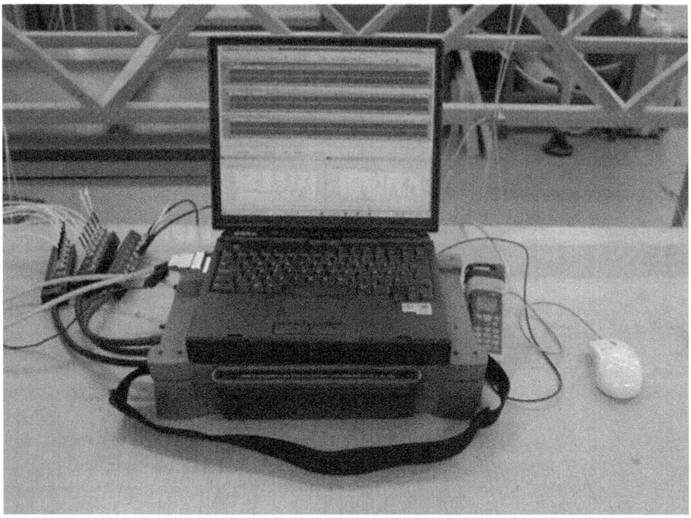

Figure 7. Data acquisition system with an alarm transmission using SMS.

of damage.)

This section gave an overview of different functions of an SHM system. Several functions are discussed in detail in later sections.

3 Sensor Validation

Structures equipped with sensors are becoming common due to the development of the sensor technology and different applications that exploit the sensor information. Vibration-based structural health monitoring is heavily based on the measurement data recorded during a long period. It is most important to detect possible sensor faults to maintain the reliability of the system. Monitoring systems typically include several sensors in order to obtain information from different locations of the structure. Such a sensor network is typically redundant, and can then be utilised to detect sensor malfunction or failure and to identify and correct the faulty sensor. These topics are discussed in this section.

Identification and reconstruction of a faulty sensor has been studied e.g. by Dunia et al. (1996) and by Kerschen et al. (2004) using principal component analysis (PCA). One disadvantage of PCA is that the number of principal components must be determined, which can be difficult. In this section, a more straightforward method is introduced, which uses direct estimation.

It is assumed as in Kerschen et al. (2004) that the number of sensors is higher than the number of excited modes. If the number of active modes is larger than the number of sensors, there is a lack of redundancy; in that case, the time-histories must be filtered to a limited frequency bandwidth.

This section focuses on sensor validation using the correlation or redundancy of the sensors. The minimum mean square (MMSE) estimation model is derived, which can be used for detection, identification, and reconstruction of a faulty sensor. It is assumed here that only one sensor is faulty at a given time. Different fault types: bias, complete failure, drifting, and precision degradation (Dunia et al., 1996) are studied using experimental acceleration data and modifying one sensor to represent the faulty sensor. Also, the effects of the environment and minor damage are present.

3.1 Minimum Mean Square Error (MMSE) Estimation

In the measurement some variables may be missing or incorrectly measured. These variables can be reproduced using *minimum mean square (MMSE) estimation*. The method uses correlation between the variables. The model is created using the training data with all variables correctly measured. The data can be time signals or static features. The MMSE model is derived in the following.

With a multivariable feature vector, a subset of variables can be estimated using the remaining variables. Each observation is divided into observed variables $\mathbf{v}$ and missing variables $\mathbf{u}$,

$$\mathbf{x} = \begin{Bmatrix} \mathbf{u} \\ \mathbf{v} \end{Bmatrix} \tag{2}$$

There are two cases in which the estimator of $\mathbf{u}$ is linear (Sorenson, 1980): (1) The data $\mathbf{x}$ are multinormally distributed, and (2) the estimator $\hat{\mathbf{u}}$ is constrained to be a linear function of the observed data,

$$\hat{\mathbf{u}} \equiv E(\mathbf{u}|\mathbf{v}) = \mathbf{K}\mathbf{v} \tag{3}$$

where $E(\cdot)$ is the expectation operator, $\mathbf{u}|\mathbf{v}$ is a conditional random variable $\mathbf{u}$ given $\mathbf{v}$. $\mathbf{K}$ is an unknown matrix, and $\mathbf{u}$ and $\mathbf{v}$ are assumed to be random vectors with zero means and known covariances, but the joint distribution need not be Gaussian. Both cases result in identical estimators. The derivation of case 1 can be found for example in Sorenson (1980) and Kullaa (2006). Here, only case 2 is presented, as the assumption of a Gaussian distribution is not generally valid, e.g. for harmonic vibration.

The optimal linear estimator $\hat{\mathbf{u}}$ is obtained by minimising the mean-square error (MSE),

$$\begin{aligned}\varepsilon_{MS} &= E[(\mathbf{u}-\hat{\mathbf{u}})^T(\mathbf{u}-\hat{\mathbf{u}})] \\ &= \mathrm{tr}E[(\mathbf{u}-\hat{\mathbf{u}})(\mathbf{u}-\hat{\mathbf{u}})^T]\end{aligned} \tag{4}$$

where tr is the trace operator. Also,

$$\begin{aligned}\mathrm{cov}(\mathbf{u}|\mathbf{v}) &\equiv E[(\mathbf{u}-\hat{\mathbf{u}})(\mathbf{u}-\hat{\mathbf{u}})^T] \\ &= E[(\mathbf{u}-\mathbf{Kv})(\mathbf{u}-\mathbf{Kv})^T] \\ &= E[\mathbf{uu}^T] - E[\mathbf{uv}^T]\mathbf{K}^\mathbf{T} - \mathbf{K}E[\mathbf{vu}^T] + \mathbf{K}E[\mathbf{vv}^T]\mathbf{K}^T \\ &= \mathbf{\Sigma}_{uu} - \mathbf{\Sigma}_{uv}\mathbf{K}^T - \mathbf{K}\mathbf{\Sigma}_{vu} + \mathbf{K}\mathbf{\Sigma}_{vv}\mathbf{K}^T\end{aligned} \tag{5}$$

By completing the square with respect to $\mathbf{K}$, equation 5 can be written in the form,

$$\mathbf{\Sigma}_{uu} - \mathbf{\Sigma}_{uv}\mathbf{K}^T - \mathbf{K}\mathbf{\Sigma}_{vu} + \mathbf{K}\mathbf{\Sigma}_{vv}\mathbf{K}^T = (\mathbf{K}+\mathbf{Q})\mathbf{P}(\mathbf{K}+\mathbf{Q})^T + \mathbf{R} \tag{6}$$

from which the unknown matrices can be obtained,

$$\begin{aligned}\mathbf{P} &= \mathbf{\Sigma}_{vv} \\ \mathbf{Q} &= -\mathbf{\Sigma}_{uv}\mathbf{\Sigma}_{vv}^{-1} \\ \mathbf{R} &= \mathbf{\Sigma}_{uu} - \mathbf{\Sigma}_{uv}\mathbf{\Sigma}_{vv}^{-1}\mathbf{\Sigma}_{vu}\end{aligned}$$

resulting in,

$$\mathrm{cov}(\mathbf{u}|\mathbf{v}) = (\mathbf{K} - \mathbf{\Sigma}_{uv}\mathbf{\Sigma}_{vv}^{-1})\mathbf{\Sigma}_{vv}(\mathbf{K} - \mathbf{\Sigma}_{uv}\mathbf{\Sigma}_{vv}^{-1})^T + \mathbf{\Sigma}_{uu} - \mathbf{\Sigma}_{uv}\mathbf{\Sigma}_{vv}^{-1}\mathbf{\Sigma}_{vu} \tag{7}$$

The gain matrix $\mathbf{K}$ appears only in the first term on the right hand side. To minimise the mean-square error (equation 4), each diagonal element of equation 7 is minimised by choosing,

$$\mathbf{K} = \mathbf{\Sigma}_{uv}\mathbf{\Sigma}_{vv}^{-1} \tag{8}$$

and finally,

$$\hat{\mathbf{u}} = \mathbf{Kv} = \mathbf{\Sigma}_{uv}\mathbf{\Sigma}_{vv}^{-1}\mathbf{v} \tag{9}$$

The error covariance matrix is directly obtained by substituting equation 8 into equation 7,

$$\mathrm{cov}(\mathbf{u}|\mathbf{v}) = \mathbf{\Sigma}_{uu} - \mathbf{\Sigma}_{uv}\mathbf{\Sigma}_{vv}^{-1}\mathbf{\Sigma}_{vu} \quad (10)$$

The results extend readily to variables with nonzero means (Sorenson, 1980),

$$\hat{\mathbf{u}} = \mu_u + \mathbf{\Sigma}_{uv}\mathbf{\Sigma}_{vv}^{-1}(\mathbf{v} - \mu_v) \quad (11)$$

where μ_u and μ_v are the mean values of $\mathbf{u}$ and $\mathbf{v}$, respectively. The error covariance matrix in equation 10 remains unchanged.

Sensor fault detection. The first step in sensor validation is sensor fault detection. The correlation model is built using the training data from the full undamaged sensor set. Then, each sensor is removed in turn and the sensor is estimated using the remaining sensors. The estimated sensor data is then subtracted from the measured sensor data and the resulting residual is used for fault detection. Each residual is normalised according to the training data, because the magnitudes of the residuals are generally different due to different levels of correlation. The number of variables is equal to the number of sensors. This number is usually too large for a statistical analysis and a dimensionality reduction is made using principal component analysis (PCA) of the whole data range including the training data and the test data. If one variable of the test data has deviation from the natural variability, the largest principal component would be in this direction in the multidimensional space. The largest principal value is finally used to design the control chart to detect sensor faults. For different types of sensor faults, the best fault indicator depends on whether the fault affects the mean value or the variance.

MMSE estimation is applied to sensor fault detection using the following procedure.

1. Define the training data with no sensor faults.
2. Estimate the mean μ and covariance matrix $\mathbf{\Sigma}$ of the training data.
3. Select a sensor and set it as a missing variable $\mathbf{u}$.
4. Form matrices μ_u, μ_v, $\mathbf{\Sigma}_{uu}$, $\mathbf{\Sigma}_{vv}$, $\mathbf{\Sigma}_{uv}$, and $\mathbf{\Sigma}_{vu}$ by partitioning μ and $\mathbf{\Sigma}$.
5. Compute the mean of $\mathbf{u}|\mathbf{v}$ for each sample using equation 11.
6. Compute the residual $\mathbf{r} = \mathbf{u} - E(\mathbf{u}|\mathbf{v})$.
7. Return to 3 until the residuals of all sensors have been evaluated.
8. Scale each residual according to the residual of the training data.
9. Perform a dimensionality reduction using principal component analysis.

10. Define the in-control samples with no sensor faults.
11. Perform a statistical analysis using control charts.
12. If the control chart raises an alarm, it is an indication of a sensor fault. Move on to fault isolation and reconstruction.

The variance of $\mathbf{u}|\mathbf{v}$ (equation 10) can be used to determine the confidence limits, but it is not used here, because a further dimensionality reduction is performed using the PCA and new control limits are set for the new features.

Faulty sensor isolation. Once a sensor fault is detected, the faulty sensor must be identified. It is assumed that one sensor only is faulty. The algorithm is as follows.

One sensor is removed in turn and a model is built according to the training samples of the remaining sensors. If the faulty sensor is present, the test data should give large residuals. On the other hand, if the faulty sensor is missing, the residual would be at the same level as for the training data. The residuals are scaled according to the mean residual of the training data; then the minimum rms value of the scaled residuals indicates the faulty sensor. The faulty sensor is the missing sensor in the data set with the minimum rms value.

The algorithm has an outer loop and an inner loop resulting in a higher computation time. While in structural health monitoring the largest residual from the full sensor set may indicate the damage location, the same approach cannot be used here. As one sensor gives faulty information, it propagates also to other sensors; as a consequence it could result in an erroneous identification of the faulty sensor.

MMSE estimation is applied to sensor identification using the following procedure.

1. Define the training data with no sensor faults.
2. Set $i = 1$.
3. Select sensor i and remove it from the data.
4. Estimate mean μ and covariance matrix $\mathbf{\Sigma}$ of the training data without sensor i.
5. Select sensor $j \neq i$ and set it as a missing variable $\mathbf{u}$.
6. Form matrices μ_u, μ_v, $\mathbf{\Sigma}_{uu}$, $\mathbf{\Sigma}_{vv}$, $\mathbf{\Sigma}_{uv}$, and $\mathbf{\Sigma}_{vu}$ by partitioning μ and $\mathbf{\Sigma}$.
7. Compute the mean of $\mathbf{u}|\mathbf{v}$ for each sample using equation 11.
8. Compute the residual $\mathbf{r} = \mathbf{u} - E(\mathbf{u}|\mathbf{v})$.
9. Return to 5 until the residuals of all sensors has been evaluated.
10. Scale each residual according to the residual of the training data.
11. Compute a single rms(i) value of all residuals.

12. Next i. Return to 3 until all sensors have been evaluated.
13. The lowest rms(i) value indicates the faulty sensor i.
14. Move on to sensor correction.

Sensor correction. The sensor correction can be made once the faulty sensor has been isolated. The procedure is very similar to that in fault detection except that only the faulty sensor is studied. The model is built using the training data from all sensors without faults. The faulty sensor is corrected by removing the sensor from the set and estimating it using the remaining sensors. The procedure is as follows.

1. Define the training data with no sensor faults.
2. Estimate mean μ and covariance matrix $\mathbf{\Sigma}$ of the training data.
3. Set the faulty sensor as a missing variable $\mathbf{u}$.
4. Form matrices μ_u, μ_v, $\mathbf{\Sigma}_{uu}$, $\mathbf{\Sigma}_{vv}$, $\mathbf{\Sigma}_{uv}$, and $\mathbf{\Sigma}_{vu}$ by partitioning μ and $\mathbf{\Sigma}$.
5. Compute the mean $E(\mathbf{u}|\mathbf{v})$ for each sample using equation 11. It is the reconstructed sensor.
6. Compute the variance $\text{cov}(\mathbf{u}|\mathbf{v})$ using equation 10. It can be used to compute the estimation error.

3.2 Experimental Results

Sensor faults were investigated with a monitoring system built in the laboratory. The structure was the wooden model bridge shown in Figure 8. The monitoring system was described in Section 2.2. See also Figure 2.

The modal parameters were identified from the output-only data, which resulted in 20 natural modes below 100 Hz. As this number was larger than the number of sensors, there was a lack of redundancy; therefore, the signals were low-pass filtered below 50 Hz. In this frequency range, the structure contained 14 natural modes.

In order to study the sensor fault identification, the training data and the test data were selected. The training data were three subsequent measurement sets from the undamaged structure (12226 samples per channel). The test data were a measurement set with an added mass of size 197 g (4076 samples per channel). Also the environmental effects were different to those of the training data. (The damage scenarios and the environmental effects will be discussed in more detail in Section 5.3.) The test data were scaled according to the training data resulting in equal rms values for both data sets. The relative magnitude between sensors was not changed. Test samples 5001–10000 were used to build the model, whereas samples 1–5000 were used as in-control samples to design the control charts. Different fault

Figure 8. Wooden bridge.

types in different sensors are studied in the following sections: bias, complete failure, drifting, and precision degradation (Dunia et al., 1996). In sensor fault detection, MMSE estimation resulted in residual accelerations for each sensor. The dimensionality of the feature vector was 15, and further dimensionality reduction was made using PCA. The largest principal value was then used in sensor fault detection using control charts. Once a fault was detected, more detailed procedures were triggered to identify and correct the faulty sensor.

No sensor fault. First, the test data without a faulty sensor is studied. This analysis is made to study the effect of the added mass and the environmental changes to the model. The Hotelling T control chart (Montgomery, 1997) was used to detect sensor failure as shown in Figure 9; Note the logarithmic scale. It can be seen that the chart shows no indication of abnormal operation and it can be concluded that all sensors are working.

The effect of the added mass was detected with the structural health monitoring system using the modal parameters as features. The influences of the environment were eliminated using factor analysis (Kullaa, 2002). The sensor fault detection was insensitive to these variations, validating the proposed approach in distinguishing between the sensor fault and structural damage or environmental changes.

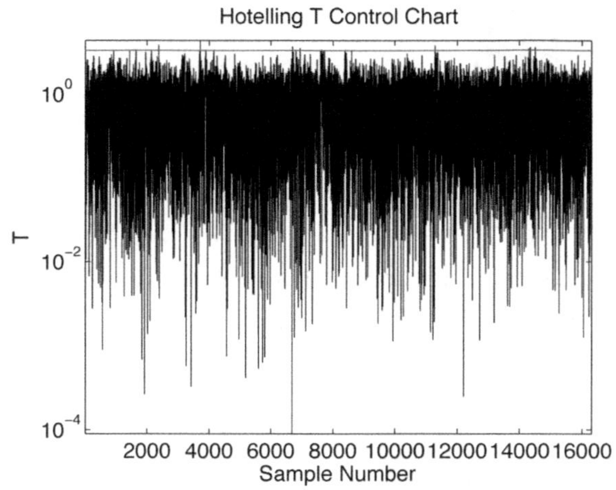

Figure 9. Control chart without sensor fault.

Bias. Bias gives values that are shifted by a constant from the true value. In the test data, sensor 3 was biased by a value of 0.01. The standard deviation of the sensor data was 0.02. The control chart for the fault detection is shown in Figure 10 (left; note the logarithmic scale) and the faulty sensor identification is shown in Figure 10 (right) plotting the rms value of the residuals for each sensor. The sensor fault was perfectly detected and the faulty sensor was correctly identified by the minimum residual.

Complete failure. In the case of complete failure the sensor gives a constant value. In the test data, sensor 13 was given a value of 0.01. The control chart for the fault detection is shown in the left of Figure 11 and the faulty sensor identification is shown in the right of Figure 11. The sensor fault could be detected using the S chart (Montgomery, 1997) monitoring the change in variance; Note the logarithmic scale. Also, the faulty sensor was correctly identified.

Drifting. Drifting of sensor 1 was simulated by adding a linearly varying value to the test data. The value varied from 0 to 0.1 over the measurement range. The control chart for the fault detection is shown in the left of Figure

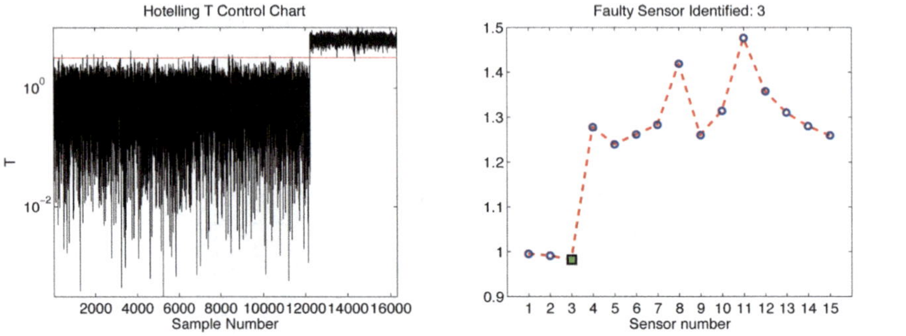

Figure 10. Left: control chart for the biased sensor 3. Right: faulty sensor isolation from the minimum rms of the residuals.

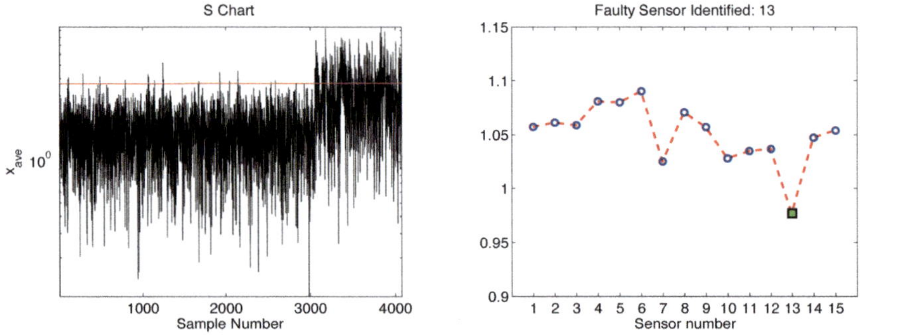

Figure 11. Left: control chart for the complete failure of sensor 13. Right: faulty sensor isolation from the minimum rms of the residuals.

12 and the faulty sensor identification is shown in the right of Figure 12. The sensor fault was perfectly detected; Note the logarithmic scale. Also, the faulty sensor was correctly identified.

Precision degradation. In the case of precision degradation, sensor 4 was multiplied by a random value between 0.5 and 1.5. This resulted in a change in variance and the S chart was the best indicator to detect the sensor failure and is shown in Figure 13 (left). The faulty sensor identification is shown in Figure 13 (right). The sensor fault was perfectly detected; Note the logarithmic scale. Also, the faulty sensor was correctly identified. The rms values for sensors 4, 5, and 6 were close to each other and made sensor identification difficult. In particular, when sensor 5 was faulty, sensor 4 was

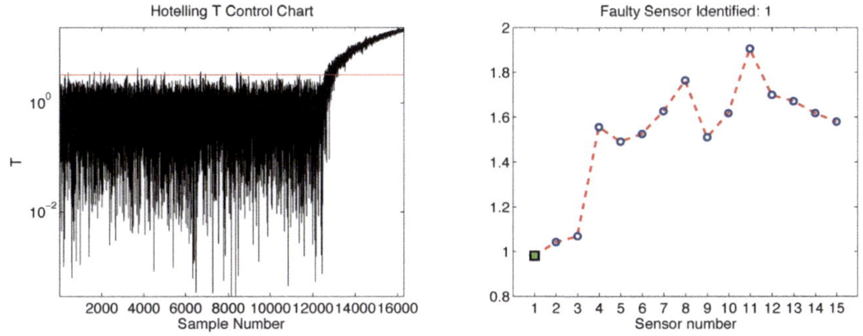

Figure 12. Left: control chart for the drifting sensor 1. Right: faulty sensor isolation from the minimum rms of the residuals.

incorrectly identified as the faulty sensor; however, fault detection caused no difficulties.

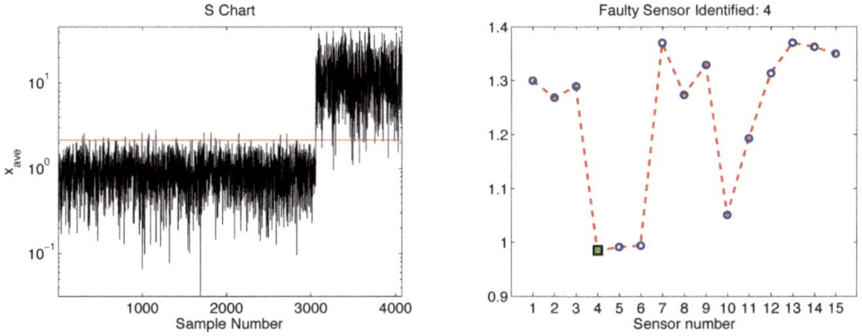

Figure 13. Left: control chart for sensor 4 with precision degradation. Right: faulty sensor isolation from the minimum rms of the residuals.

Sensor correction. Sensor 3 with the bias studied in section 3.2 was reconstructed. The true signal (full line) and the estimated signal (dashed line) are shown in Figure 14, in which it is difficult to see much difference between the two. It was therefore possible to reconstruct a sensor when enough redundancy existed in the measured data.

Failure in fault detection, isolation, and reconstruction. The previous sections showed successful cases in sensor fault detection, isolation, and

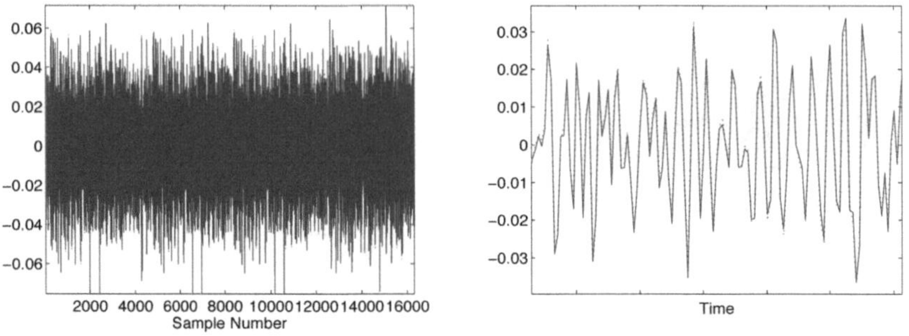

Figure 14. Correction of sensor 3. Left: whole range. Right: detail. Full line: true signal; Dashed line: estimated signal.

reconstruction. In the following, a case with erroneous results is discussed.

Sensor 14 was modified by the same amount of precision degradation as in Section 3.2. The results are shown in Figure 15. The sensor fault was not detected and the faulty sensor was incorrectly identified as sensor 15. The rms value for sensor 14 was only the third lowest. In sensor reconstruction (Figure 15 right) it can be seen that for sensor 14 there is not enough redundancy, resulting in poor reconstruction. The problem would probably have not existed if more sensors had been used.

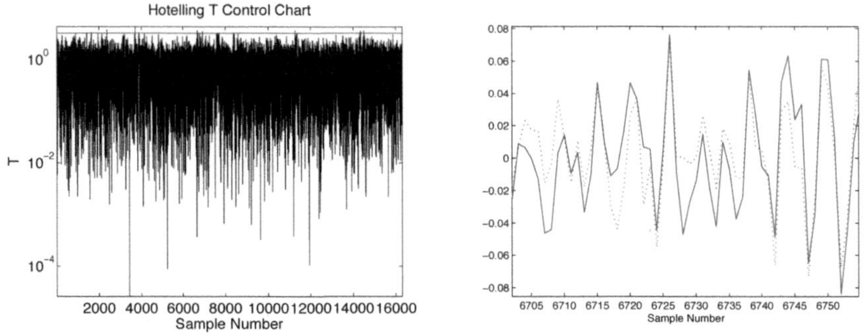

Figure 15. Left: control chart for sensor 14 with precision degradation. Right: sensor correction, detail. Full line: true signal; Dashed line: estimated signal.

3.3 Summary

A straightforward method for sensor fault detection, identification, and correction was introduced. All steps were based on minimum mean square error estimation using the correlation of the variables to estimate the removed sensor. The main assumption is that initially no sensor is faulty so that training data can be acquired to build the model. The method was validated using experimental vibration data. If the sensor set provided enough redundancy, the results were excellent. For some sensors, however, this was not the case and erroneous results were obtained: a) no fault was detected; b) a fault was detected but an incorrect sensor was identified. In these cases the reconstruction was also inaccurate. The number of sensors (15) was close to the number of modes (14), which probably caused a lack of redundancy. The results would probably have been better if more sensors had been used or if the frequency bandwidth had been reduced. It was assumed that a single sensor only was faulty; the method can be further developed to allow more faulty sensors. Also, the faulty sensor isolation may be too slow in practice with several sensors, and a faster method should be developed. It is anticipated that in the future more structures will be equipped with a sensor array having numerous sensors; this fact makes the proposed method valuable and available for practical applications.

4 Damage Detection Using Control Charts

In structural health monitoring, each measurement is classified into two categories according to the state of the structure: undamaged or damaged. This classification must usually be done using unsupervised learning methods, because data from the damaged structure are not available in the training phase. It is assumed that the vibration characteristics of the structure change due to damage, and by identifying the appropriate features and monitoring their variation in time, the existence of damage can be observed. However, no measurement set results in exactly the same numbers, and the features have natural variability. The objective is therefore to decide if the change of the features is statistically significant. In an outlier analysis, the distribution of the training data is estimated and when new data arrive, the fit in the distribution is tested. In this section, damage detection is studied using control charts (Sohn et al., 2000a; Kullaa, 2003).

Damage detection can be performed either on-line or off-line; the ultimate goal is a monitoring system with on-line damage detection. There are however still problems to be solved: (1) Which variables should be selected for the damage detection? (2) Which damage detection algorithm or control chart would observe the possible damage? (3) How should false alarms be handled? In off-line damage detection, one can experiment with different variables and algorithms and study their sensitivity. Moreover, in an interactive study the final decision can be made by a human, not by the computer, facilitating the solution considerably.

A comparison of different control charts is performed here using numerical modal data. The features include natural frequencies and modeshapes. The following control charts are studied: univariate Shewhart, x, CUSUM, and EWMA charts, and multivariate Shewhart T, Hotelling T, MCUSUM, and MEWMA control charts. The reliability and sensitivity of the control charts are investigated. Differences between off-line and on-line results are discussed. Dimensionality reduction is discussed and applied using principal component analysis.

4.1 Control Charts

Statistical methods must be used for damage detection. A control chart (Montgomery, 1997) is one of the primary techniques of statistical process control and is a very useful process monitoring technique. It plots a quality characteristic as a function of the sample number. The chart has lower and upper control limits, which are computed from those samples when the process is assumed to be in control. When unusual sources of variability are present, sample statistics will plot outside the control limits; in that case an alarm signal will be produced. An advantage of control charts is that they can be automated for on-line structural health monitoring. Different control charts are studied including univariate and multivariate Shewhart, CUSUM, and EWMA control charts. They are described briefly below.

Shewhart chart. In the Shewhart chart, or xbar chart (Montgomery, 1997) the subgroup mean $\overline{x}$ of a variable is plotted. The upper and lower control limits, UCL and LCL, respectively, are,

$$\left.\begin{array}{c}\mathrm{UCL}\\ \mathrm{LCL}\end{array}\right\} = \overline{\overline{x}} \pm A_3\overline{S} \tag{12}$$

where $\overline{\overline{x}} = \mathrm{mean}(\overline{x})$ is the average of the subgroup means, $\overline{S}$ is the average of the subgroup standard deviations, and A_3 is a constant depending on the subgroup size, which is typically 4, 5, or 6.

Shewhart T chart. The multivariate counterpart of the Shewhart chart is the Shewhart T, or Hotelling T control chart (Montgomery, 1997), where the plotted characteristic is,

$$T^2 = n(\overline{\mathbf{x}} - \overline{\overline{\mathbf{x}}})^T \mathbf{S}^{-1} (\overline{\mathbf{x}} - \overline{\overline{\mathbf{x}}}) \tag{13}$$

where $\overline{\mathbf{x}}$ is the subgroup average, $\overline{\overline{\mathbf{x}}}$ is the process average, which is the mean of the subgroup averages when the process is in control, and $\mathbf{S}$ is the matrix consisting of the grand average of the subgroup variances and covariances. The upper control limit is,

$$\mathrm{UCL} = \frac{p(m+1)(n-1)}{mn - m - p + 1} F_{\alpha,p,mn-m-p+1} \tag{14}$$

where p is the dimension of the variable, n is the subgroup size, m is the number of subgroups when the process is assumed to be in control, and $F_{\alpha,p,mn-m-p+1}$ denotes the α percentage point of the F distribution with p and $mn - m - p + 1$ degrees of freedom.

x **chart.** Control charts for individual measurements include the x chart, cumulative sum, and exponentially weighted moving average control charts. On the x chart (Montgomery, 1997) the variable x should fall between the control limits,

$$\left.\begin{matrix}\mathrm{UCL}\\ \mathrm{LCL}\end{matrix}\right\} = \overline{x} \pm 3\frac{\overline{\mathrm{MR}}}{d_2} \tag{15}$$

where $\overline{x}$ is the sample average and $\overline{\mathrm{MR}}$ is the average of the moving ranges of two successive observations $\mathrm{MR}_i = |x_i - x_{i-1}|$ and $d_2 = 1.128$.

There is a well-known relationship between the range of a sample from a normal distribution and the standard deviation of that distribution (Montgomery, 1997). If $R = x_{\max} - x_{\min}$ is the range of the sample of size n and $\overline{R}$ is the average range, the quantity $\overline{R}/d_2$ is an estimate of the standard deviation σ, where d_2 depends on the sample size n. For an x chart, $n = 2$ and equation 15 with $\overline{x} \pm 3\sigma$ limits follows.

Hotelling T chart with $n = 1$. The multivariate counterpart of the x chart is the Hotelling T chart with the subgroup size $n = 1$ (Montgomery, 1997). The plotted statistic T^2 (Mahalanobis distance) and the upper control limit are defined as,

$$T^2 = (\mathbf{x} - \overline{\mathbf{x}})^T \mathbf{S}^{-1} (\mathbf{x} - \overline{\mathbf{x}}) \tag{16}$$

$$\mathrm{UCL} = \frac{p(m+1)(m-1)}{m^2 - mp} F_{\alpha,p,m-p} \tag{17}$$

where $\overline{\mathbf{x}}$ is the mean of $\mathbf{x}$, and $\mathbf{S}$ the data covariance matrix, when the process is in control.

CUSUM chart. The cumulative sum (CUSUM) and the exponentially weighted moving average (EWMA) control charts are good alternatives to those above, when detection of small shifts is important. While the control charts above use the information of the last sample only, the CUSUM and EWMA charts incorporate all the information in the sequence of sample values. The tabular cusum (Montgomery, 1997) works by accumulating deviations from $\overline{x}$ that are above target with one statistic C^+ and accumulating deviations from $\overline{x}$ that are below target with another statistic C^-. The statistics C^+ and C^- are the one-sided upper and lower cusums, respectively. They are computed as follows,

$$\begin{aligned} C_i^+ &= \max\{0, x_i - (\overline{x} + k\sigma) + C_{i-1}^+\} \\ C_i^- &= \max\{0, (\overline{x} - k\sigma) - x_i + C_{i-1}^-\} \end{aligned} \tag{18}$$

where $C_0^+ = C_0^- = 0$, $\overline{x}$ is the target value for the characteristic x, and k is called the reference value and is often chosen halfway between the target and the out-of-control value that one is interested in detecting quickly. σ is the standard deviation of the variable used in forming the cusum. The process is considered to be out of control, if either C^+ or C^- exceeds the decision interval $H = h\sigma$, where h depends on the shift to be detected. Notice that C_i^+ and C_i^- accumulate deviations from the target value $\overline{x}$ that are greater than $k\sigma$, with both quantities reset to zero upon becoming negative (Montgomery, 1997).

MCUSUM chart. A vector-valued CUSUM (Crosier, 1988), or MCUSUM is a multivariate counterpart of the CUSUM chart. The multivariate CUSUM chart signals when the plotted characteristic $Y_i > h$, where the constant h depends e.g. on the dimension of the feature vector.

$$Y_i^2 = \mathbf{s}_i^T \mathbf{S}^{-1} \mathbf{s}_i \tag{19}$$

where,

$$\mathbf{s}_i = \begin{cases} \mathbf{0}, & C_i \leq k \\ (\mathbf{s}_{i-1} + \mathbf{x}_i - \overline{\mathbf{x}})(1 - k/C_i), & C_i > k \end{cases} \tag{20}$$

$$C_i^2 = (\mathbf{s}_{i-1} + \mathbf{x}_i - \overline{\mathbf{x}})^T \mathbf{S}^{-1} (\mathbf{s}_{i-1} + \mathbf{x}_i - \overline{\mathbf{x}}) \tag{21}$$

where $\mathbf{s}_0 = 0$ and $k > 0$ is the same as for the CUSUM chart.

EWMA chart. The exponentially weighted moving average (EWMA) chart (Montgomery, 1997) is also a good alternative when one is interested in small shifts. It is defined as,

$$z_i = \lambda x_i + (1 - \lambda) z_{i-1} \tag{22}$$

where $0 < \lambda \leq 1$. The control limits are as follows,

$$\left.\begin{matrix} \mathrm{UCL} \\ \mathrm{LCL} \end{matrix}\right\} = \overline{x} \pm L\sigma \sqrt{\frac{\lambda}{2-\lambda}\left[1 - (1-\lambda)^{2i}\right]} \tag{23}$$

where L is the design parameter of the chart.

MEWMA chart. A multivariate exponentially weighted moving average (MEWMA) chart (Lowry et al., 1992) is a multivariate counterpart of the EWMA chart. It is defined analogously by,

$$\mathbf{Z}_i = \lambda \mathbf{x}_i + (1 - \lambda)\mathbf{Z}_{i-1} \tag{24}$$

where $0 < \lambda \leq 1$ and $\mathbf{Z}_0 = \mathbf{0}$. The quantity plotted on the control chart is,

$$T_i^2 = \mathbf{Z}_i^T \mathbf{S}_{Z_i}^{-1} \mathbf{Z}_i \tag{25}$$

where the covariance matrix is,

$$\mathbf{S}_{Z_i} = \frac{\lambda}{2-\lambda}\left[1 - (1-\lambda)^{2i}\right] \mathbf{S} \tag{26}$$

The MEWMA chart signals when $T_i^2 > h$, where the constant h depends e.g. on the dimension of the feature vector.

4.2 Numerical Example

Description of structure and data. The application here is to a finite element model of a vehicle crane (Figure 16) at varying joint positions and tip loads. Experimental results of the same structure will be given in Section 7. The structure was modelled with planar beam and plane stress elements. The hydraulic cylinders were modelled with spring and dashpot elements. A point mass represented the load at the tip. The bottom of the fixed link was clamped and constraint equations were used in the modelling of the joints.

The natural frequencies and complex modeshapes were first chosen as the damage-sensitive features. They were influenced by four latent variables corresponding to the configuration: three cylinder strokes and the load at the tip, which were assumed unknown. In the simulations they varied randomly in a limited range. Figure 17 shows the limit configurations. The load varied randomly between 0 and 50 kg. In addition, the relationship between the cylinder stiffness and stroke was nonlinear.

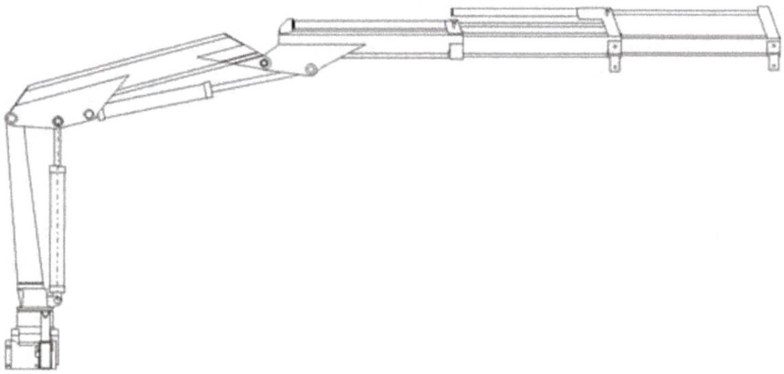

Figure 16. Vehicle crane HIAB-022-2L.

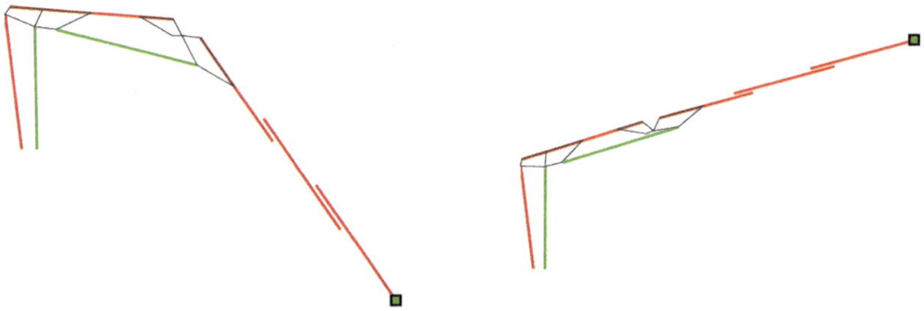

Figure 17. Finite element models of the crane in the limit configurations.

Damage was modelled as a stiffness reduction in the bottom beam element of the fixed link. Different sizes of damage were studied, ranging from 1 to 10 percent of the original stiffness. The undamaged structure was first monitored with 2000 measurements; then, damage was gradually increased, and each case was monitored with 250 measurements. Noise with five percent standard deviation was added to all features.

The modeshapes were recorded at four different positions of the structure in the links' transverse direction. The five lowest modes were used for damage detection, and after the normalisation of the modeshapes, this resulted in 35 features.

The first 700 observations were used as training data to remove the operational effects using minimum mean square error (MMSE) estimation. Next 700 observations were used as the in-control samples to compute the control limits for the control chart. After MMSE estimation, the dimensionality of the multivariate residual vector was reduced using principal component analysis and the remaining variables were used for damage detection using different control charts.

Dimensionality reduction. Feature extraction from measured time histories into modal parameters compresses the data considerably. However, the dimensionality of the feature vector is still often too high for a reliable statistical analysis. For example, when five natural frequencies and the corresponding modeshapes from four sensors were used for the damage detection, the dimension of the feature vector was 35, being too high with 700 samples only from the undamaged structure. Therefore, a further dimension reduction was required.

Several dimensionality reduction techniques have been proposed, the most popular being principal component analysis (PCA) (Sharma, 1996). Suppose that instead of using all the original variables one only uses a few new variables to represent most of the information contained in the data. Principal component analysis projects the original data into new orthogonal axes, which are the eigenvectors of the data covariance matrix. The new variables, called principal component scores, are ordered so that the first variable accounts for the maximum variance in the data. The number of variables selected for subsequent analysis depends on the application. In this study, the first principal component only is used in the univariate control charts, whereas for the multivariate control charts, the number of principal components was determined so that the remaining variables explained at least 90% of the variance.

Results. In all control charts of this study the number of in-control samples used was 700. In the Shewhart and Shewhart T control charts the subgroup size was 4. The dimension of the feature vector was 35, being too high for a reliable statistical analysis. Therefore, those first principal components were selected which explained 90% of the variance in the features. The resulting number of principal components was 16. The multivariate control charts for the features are shown in Figure 18. The initialisation of each damage size is also shown. Because the damage cases were known, the sensitivity and reliability of different control charts could be assessed. The number of outliers in different control charts at each damage level is shown in Table 1. It can be seen that damage levels 2–4 were clearly detected with all multivariate charts, except that the Hotelling T control chart was relatively insensitive to damage level 2. The MCUSUM chart was the most sensitive as expected, detecting also the level 1 damage. It is interesting that the Hotelling T control chart resulted in most false indications of damage together with the MEWMA chart and that the MCUSUM chart produced no false indications of damage.

For the univariate control charts, the first principal component only was used. It was seen to explain 53% of the variance in the features and resulted in the control charts shown in Figure 19 and the number of outliers at each damage level listed in Table 1. All control charts worked perfectly for damage levels 2–4; only the x chart classified half of the data of damage level 2 as undamaged. The CUSUM chart was the most sensitive for the damage level 1. All charts had some false indications of damage, but their number was statistically acceptable.

Comparison between the univariate and multivariate control charts is, however, difficult, because the variables in the charts were different. Generally, multivariate control charts offer more flexibility, as they can always be applied regardless of the dimensionality of the feature vector. For a higher accuracy, the dimensionality should, however, be reduced below five, preferably even lower. The choice of the control chart depends also on the application. Generally, the Shewhart and Shewhart T charts are quite robust, because they are based on sample averages. The main disadvantage is that a higher amount of training is needed. If high sensitivity is desired to detect small faults, then the MCUSUM control chart would be justified.

On-line damage detection. In the previous section, the control charts were plotted after all data had arrived with information on all four levels of damage. In a real monitoring application, the control charts are applied on-line after each measurement. As the new sample has arrived, it can change the direction of the principal axes and consequently the control chart. Also, the number of remaining principal components may vary. On the other hand, the control charts are identical in the on-line or off-line mode, if no dimensionality reduction is applied.

The results in the previous section hold essentially only for that particular condition of the structure. With information on the lowest damage levels only, the directions of the principal axes are probably not yet established. In this section, different control charts are studied in an on-line mode

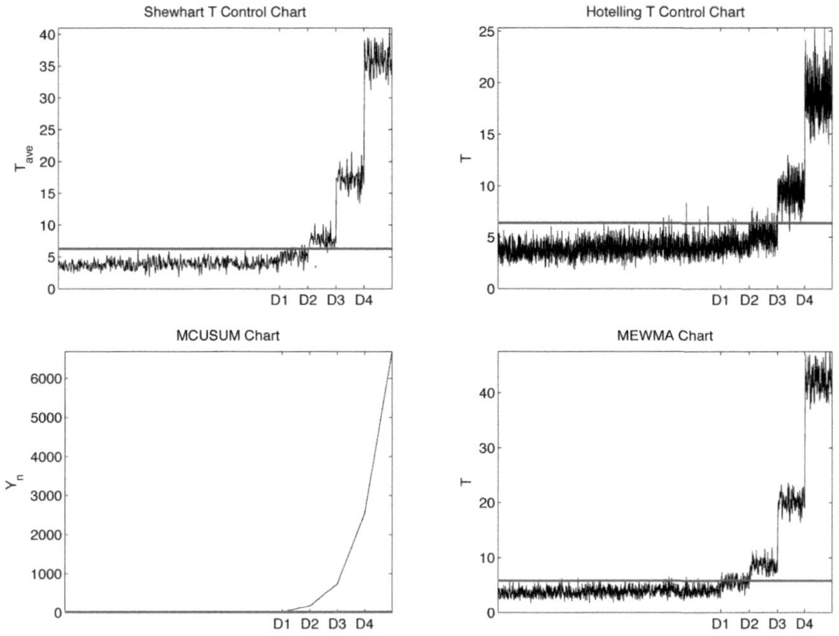

Figure 18. Multivariate control charts for the first 16 principal components.

when data arrive sequentially. The objective is to study the performance of early damage detection. When the last measurement has arrived, the results agree with those in the previous section.

Multivariate and univariate control charts were applied to on-line damage detection. The modified on-line control charts are shown in Figures 20 and 21, where the statistics of the most recent measurement only and the corresponding control limits are plotted. The number of principal components in the multivariate control charts varied between 16 and 20 depending on the damage size. The lowest number corresponded to the largest damage level as expected.

The modified on-line control charts can be applied also in practice when information of damage is not available. In this example, damage is known and the sensitivity and reliability can be analysed as follows. Figures 22 and 23 show the proportion of outliers at each damage level after the most recent sample. The figures show that the Shewhart T and MEWMA control charts detect damage at level 2, the Hotelling T control chart only from level 3, and the MCUSUM chart from level 1. Reasonably few false indications

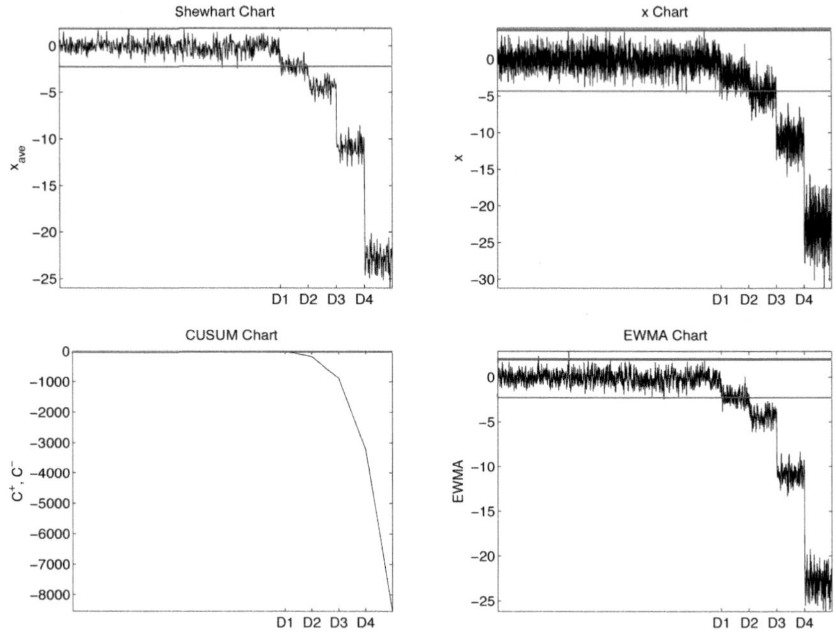

Figure 19. Univariate control charts for the first principal component.

Table 1. Number of outliers in different control charts. The first four charts are for the first 16 principal components, while the last four charts are for the first principal component.

Control chart	Undamaged	Damage 1	Damage 2	Damage 3	Damage 4
Shewhart T	0/325	8/62	57/63	62/62	63/63
Hotelling T	6/1300	3/250	23/250	246/250	250/250
MCUSUM	0/1300	234/250	250/250	250/250	250/250
MEWMA	9/1300	89/250	249/250	250/250	250/250
Shewhart	2/325	32/62	63/63	62/62	63/63
x chart	3/1300	13/250	126/250	250/250	250/250
CUSUM	5/1300	244/250	250/250	250/250	250/250
EWMA	3/1300	127/250	250/250	250/250	250/250

of damage occur in each chart at different levels of damage. The univariate control charts show similar behaviour. The Shewhart chart, CUSUM, and EWMA charts detected damage at level 2, whereas detection in the x chart occurred clearly only at level 3. Also here the number of false indications of damage was not excessive.

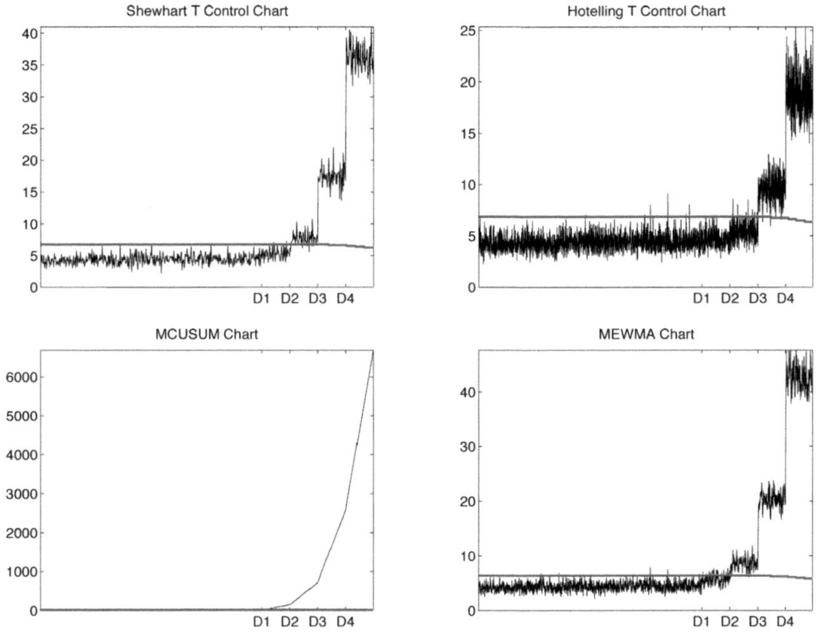

Figure 20. Multivariate on-line control charts for the first 16–20 principal components. The charts represent the most recent sample and the corresponding instantaneous limits.

4.3 Summary

Damage detection is a statistical problem to find changes in a structure through changes in damage-sensitive features. Control charts are powerful and visual tools, and also easy to apply to damage detection. Different univariate and multivariate control charts were studied to monitor the change in the mean. Also the change in the variability can be studied using the S charts (Montgomery, 1997). They are useful e.g. for detecting multiplicative sensor faults.

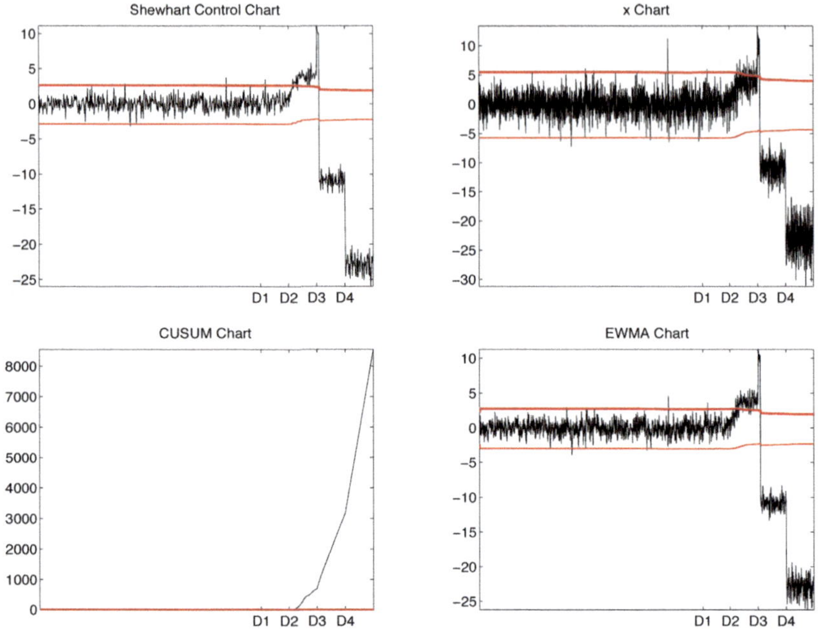

Figure 21. Univariate on-line control charts for the first principal component. The charts represent the most recent sample and the corresponding instantaneous limits.

The control charts were applied to detect damage in a vehicle crane. All control charts performed well. The multivariate control charts offer more flexibility as they can be applied to any number of features. However, in practice the maximum number of features is restricted to around five, because the statistical accuracy decreases with an increasing number of variables. Dimensionality reduction is therefore an essential step before applying control charts. Principal component analysis was applied to the whole data range including the training data and the test data; it finds the directions of maximum variability indicating damage. Projecting the features onto these directions, the dimensionality can be considerably reduced without a loss of relevant information needed for damage detection. Experience has shown that usually one or two largest principal components are sufficient. In this study, however, the applied criterion resulted in a larger number of principal components.

The control chart can be chosen based on the size of damage that must

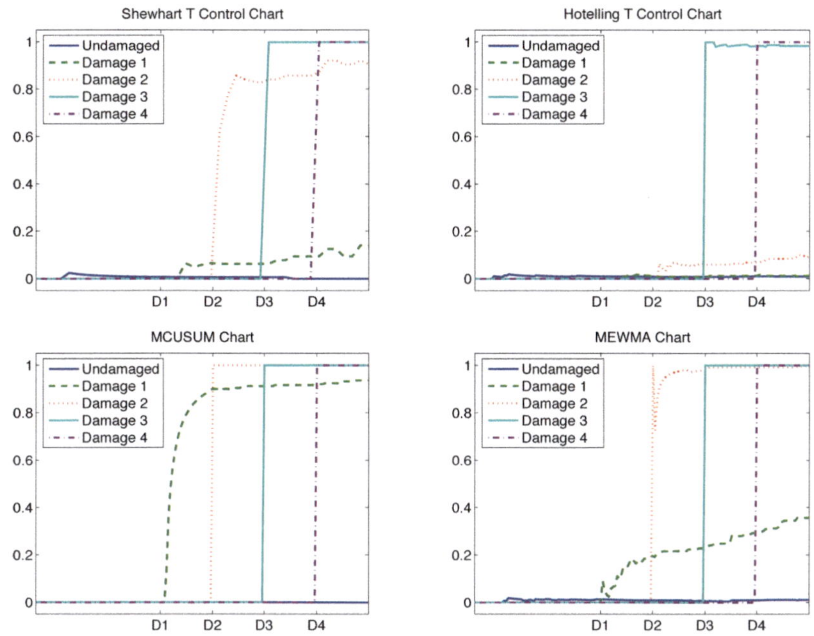

Figure 22. The proportion of outliers in multivariate on-line control charts at each damage level $(D_1, \ldots, D_4)$ after the most recent sample.

be detected. To detect small changes, CUSUM or EWMA charts can be applied. Increasing the sensitivity to detect small changes also increases the sensitivity to false alarms. Shewhart and x charts can detect moderate changes, but they are more insensitive to false alarms. One suggestion to eliminate false alarms is to state that a certain percentage of the last N samples must be out of control. This decision strategy would be more reliable, but the alarm may be received after a delay. For the CUSUM and EWMA charts an equivalent strategy would be to extend the control limits. Another strategy would be to disregard the sample indicating the first out-of-control condition and to make another measurement.

Usually the statistic plotted is assumed to be normally distributed; in many cases this assumption cannot be guaranteed. The statistic in the Shewhart charts is an average of 4–6 successive variables, which makes the distribution of the plotted statistic close to normal. However, if the features are autocorrelated, the features must be pre-processed to remove the autocorrelation. For example, the autocorrelation can be modelled with

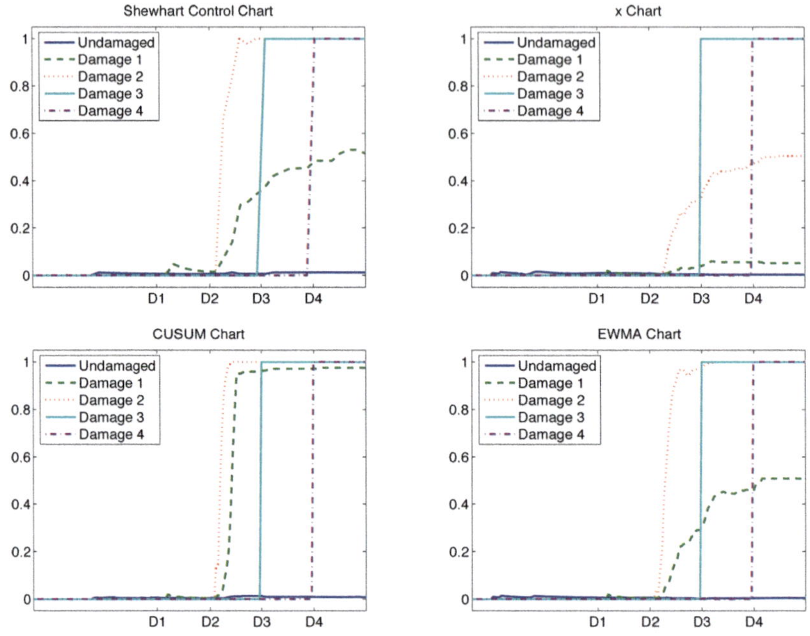

Figure 23. The proportion of outliers in univariate on-line control charts at each damage level ($D_1, \ldots, D_4$) after the most recent sample.

appropriate time series models, and one then applies control charts to the residuals.

As the SHM systems work on-line trying to detect damage once it occurs, it is necessary to apply on-line methods to assess different control charts. If PCA is applied, the plotted statistic is different after each measurement and the control chart must be re-designed. Off-line methods with all data available tend to give too-optimistic results suggesting that also small damage levels can be detected. However, an on-line study revealed that small damage is often invisible, but is detected only after it has grown to a certain level.

The visual effect of the control chart cannot be underestimated. A control chart gives a picture of the damage evolution with time at a glance, and makes it also possible to subjectively assess the reliability of the analysis. Control charts can be easily automated and different reports can be generated from the results.

5 SHM Under Changing Environmental or Operational Conditions: Linear Models

Structural health monitoring of machines and structures often utilises information extracted from vibration measurements. Changes in the selected features extracted from the measurements compared with those of the healthy structure can be interpreted as an indication of damage. However, the natural variation in environmental or operational conditions may also influence the same features and thus yield false indications of damage (Wahab and Roeck, 1997; Farrar et al., 1997; Alampalli, 1998; Cornwell et al., 1999; Peeters and De Roeck, 2000; Rohrmann et al., 2000; Sohn et al., 2000b, 2001, 2003; Manson, 2002; Kullaa, 2002, 2003). For example, temperature, humidity, wind, traffic, etc. can cause such natural variability; if their effects are not taken into account, they can make vibration-based health monitoring unreliable.

In spite of the importance of the problem, there are only a few studies investigating the normal variation of dynamic response of structures due to changes in environmental or operational conditions. Problems in health monitoring due to these effects include:

1. Although temperature measurements are relatively easy to perform, the sensor locations may be difficult to determine.
2. Other environmental or operational quantities may affect the data. Some quantities may be ignored or can be difficult to measure.
3. Each structure is individual. Models accounting for environmental effects may be complicated and need a lot of effort to develop separately for each structure.
4. Long monitoring is needed before automatic damage detection can be initiated.
5. Only part of the measurements can be utilised, for example data from similar conditions (Sohn et al., 2000b) or features for which a model has been developed (Peeters and De Roeck, 2000).

This study introduces a technique, factor analysis, that attempts to eliminate most of the problems above. The measurement of environmental variables is not needed; affecting quantities need not be known; there can be several environmental variables acting simultaneously; the same model can be used for different structures; and all monitoring data can be utilised regardless of current conditions. The restriction of the model is the assumption of a linear relationship between the features and the unknown latent variables.

5.1 Factor Analysis

Let it be assumed that the environmental changes (temperature, humidity, etc.) have an effect on the measured variables (e.g. natural frequencies). The environmental variables are not measured but their effects are merely observed from the measured variables. The objective of factor analysis is to search for or identify the underlying latent factors that can explain the correlation among the variables (Sharma, 1996). The measured variables are functions of these factors, and they reflect the presence of the latent constructs. Although factor analysis is typically labelled as a data-reduction technique, this is not the case in this application. The amount of data is not changed during factor analysis, because the factors themselves are not used, but their influences are only removed from the observations.

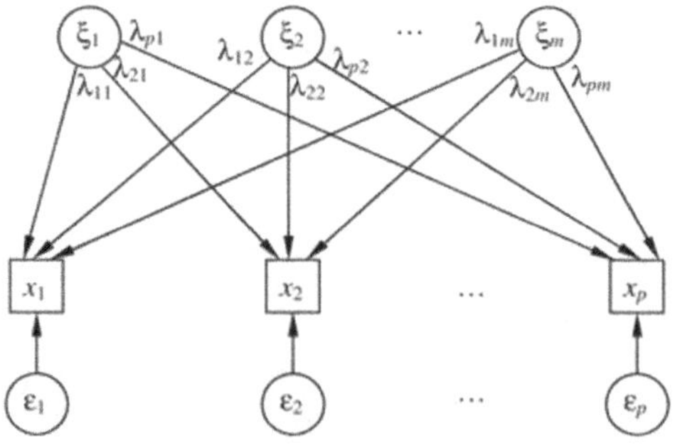

Figure 24. Factor model. The measured variables x_i are also called indicators, ξ_i are common factors, ε_i are unique factors, and λ_{ij} are factor loadings.

The underlying model is sketched in Figure 24. Mathematically it can be written as,

$$\mathbf{x} = \mathbf{\Lambda}\xi + \varepsilon \tag{27}$$

where $\mathbf{x}$ is a $p \times 1$ vector of the measured variables, $\mathbf{\Lambda}$ is a $p \times m$ matrix of *factor loadings* ($m < p$), ξ is an $m \times 1$ vector of unobservable *factors*, and ε is a $p \times 1$ vector of *unique factors*. Equation 27 is the basic factor analysis equation. It is assumed that the factors are not correlated with the unique factors, and without loss of generality it is assumed that the means and variances of the factors are zero and one, respectively.

In a structural health monitoring application, the measured variables $\mathbf{x}$ can be, for example, the lowest natural frequencies of the structure; the factors ξ are the latent variables affecting those natural frequencies, e.g. temperature or the joint positions of a manipulator. Usually the objective of factor analysis is to search for or identify the underlying latent factors. In structural health monitoring, however, one is not usually interested in the quantitative values of the factors, but merely in their effects on the measured variables.

The factor model is constructed using the measurements from the undamaged structure under different conditions. This model is then used to estimate the underlying factors and finally to eliminate their effects from the data. What remains are the unique factors ε, which now should be independent variables and insensitive to the environmental or operational conditions. These unique factors are then used for damage detection. If the structure deteriorates so that the measured variables change, the previously trained factor model cannot explain their changes. These changes would therefore remain and result in an alarm. The learning phase should contain a full range of operational conditions. It should also be emphasised that the model is solely based on measurement data; no mathematical model of the structure is needed.

It is assumed that the factors are mutually independent and normally distributed with zero mean and unit variance: $\xi \sim \mathcal{N}(\mathbf{0}, \mathbf{I})$. The vector of unique factors ε is normally distributed with zero mean and a diagonal covariance matrix $\mathbf{\Psi}$: $\varepsilon \sim \mathcal{N}(\mathbf{0}, \mathbf{\Psi})$. The diagonality of $\mathbf{\Psi}$ is one of the key assumptions in factor analysis. According to the model (equation 27), $\mathbf{x}$ is therefore distributed with zero mean as in,

$$\begin{aligned} E(\mathbf{x}) &= E(\mathbf{\Lambda}\xi + \varepsilon) \\ &= \mathbf{\Lambda}E(\xi) + E(\varepsilon) \\ &= \mathbf{0} \end{aligned} \tag{28}$$

and with covariance matrix, $\mathbf{R}$,

$$\begin{aligned} \mathbf{R} &= E(\mathbf{x}\mathbf{x}^T) \\ &= E[(\mathbf{\Lambda}\xi + \varepsilon)(\mathbf{\Lambda}\xi + \varepsilon)^T] \\ &= E[(\mathbf{\Lambda}\xi + \varepsilon)(\xi^T\mathbf{\Lambda}^T + \varepsilon^T)] \\ &= E(\mathbf{\Lambda}\xi\xi^T\mathbf{\Lambda}^T) + E(\varepsilon\varepsilon^T) \\ &= \mathbf{\Lambda}\mathbf{\Lambda}^T + \mathbf{\Psi} \end{aligned} \tag{29}$$

The measured variables are then also normally distributed: $\mathbf{x} \sim \mathcal{N}(\mathbf{0}, \mathbf{\Lambda\Lambda}^T + \mathbf{\Psi})$. In equation 29 it was assumed that $E(\xi\varepsilon^T) = 0$.

The objective of factor analysis is to estimate the parameter matrices $\mathbf{\Lambda}$ and $\mathbf{\Psi}$ that best model the covariance structure of $\mathbf{x}$. This can be done using e.g. principal factor analysis. After studying the algorithm, the factor score estimates $\hat{\xi}$ are discussed.

The real, symmetric and non-singular data covariance matrix $\mathbf{R}$ can be decomposed as in PCA,

$$\begin{aligned}\mathbf{R} &= \mathbf{P\Gamma P}^T \\ &= \begin{bmatrix}\mathbf{P}_1 & \mathbf{P}_2\end{bmatrix}\begin{bmatrix}\mathbf{\Gamma}_1 & \mathbf{0} \\ \mathbf{0} & \mathbf{\Gamma}_2\end{bmatrix}\begin{bmatrix}\mathbf{P}_1 \\ \mathbf{P}_2\end{bmatrix} \\ &= \mathbf{P}_1\mathbf{\Gamma}_1\mathbf{P}_1^T + \mathbf{P}_2\mathbf{\Gamma}_2\mathbf{P}_2^T \\ &= \mathbf{\Lambda\Lambda}^T + \mathbf{\Psi}\end{aligned} \tag{30}$$

where $\mathbf{P}$ is an orthogonal matrix containing the eigenvectors of matrix $\mathbf{R}$. $\mathbf{\Gamma}$ is a diagonal matrix containing the eigenvalues of $\mathbf{R}$ in a decreasing order. In the third row of equation 30, the eigenvectors $\mathbf{P}$ are partitioned into $\mathbf{P}_1$ and $\mathbf{P}_2$ corresponding to the largest and lowest eigenvalues, respectively. If $\mathbf{P}_1$ consists of the m largest eigenvalues, where m is the number of factors, the last row of equation 30 follows provided that,

$$\mathbf{\Lambda} = \mathbf{P}_1\mathbf{\Gamma}_1^{1/2} \tag{31}$$

and,

$$\mathbf{\Psi} = \mathbf{P}_2\mathbf{\Gamma}_2\mathbf{P}_2^T \tag{32}$$

However, the model assumptions are violated, because $\mathbf{\Psi}$ in equation 32 is a full matrix. An iterative algorithm is needed to force $\mathbf{\Psi}$ diagonal. The ith iteration includes the following two steps. First, $\mathbf{\Lambda}$ is estimated assuming $\mathbf{\Psi}$ is known and assuming $\mathbf{\Psi}^{(0)} = \mathbf{0}$,

$$\begin{aligned}\mathbf{R} - \mathbf{\Psi}^{(i-1)} &= \mathbf{P}_1^{(i)}\mathbf{\Gamma}_1^{(i)}\mathbf{P}_1^{(i)^T} + \mathbf{P}_2^{(i)}\mathbf{\Gamma}_2^{(i)}\mathbf{P}_2^{(i)^T} \\ &= \mathbf{\Lambda}^{(i)}\mathbf{\Lambda}^{(i)^T} + \mathbf{P}_2^{(i)}\mathbf{\Gamma}_2^{(i)}\mathbf{P}_2^{(i)^T}\end{aligned} \tag{33}$$

The error covariance matrix $\mathbf{\Psi}$ is updated in the second step,

$$\mathbf{\Psi}^{(i)} = \text{diag}\left[\mathbf{R} - \mathbf{\Lambda}^{(i)}\mathbf{\Lambda}^{(i)^T}\right] \tag{34}$$

The iteration is continued until $\mathbf{\Psi}$ converges.

Given $\mathbf{\Lambda}$ and $\mathbf{\Psi}$, the estimate of the *factor scores* $\hat{\xi}$ can be obtained. Different estimates have been suggested; their comparison in damage detection can be found in Kullaa (2004). The Thomson's factor score (Mardia et al., 1979) is obtained using the Bayesian approach. The posterior probability distribution can be computed by Bayes' rule,

$$p(\xi|\mathbf{x}) = \frac{p(\mathbf{x}|\xi)p(\xi)}{p(\mathbf{x})} = \frac{\mathcal{N}(\mathbf{x}|\mathbf{\Lambda}\xi, \mathbf{\Psi})\mathcal{N}(\xi|\mathbf{0}, \mathbf{I})}{\mathcal{N}(\mathbf{x}|\mathbf{0}, \mathbf{\Lambda}\mathbf{\Lambda}^T + \mathbf{\Psi})} \tag{35}$$
$$= c_1 \exp\left[-\frac{1}{2}(\mathbf{x} - \mathbf{\Lambda}\xi)^T\mathbf{\Psi}^{-1}(\mathbf{x} - \mathbf{\Lambda}\xi) - \frac{1}{2}\xi^T\xi + \frac{1}{2}\mathbf{x}^T(\mathbf{\Lambda}\mathbf{\Lambda}^T + \mathbf{\Psi})^{-1}\mathbf{x}\right]$$

which is a multinormal density and c_1 is a constant. The mean and covariance of ξ are obtained by completing the square of the exponent with respect to ξ,

$$\begin{aligned}
&-\frac{1}{2}(\mathbf{x} - \mathbf{\Lambda}\xi)^T\mathbf{\Psi}^{-1}(\mathbf{x} - \mathbf{\Lambda}\xi) - \frac{1}{2}\xi^T\xi + \frac{1}{2}\mathbf{x}^T(\mathbf{\Lambda}\mathbf{\Lambda}^T + \mathbf{\Psi})^{-1}\mathbf{x} \\
&= -\frac{1}{2}\mathbf{x}^T\mathbf{\Psi}^{-1}\mathbf{x} + \xi^T\mathbf{\Lambda}^T\mathbf{\Psi}^{-1}\mathbf{x} - \frac{1}{2}\xi^T\mathbf{\Lambda}^T\mathbf{\Psi}^{-1}\mathbf{\Lambda}\xi - \frac{1}{2}\mathbf{x}^T(\mathbf{\Lambda}\mathbf{\Lambda}^T + \mathbf{\Psi})^{-1}\mathbf{x} \\
&= -\frac{1}{2}\xi^T(\mathbf{I} + \mathbf{\Lambda}^T\mathbf{\Psi}^{-1}\mathbf{\Lambda})\xi + \xi^T\mathbf{\Lambda}^T\mathbf{\Psi}^{-1}\mathbf{x} - \frac{1}{2}\mathbf{x}^T[\mathbf{\Psi}^{-1} - (\mathbf{\Lambda}\mathbf{\Lambda}^T + \mathbf{\Psi})^{-1}]\mathbf{x} \\
&= -\frac{1}{2}\left[\xi - (\mathbf{I} + \mathbf{\Lambda}^T\mathbf{\Psi}^{-1}\mathbf{\Lambda})^{-1}\mathbf{\Lambda}^T\mathbf{\Psi}^{-1}\mathbf{x}\right]^T(\mathbf{I} + \mathbf{\Lambda}^T\mathbf{\Psi}^{-1}\mathbf{\Lambda}) \\
&\quad \times \left[\xi - (\mathbf{I} + \mathbf{\Lambda}^T\mathbf{\Psi}^{-1}\mathbf{\Lambda})^{-1}\mathbf{\Lambda}^T\mathbf{\Psi}^{-1}\mathbf{x}\right] + c_2
\end{aligned} \tag{36}$$

where c_2 is a constant. The Thomson's factor score is the mean of the distribution and can be readily found from above,

$$\hat{\xi} = (\mathbf{I} + \mathbf{\Lambda}^T\mathbf{\Psi}^{-1}\mathbf{\Lambda})^{-1}\mathbf{\Lambda}^T\mathbf{\Psi}^{-1}\mathbf{x} \tag{37}$$

and the covariance of ξ given $\mathbf{x}$ is $(\mathbf{I} + \mathbf{\Lambda}^T\mathbf{\Psi}^{-1}\mathbf{\Lambda})^{-1}$.

For damage detection, the unique factors ε can be computed from equation 27 once the factor scores ξ have been estimated,

$$\hat{\varepsilon} = \mathbf{x} - \mathbf{\Lambda}\hat{\xi} \tag{38}$$

These unique factor scores are then used for damage detection. Applications of the factor analysis are given below.

The dimensionality of the feature vector is not affected by the factor analysis; therefore, it is usually necessary to perform a dimensionality reduction before using statistical methods to assess damage. Principal component analysis (Sharma, 1996) is again used in this study. Moreover, if the multivariate Hotelling T statistic is used in damage detection, factor analysis alone is not sufficient, because the Hotelling T statistic of the original variables $\mathbf{x}$ and the unique factor score estimates $\hat{\varepsilon}$ are identical (Fukunaga, 1990).

5.2 Numerical Analysis

The numerical data here simulates four natural frequencies monitored once a day during nine years under varying temperature. The variation of the natural frequencies is considered to be a consequence of temperature, damage, and normal statistical fluctuation,

$$f_i = f_i(T_0) + k_i(T - T_0) + e_i \tag{39}$$

$$f_{iD} = f_{iD}(T_0) + k_i(T - T_0) + e_i \tag{40}$$

where f_i and f_{iD} are the natural frequencies ($i = 1, 2, 3, 4$) of the undamaged and damaged structure, respectively; T is the current temperature; T_0 is the reference temperature; coefficient k_i defines the linear relationship between temperature and frequency i; and $e_i \sim \mathcal{N}(0, \sigma_i^2)$ is the natural variability, or the error term. Notice that the damage was assumed to affect the mean value of the frequency only. The second and third terms in equation 40 were unaffected by damage.

The structure was undamaged during the first two years. Damage was introduced in the beginning of the third year, causing the natural frequencies to decrease. Different magnitudes of damage were defined and the influence of damage level D on the natural frequencies was modelled using a coefficient d_{iD} for frequency i:

$$f_{iD}(T_0) = f_i(T_0) - d_{iD}\sigma_i \tag{41}$$

The damage level increased in the beginning of each successive year and remained constant until the end of the year. The coefficient d_{iD} varied between 0 and 4. The reference temperature was $T_0 = 14°C$. The simulation parameters are shown in Table 2.

Temperature T was simulated by random sampling given the minimum and maximum average values in each month. The temperature and frequency variations during nine years are shown in Figure 25. It should be noted that the simulation might not be physically realistic for structures

Table 2. Simulation parameters.

i	$f_i(T_0)$ Hz	k_i Hz/K	σ_i Hz	d_{i1}	d_{i2}	d_{i3}	d_{i4}	d_{i5}	d_{i6}	d_{i7}
1	4.0	−0.01	0.02	0	0.5	1	1.5	2	2.5	3
2	9.0	−0.015	0.06	0.33	0.83	1.33	1.83	2.33	2.83	3.33
3	10.0	−0.02	0.06	0.67	1.17	1.67	2.17	2.67	3.17	3.67
4	12.0	−0.03	0.1	1	1.5	2	2.5	3	3.5	4

including materials that tend to behave nonlinearly below freezing point. nonlinear effects are considered in Section 6.

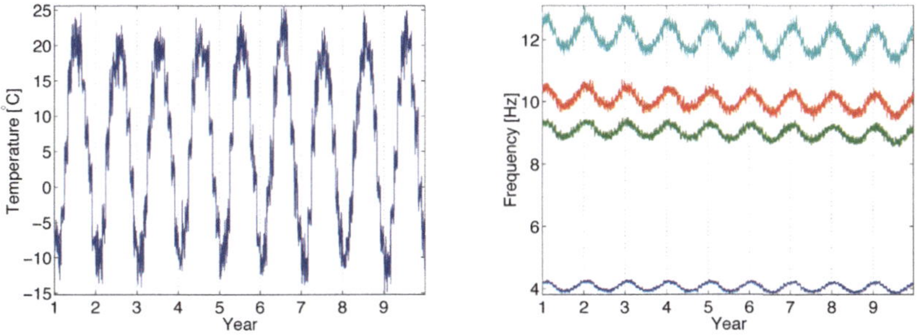

Figure 25. Temperature (left) and the four natural frequencies (right) during nine years.

It was assumed that temperature T was not measured, but the four frequencies only were available. The factor model was estimated using the data from the first year of monitoring. The first year was also considered as the in-control period, the data of which were used to design the control charts for damage detection.

Multivariate Shewhart charts (Montgomery, 1997) for all four natural frequencies before and after factor analysis are shown in Figure 26. The day of initiation of each damage scenario is also shown. The improvement using factor analysis for damage detection can be clearly observed. Factor analysis clearly eliminated the temperature effect. All damage scenarios were detected and no false indication of damage was obtained.

One of the remaining problems is how many factors should be chosen. In the simulated data, there was only one underlying factor, but in reality the number of latent factors is usually not known. The effect of the number of factors was tested using 1, 2, and 3 latent factors. All produced excellent results and no final conclusion could be made. The number of factors is also studied with real measurement data as reported in the following section.

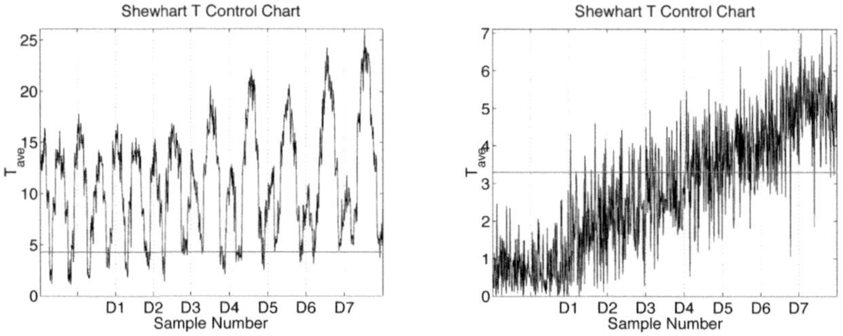

Figure 26. Control charts for Hotelling T statistic for original data (left), and after factor analysis followed by PCA (right).

5.3 Wooden Bridge

The structure in the laboratory was the wooden model bridge shown in Figures 2 and 8. The experimental setup was described in Section 2.2.

The healthy structure was monitored for several days to acquire training data at different environmental conditions. Five artificial damage scenarios were then introduced by adding small point masses of different size to the structure. They were attached on the top flange, 600 mm left from the midspan. The damage scenarios are shown in Table 3, in which damage number 0 means the healthy structure. The last measurements were again from a healthy structure. The added mass was very small compared to the total weight of the structure, even the highest mass increase was only half a percent.

The environmental effects in the laboratory were not anticipated, but it was noticed that the identified features of the bridge varied significantly. The reason for the environmental changes was assumed to be due to the fan, running during working hours and being turned off for the night. Because the structure was made of wood, it was assumed that the variation of the modal properties was mostly due to temperature and humidity variations.

The features used in damage detection were the natural frequencies and modal co-ordinates of modes 6–8, 10, and 12–16. The first 800 samples were used as the in-control samples. Control charts for the natural frequency of mode 6 before and after factor analysis are shown in Figure 27. The number of factors used was 87. Factor analysis resulted in much better performance

Table 3. Increase of the structure's mass at each damage scenario and the corresponding measurement numbers.

Damage case	0	1	2	3	4	5	0
Added mass (g)	0	23.5	47.0	70.5	123.2	193.7	0
Proportional increase (%)	0	0.065	0.13	0.20	0.34	0.54	0
First measurement number	1	1881	1901	1924	1946	1966	1986
Last measurement number	1880	1900	1923	1945	1965	1985	2008

and different damage levels can be observed. However, several outliers are present also in the in-control stage.

The effect of the number of underlying factors in the model was studied using 0, 1, 2, 3, 20, and 87 factors. The Hotelling T control charts are shown in Figure 28. The dimensionality of the feature vector was $p = 261$, and dimensionality reduction was made using PCA. Samples 801–1600 were used for factor analysis, whereas samples 1–800 were used as in-control samples to design the control charts. This distinction is important, because if the same data were used for both the factor model construction and control chart design, too-tight control limits may result. It can be seen that there were probably two significant factors affecting the modal properties, because using two factors produced superior results compared to those with 0, 1, or 3 factors. However, increasing the number of factors to 20 or higher also resulted in reliable damage detection. The number of factors is usually not known in advance, and the results may be sensitive to the choice, as seen in Figure 28. It is therefore suggested to use quite a few factors, because it was found that the results stayed stable with an increasing number of factors. In this study, the number of factors was determined to account for at least 99.9% of the variance in the data, resulting in 87 factors. All damage configurations were detected. Occasional false indications of damage were also present, suggesting that at least two successive out-of-control samples should be present before an alarm is generated.

5.4 Summary

A promising attempt was made to eliminate environmental and operational effects from a structural health monitoring system using factor analysis. Factor analysis offers many advantages compared to existing techniques: the underlying quantities need not be measured, they need not be

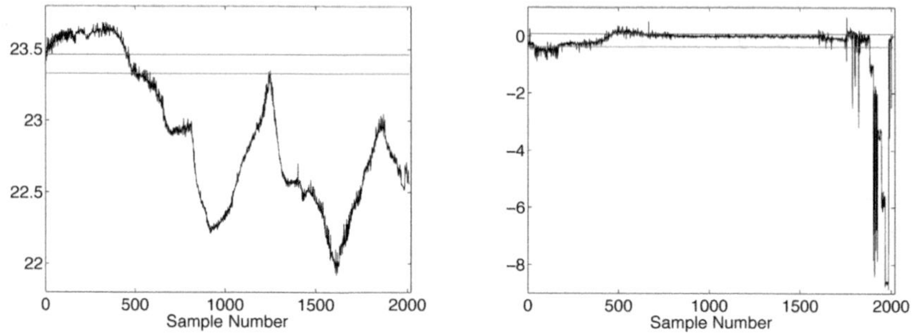

Figure 27. Control charts for the 6th natural frequency, before (left), and after factor analysis (right).

even known, there can be several factors affecting simultaneously, and the same model can be used for different structures. The technique assumes a linear relationship between the features and latent factors. The training data must include at least one complete cycle of in-service conditions to build the factor model.

Because the number of factors is not usually known, it is suggested to use quite a few factors. This was also found to produce the most reliable results. It was found that Hotelling T control charts, both for the original features and features after factor analysis, produced identical results. Therefore, in order to exploit the advantageous influence of factor analysis in removing environmental and operational effects, the factor analysis was followed by principal component analysis. It was also found with the laboratory experiments that not all features individually were in control after the factor analysis, whereas the multivariate Hotelling T control chart worked well.

It is recommended to use different data segments for the factor model construction and for the control chart design. It is also suggested to use a high-dimensional feature vector, letting PCA decide which variables are affected by damage. This approach can be utilised in the SHM system design, since it is usually difficult to tell in advance, which features are sensitive to damage.

6 SHM Under Changing Environmental or Operational Conditions: Nonlinear Models

Damage in a structure can be indirectly detected from changes in features extracted from the measurement data. However, the influence of environ-

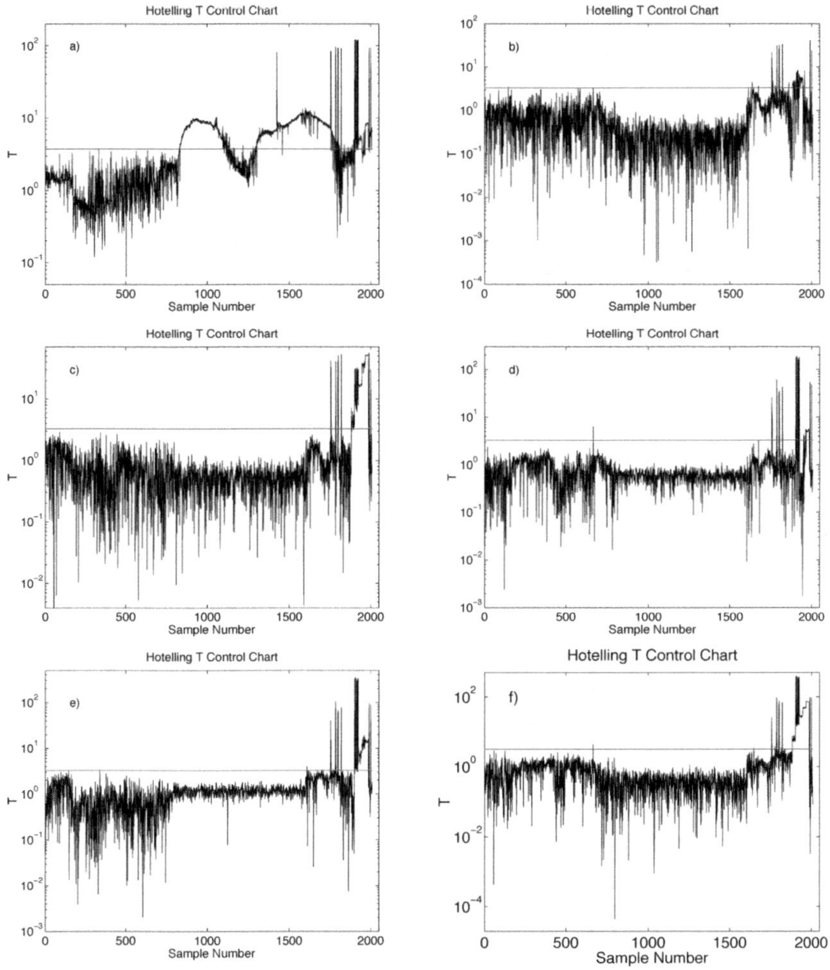

Figure 28. Effects of the number of factors: a) 0, b) 1, c) 2, d) 3, e) 20, and f) 87.

mental or operational variations on the same features is often stronger than that of damage, making damage detection difficult and unreliable. If this effect is not taken into account, it can result in false identifications of damage or incapability to detect minor damage.

The effects of environmental or operational variations on the damage-sensitive features can be eliminated using latent variable models. Such methods which are able to compensate linear influences on structural features have been developed, including factor analysis (Kullaa, 2002), principal component analysis (PCA) (Manson, 2002; Yan et al., 2005a), robust singular value decomposition (Vanlanduit et al., 2004), and minimum mean square error (MMSE) estimation (Kullaa, 2004, 2005). All the methods are based on the correlation structure between the features. The main advantage of latent variable models is that the environmental or operational variables need not be measured.

However, the environmental or operational variations have often a nonlinear effect on the features. For example, as the temperature falls below zero, its influence on the features can change abruptly. Usually this results also in a nonlinear correlation structure between the features. There are only a few studies of nonlinear models. Kullaa (2003) used a piecewise linear model to compensate the nonlinear effects. A similar approach was used by Yan et al. (2005b) having local PCA models for local regions in the data space. Sohn et al. (2003) used an autoassociative neural network that can be thought of as a nonlinear PCA (Kramer, 1991).

The mixture of factor analysers model (Ghahramani and Hinton, 1996) is used here to compensate the nonlinear effects. It is based on piecewise linear factor analysers each modelling a region in the input space. An advantage of the algorithm is that the clustering and factor analysis are performed simultaneously using the expectation maximisation (EM) algorithm.

In the following sections, the model is introduced, the model parameters identified, and the method is subsequently applied to a numerical study. The second application is the Z24 Bridge, in which the natural frequencies vary due to the temperature. The method is also compared to factor analysis and nonlinear principal component analysis.

6.1 Mixture of Factor Analysers

Ghahramani and Hinton (1996) developed a nonlinear model, which is an extended factor analysis allowing different local linear factor models for different regions in the input space. For each factor analyser j, the model is,

$$\mathbf{x} = \mu_j + \mathbf{\Lambda}_j \xi + \varepsilon \tag{42}$$

where $\mathbf{x}$ is a $p \times 1$ vector of the measured variables, μ_j is the mean vector, $\mathbf{\Lambda}_j$ is a $p \times m$ matrix of *factor loadings*, ξ is an $m \times 1$ vector of unobservable *factors*, and ε is a $p \times 1$ vector of *unique factors*. The factors are assumed to be $\mathcal{N}(\mathbf{0}, \mathbf{I})$ distributed (independent of the factor analyser j), and ε is $\mathcal{N}(\mathbf{0}, \mathbf{\Psi}_j)$ distributed, with a diagonal $\mathbf{\Psi}_j$.

In the mixture of factor analysers (MFA) model having m factor analysers, an m-dimensional binary random variable $\mathbf{w}$ is introduced, having a 1-of-m representation in which a particular element w_j is equal to 1 and all other elements are equal to 0 (Bishop, 2006). The marginal distribution over $\mathbf{w}$ is specified in terms of the mixing coefficients π_j, such that,

$$\pi_j = p(w_j = 1) \tag{43}$$

in which,

$$0 \leq \pi_j \leq 1 \tag{44}$$

and,

$$\sum_j \pi_j = 1 \tag{45}$$

The unknown parameters θ of the model are π_j, μ_j, $\mathbf{\Lambda}_j$, and $\mathbf{\Psi}_j$ for each factor analyser ($j = 1, \ldots, m$). Ghahramani and Hinton (1996) assumed that each mixture component has the same $\mathbf{\Psi}$. The analysis here, however, is not restricted to this assumption.

For every observed data point $\mathbf{x}_i$, there are corresponding latent variables ξ_i and $\mathbf{w}_i$, where $i = 1, \ldots, n$ and n is the number of observations.

Consider a joint distribution of the complete data given the model parameters θ,

$$p(\mathbf{x}, \xi, \mathbf{w}) = p(\mathbf{x}|\xi, \mathbf{w}) p(\xi|\mathbf{w}) p(\mathbf{w}) \tag{46}$$

The distributions in (46) are,

$$p(\mathbf{x}|\xi, \mathbf{w}) = \prod_j \left[\mathcal{N}(\mathbf{x}|\mu_j + \mathbf{\Lambda}_j \xi, \mathbf{\Psi}_j)\right]^{w_j} \tag{47}$$

$$p(\xi|\mathbf{w}) = p(\xi) = \mathcal{N}(\xi|\mathbf{0}, \mathbf{I}) \tag{48}$$

$$p(\mathbf{w}) = \prod_j \pi_j^{w_j} \tag{49}$$

resulting in,

$$p(\mathbf{x}, \xi, \mathbf{w}) = \prod_j \left[\pi_j \mathcal{N}(\mathbf{x}|\mu_j + \mathbf{\Lambda}_j\xi, \mathbf{\Psi}_j)\right]^{w_j} \mathcal{N}(\xi|\mathbf{0}, \mathbf{I}) \tag{50}$$

The likelihood function for the complete data set to be maximised is,

$$p(\mathbf{X}, \Xi, \mathbf{W}) = \prod_i \prod_j \left[\pi_j \mathcal{N}(\mathbf{x}_i|\mu_j + \mathbf{\Lambda}_j\xi_i, \mathbf{\Psi}_j)\right]^{w_{ij}} \mathcal{N}(\xi_i|\mathbf{0}, \mathbf{I}) \tag{51}$$

The model parameters can be obtained by maximising the likelihood function. The last conditional distribution can be left out as it contains no unknown parameters. However, the maximisation of the likelihood function above cannot be done, because the latent variables are not available, only the incomplete data $\mathbf{X}$.

The expectation maximisation (EM) algorithm can be used to estimate the unknown parameters of a probabilistic model that maximise the likelihood. The algorithm starts with an initial estimate for the parameters and iteratively modifies them to increase the likelihood of the observed data. The EM-algorithm has two steps: the expectation step (E-step) and the maximisation step (M-step). The E-step computes expectations given parameters, and the M-step provides new estimates of the parameters. The algorithm guarantees that the log-likelihood will increase monotonically in each iteration.

The expected values of the latent variable functions given the model parameters are computed in the E step of the EM algorithm. The expectation is taken from the log-likelihood function, resulting in the following function,

$$\begin{aligned}
Q &= E\left[\log \prod_i \prod_j \left[\pi_j \mathcal{N}(\mathbf{x}_i|\mu_j + \mathbf{\Lambda}_j\xi_i, \mathbf{\Psi}_j)\right]^{w_{ij}}\right] \\
&= E\left[\log \prod_i \prod_j \left\{\pi_j (2\pi)^{-p/2} |\mathbf{\Psi}_j|^{-1/2}\right.\right. \\
&\quad \left.\left. \times \exp\left[-\frac{1}{2}(\mathbf{x}_i - \mu_j - \mathbf{\Lambda}_j\xi_i)^T \mathbf{\Psi}_j^{-1} (\mathbf{x}_i - \mu_j - \mathbf{\Lambda}_j\xi_i)\right]\right\}^{w_{ij}}\right] \\
&= E\left[\log \prod_i \prod_j \left\{\pi_j (2\pi)^{-p/2} |\mathbf{\Psi}_j|^{-1/2}\right.\right. \\
&\quad \left.\left. \times \exp\left[-\frac{1}{2}(\mathbf{x}_i - \mathbf{\Gamma}_j\mathbf{z}_i)^T \mathbf{\Psi}_j^{-1} (\mathbf{x}_i - \mathbf{\Gamma}_j\mathbf{z}_i)\right]\right\}^{w_{ij}}\right]
\end{aligned} \tag{52}$$

in which,

$$\mathbf{z} = \begin{Bmatrix} \xi \\ 1 \end{Bmatrix} \tag{53}$$

$$\mathbf{\Gamma}_j = \begin{bmatrix} \mathbf{\Lambda}_j & \mu_j \end{bmatrix} \tag{54}$$

Apply first the logarithm,

$$\begin{aligned} Q &= c + E\left[\sum_{i,j}\left(w_{ij}\log\pi_j - \frac{1}{2}w_{ij}\log|\mathbf{\Psi}_j| - \frac{1}{2}w_{ij}\mathbf{x}_i^T\mathbf{\Psi}_j^{-1}\mathbf{x}_i\right.\right. \\ &\quad \left.\left.+\frac{1}{2}w_{ij}\mathbf{x}_i^T\mathbf{\Psi}_j^{-1}\mathbf{\Gamma}_j\mathbf{z}_i + \frac{1}{2}w_{ij}\mathbf{z}_i^T\mathbf{\Gamma}_j^T\mathbf{\Psi}_j^{-1}\mathbf{x}_i - \frac{1}{2}\mathbf{z}_i^T\mathbf{\Gamma}_j^T\mathbf{\Psi}_j^{-1}\mathbf{\Gamma}_j\mathbf{z}_i\right)\right] \\ &= c + E\left[\sum_{i,j}\left(w_{ij}\log\pi_j - \frac{1}{2}w_{ij}\log|\mathbf{\Psi}_j| - \frac{1}{2}w_{ij}\mathbf{x}_i^T\mathbf{\Psi}_j^{-1}\mathbf{x}_i\right.\right. \\ &\quad \left.\left.+w_{ij}\mathbf{x}_i^T\mathbf{\Psi}_j^{-1}\mathbf{\Gamma}_j\mathbf{z}_i - \frac{1}{2}\mathrm{tr}\left(w_{ij}\mathbf{\Gamma}_j^T\mathbf{\Psi}_j^{-1}\mathbf{\Gamma}_j\mathbf{z}_i\mathbf{z}_i^T\right)\right)\right] \end{aligned} \tag{55}$$

where c is a constant and tr is the trace operator. Next apply the expectation operator,

$$\begin{aligned} Q &= c + \sum_{i,j}\left(h_{ij}\log\pi_j - \frac{1}{2}h_{ij}\log|\mathbf{\Psi}_j| - \frac{1}{2}h_{ij}\mathbf{x}_i^T\mathbf{\Psi}_j^{-1}\mathbf{x}_i\right. \\ &\quad \left.+\mathbf{x}_i^T\mathbf{\Psi}_j^{-1}\mathbf{\Gamma}_j E\left[w_{ij}\mathbf{z}|\mathbf{x}_i\right] - \frac{1}{2}\mathrm{tr}\left(\mathbf{\Gamma}_j^T\mathbf{\Psi}_j^{-1}\mathbf{\Gamma}_j E\left[w_{ij}\mathbf{z}\mathbf{z}^T|\mathbf{x}_i\right]\right)\right) \\ &= c + \sum_{i,j}\left(h_{ij}\log\pi_j - \frac{1}{2}h_{ij}\log|\mathbf{\Psi}_j| - \frac{1}{2}h_{ij}\mathbf{x}_i^T\mathbf{\Psi}_j^{-1}\mathbf{x}_i\right. \\ &\quad \left.+h_{ij}\mathbf{x}_i^T\mathbf{\Psi}_j^{-1}\mathbf{\Gamma}_j E\left[\mathbf{z}|\mathbf{x}_i,\mathbf{w}_i\right] - \frac{1}{2}h_{ij}\mathrm{tr}\left(\mathbf{\Gamma}_j^T\mathbf{\Psi}_j^{-1}\mathbf{\Gamma}_j E\left[\mathbf{z}\mathbf{z}^T|\mathbf{x}_i,\mathbf{w}_i\right]\right)\right) \end{aligned} \tag{56}$$

where,

$$h_{ij} = E\left[w_j|\mathbf{x}_i\right] \propto \pi_j p(\mathbf{x}_i|\mathbf{w}_i) = \pi_j\mathcal{N}\left(\mathbf{x}_i|\mu_j, \mathbf{\Lambda}_j\mathbf{\Lambda}_j^T + \mathbf{\Psi}_j\right) \tag{57}$$

Equation 57 was derived using Bayes' rule and marginalising variable ξ. In the E-step, h_{ij} and the other expectations for all data points $\mathbf{x}_i$

and mixture components j must be computed given the model parameters $\pi_j, \mu_j, \mathbf{\Lambda}_j$, and $\mathbf{\Psi}_j$,

$$E\left[\mathbf{z}|\mathbf{x}_i, \mathbf{w}_i\right] = \begin{Bmatrix} E\left[\xi|\mathbf{x}_i, \mathbf{w}_i\right] \\ 1 \end{Bmatrix} \tag{58}$$

$$E\left[\mathbf{z}\mathbf{z}^T|\mathbf{x}_i, \mathbf{w}_i\right] = \begin{bmatrix} E\left[\xi\xi^T|\mathbf{x}_i, \mathbf{w}_i\right] & E[\xi|\mathbf{x}_i, \mathbf{w}_i] \\ E[\xi|\mathbf{x}_i, \mathbf{w}_i]^T & 1 \end{bmatrix} \tag{59}$$

The expectations are (Ghahramani and Hinton, 1996),

$$E\left[\xi|\mathbf{x}_i, \mathbf{w}_i\right] = \beta_j(\mathbf{x}_i - \mu_j) \tag{60}$$

$$E\left[\xi\xi^T|\mathbf{x}_i, \mathbf{w}_i\right] = \mathbf{I} - \beta_j\mathbf{\Lambda}_j + \beta_j(\mathbf{x}_i - \mu_j)(\mathbf{x}_i - \mu_j)^T\beta_j^T \tag{61}$$

where $\beta_j = (\mathbf{I} + \mathbf{\Lambda}_j^T\mathbf{\Psi}_j^{-1}\mathbf{\Lambda}_j)^{-1}\mathbf{\Lambda}_j^T\mathbf{\Psi}_j^{-1}$ (see equation 37). Equation 61 was derived as follows,

$$E\left[\xi\xi^T|\mathbf{x}_i, \mathbf{w}_i\right] = \mathrm{cov}(\xi|\mathbf{x}_i, \mathbf{w}_i) + E\left[\xi|\mathbf{x}_i, \mathbf{w}_i\right]E\left[\xi|\mathbf{x}_i, \mathbf{w}_i\right]^T \tag{62}$$

The first term in the right hand side is derived taking into account the covariance expression below equation 37,

$$\begin{aligned} \mathrm{cov}(\xi|\mathbf{x}_i, \mathbf{w}_i) &= (\mathbf{I} + \mathbf{\Lambda}_j^T\mathbf{\Psi}_j^{-1}\mathbf{\Lambda}_j)^{-1} \\ &= (\mathbf{I} + \mathbf{\Lambda}_j^T\mathbf{\Psi}_j^{-1}\mathbf{\Lambda}_j)^{-1}\left[(\mathbf{I} + \mathbf{\Lambda}_j^T\mathbf{\Psi}_j^{-1}\mathbf{\Lambda}_j) - \mathbf{\Lambda}_j^T\mathbf{\Psi}_j^{-1}\mathbf{\Lambda}_j\right] \\ &= \mathbf{I} - (\mathbf{I} + \mathbf{\Lambda}_j^T\mathbf{\Psi}_j^{-1}\mathbf{\Lambda}_j)^{-1}\mathbf{\Lambda}_j^T\mathbf{\Psi}_j^{-1}\mathbf{\Lambda}_j \\ &= \mathbf{I} - \beta_j\mathbf{\Lambda}_j \end{aligned} \tag{63}$$

In the M-step the log-likelihood is maximised by differentiation with respect to the unknown parameters (the latent variables are held constant at their expected values),

$$\frac{\partial Q}{\partial \mathbf{\Gamma}_j} = \sum_i \left(h_{ij}\mathbf{\Psi}_j^{-1}\mathbf{x}_i^T E\left[\mathbf{z}|\mathbf{x}_i, \mathbf{w}_i\right]^T - h_{ij}\mathbf{\Psi}_j^{-1}\mathbf{\Gamma}_j E\left[\mathbf{z}\mathbf{z}^T|\mathbf{x}_i, \mathbf{w}_i\right]\right) = 0 \tag{64}$$

$$\begin{aligned} \frac{\partial Q}{\partial \mathbf{\Psi}_j^{-1}} = \sum_i \bigg(&\frac{1}{2}h_{ij}\mathbf{\Psi}_j - \frac{1}{2}h_{ij}\mathbf{x}_i\mathbf{x}_i^T + h_{ij}\mathbf{\Gamma}_j E\left[\mathbf{z}|\mathbf{x}_i, \mathbf{w}_i\right]\mathbf{x}_i^T \\ &-\frac{1}{2}h_{ij}\mathbf{\Gamma}_j E\left[\mathbf{z}\mathbf{z}^T|\mathbf{x}_i, \mathbf{w}_i\right]\mathbf{\Gamma}_j^T\bigg) = 0 \end{aligned} \tag{65}$$

from which the updates are obtained,

$$\mathbf{\Gamma}_j = \left(\sum_i h_{ij} \mathbf{x}_i E\left[\mathbf{z}|\mathbf{x}_i, \mathbf{w}_i\right]^T \right) \left(\sum_l h_{lj} E\left[\mathbf{z}\mathbf{z}^T|\mathbf{x}_l, \mathbf{w}_l\right] \right)^{-1} \tag{66}$$

$$\mathbf{\Psi}_j = \frac{\operatorname{diag} \sum_{i,j} h_{ij} \left(\mathbf{x}_i - \mathbf{\Gamma}_j E\left[\mathbf{z}|\mathbf{x}_i, \mathbf{w}_i\right]\right) \mathbf{x}_i^T}{\sum_i h_{ij}} \tag{67}$$

The solution in equation 67 was obtained by substituting equation 66 for the last $\mathbf{\Gamma}_j$ in equation 65 and using the diagonal constraint (Ghahramani and Hinton, 1996).

The mixing coefficients π_j are solved using a Lagrange multiplier λ and maximising the following function (Bishop, 2006),

$$Q_1 = \sum_{i,j} (h_{ij} \log \pi_j) + \lambda \left(\sum_j \pi_j - 1 \right) \tag{68}$$

Differentiating with respect to π_j and using (45) results in,

$$\begin{aligned} \frac{\partial Q_1}{\partial \pi_j} &= \sum_i \frac{h_{ij}}{\pi_j} + \lambda = 0 \\ \Rightarrow \qquad \sum_i h_{ij} + \pi_j \lambda &= 0 \\ \Rightarrow \qquad \sum_{i,j} h_{ij} + \lambda \sum_j \pi_j &= 0 \\ \Rightarrow \qquad \lambda &= -n \end{aligned} \tag{69}$$

Substituting this into the equation on the second line results finally in,

$$\pi_j = \frac{1}{n} \sum_i h_{ij} \tag{70}$$

The factor model is constructed using the measurements from the undamaged structure as the training data. The model is then used to estimate the underlying factors. Once they have been estimated, the unique factor scores are computed by,

$$\hat{\varepsilon} = \mathbf{x} - E(\mathbf{x}|\xi) = \mathbf{x} - \sum_j \pi_j (\mu_j + \mathbf{\Lambda}_j \xi) \tag{71}$$

where ξ is computed from equation 60. The unique factor scores, or residuals, are used for damage detection. After the factor analysis, it is usually necessary to perform a dimensionality reduction before using statistical methods to assess damage. Principal component analysis (Sharma, 1996) is used again in this study.

6.2 Numerical Analysis

The mixture of factor analysers model is first applied to numerical data with two variables generated by a single factor and two factor analysers. The nonlinear correlation of the two variables is shown in Figure 29. The training data are plotted with symbol 'x'. The two clusters with a nonlinear correlation can be clearly observed. Also shown are groups of test data at different regions of the input space. These are used to test the clustering and outlier diagnosis performance of MFA. Groups 1–4 are undoubtedly outliers as they do not fit in either of the two factor analysers. On the other hand, the diagnosis of the data in groups 5–12 depends on which factor analyser they belong to. In this simple example it is easy to see that groups 5, 8, 11, and 12 expand the range of the training data, making extrapolation possible. However, with groups 6, 7, 9, and 10, extrapolation is probably not desired due to the assumed continuity of variables. Therefore, these groups preferably belong to the other factor analyser and should be diagnosed as outliers. This example shows that in the nonlinear case, clustering addresses an additional problem to structural health monitoring. A similar problem was also noticed with the data from the Z24 Bridge and is discussed in the following section.

When equation 71 was used in the residual estimation, groups 6, 7, 9, and 10 were assigned to the extension of the lines defining the factor analysers. These groups then were not diagnosed as outliers. A modified clustering criterion was used, which was based on the factor scores computed for each factor analyser. As the model assumes that the factors are $\mathcal{N}(\mathbf{0}, \mathbf{I})$ distributed, the minimum distance of each factor score from the distribution determined the corresponding factor analyser. This method assigned groups 6, 7, 9, and 10 to the other factor analysers and were diagnosed as outliers. The proposed solution to the clustering problem works only if there is one-to-one correspondence between the variables. If this is not the case, the probability criterion should be used.

The results of the MFA are shown in Figure 30. The plot on the left shows the fit of each data point to the model. On the right is the control chart with the group number indicated. All outliers were correctly detected and linear extrapolation was also possible. As a comparison, the linear factor analysis was used for the same data, and the results are plotted in Figure 31. In this case, there is a single line that best describes the training data. The variance of the residual ε in the training data was higher and

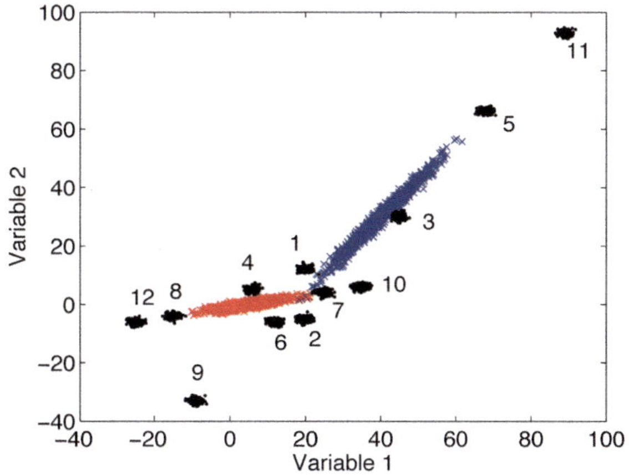

Figure 29. nonlinear correlation of variables. Training data are marked with 'x'. The numbered data clusters are test data.

the sensitivity to detect damage was decreased: from groups 1–4, group 2 only was diagnosed as an outlier. All other groups 5–12 were all diagnosed as outliers. The linear factor analysis is nevertheless a very useful tool as it works reasonably well in most cases. It is fast, automatic, and it can be easily applied to high-dimensional data. It should always be preferred as the first choice before applying any nonlinear models.

As a third method, an autoassociative neural network (Kramer, 1991; Sohn et al., 2003) was applied to the same data set and the results are shown in Figure 32. The number of neurons in the mapping layer was chosen to be 5, a result from different trials. All groups in the test data were identified as outliers. The advantage of this method is that it is fully nonlinear. However, extrapolation was not feasible. Also the training by trial and error was more time-consuming.

6.3 The Z24 Bridge

The pairwise correlation of the four lowest identified natural frequencies of the Z24 Bridge (see Peeters (2000) for details) is plotted in Figure 33. It can be seen that the frequencies varied considerably due to environmental effects and are nonlinearly correlated. The physical reason was the different behaviour of the bridge below and above the freezing point. From the figure it was decided to use two factor analysers and one factor.

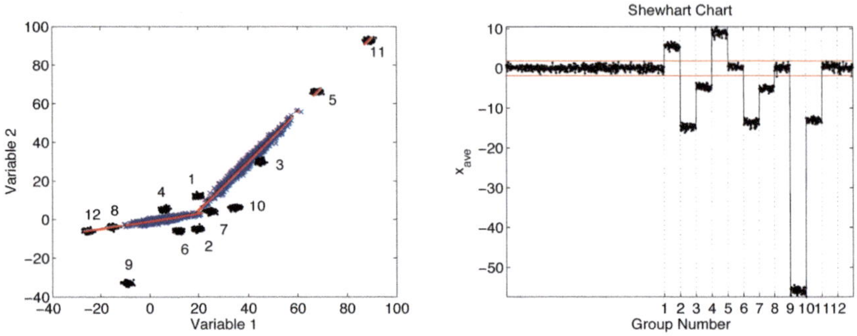

Figure 30. Model fit and diagnosis using the mixture of factor analysers.

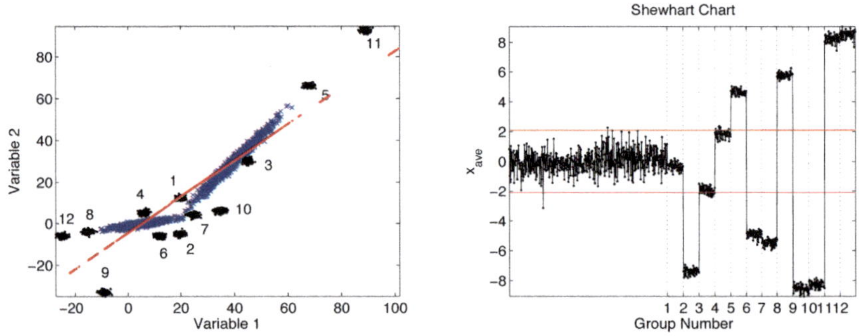

Figure 31. Model fit and diagnosis using the linear factor analysis.

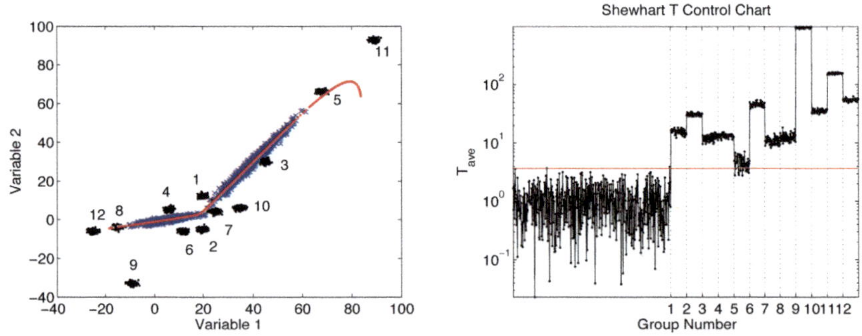

Figure 32. Model fit and diagnosis using the auto-associative neural network.

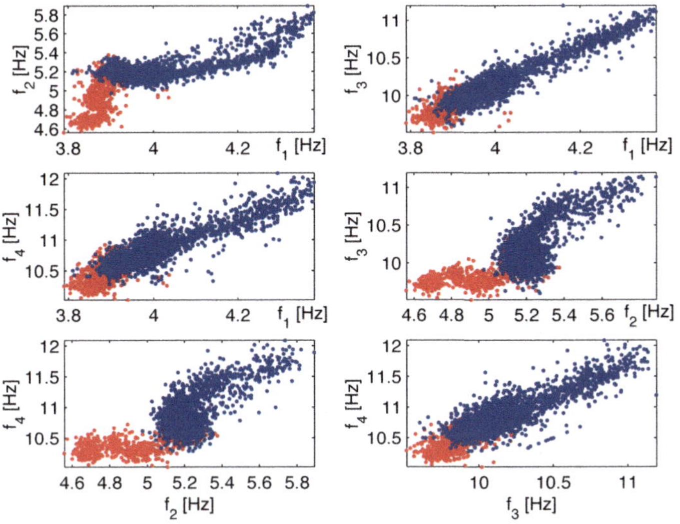

Figure 33. Correlation between the four lowest natural frequencies of the Z24 Bridge. Darker symbols indicate the training data.

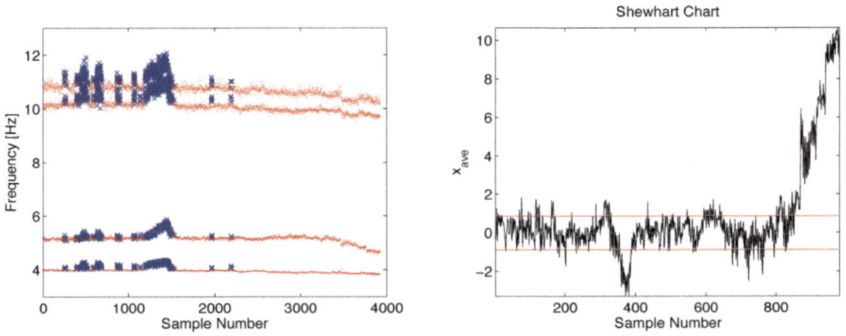

Figure 34. Left: Classification of the identified natural frequencies of the Z24 Bridge. The two different symbols indicate the two classes (factor analysers). Right: Control chart for the first principal component after MFA.

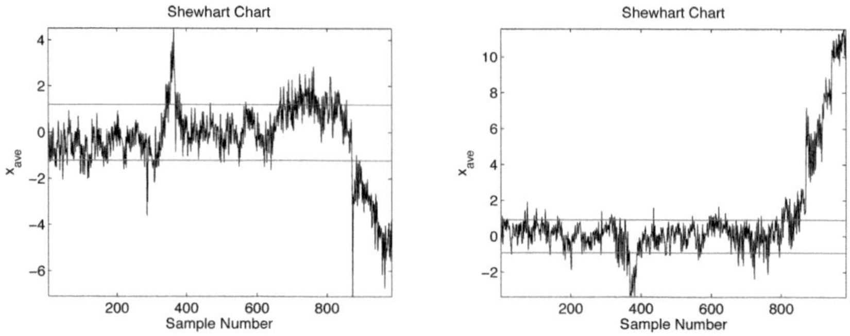

Figure 35. Control chart for the first principal component after FA (left) and NLPCA (right).

The training data were the first 3000 samples (see Figure 34 left) shown with a darker colour in Figure 33. Classifying the data from the damaged structure showed similar difficulties as with the numerical data in the previous section. This can be explained by looking at Figure 33, in which the data from the damaged structure are located at the extension of the line of the training data measured below freezing point. Therefore, the clustering algorithm decides that the data belongs to the low temperature cluster. However, the Euclidian distance of the new data to the data measured above freezing point is lower. As a solution to the clustering problem the same method as with the numerical model was used using the factor score distance as the criterion. This assigned the new data to correct factor analysers. The four lowest identified natural frequencies of the Z24 Bridge are plotted in Figure 34 left with different symbols and colours indicating the two clusters.

After the factor analysis MFA, the largest principal value of the multivariate feature vector was used for damage detection using the Shewhart control chart (Montgomery, 1997) shown in Figure 34 right. In this control chart, each point is the subgroup average of four samples. From the figure it can be seen that different sizes or types of damage could be clearly detected. However, the control limits are too tight resulting in frequent false indications of damage. Moreover, the temperature effect could not be totally eliminated. If more variables were available, for example the modeshapes, the temperature compensation would probably be better.

For a comparison, the linear factor analysis and the autoassociative neural network were applied to the same data set and the results are shown in Figure 35. In the autoassociative neural network, the number of neurons in the mapping layer was chosen to be 6. Comparing Figures 34 and 35 it can be seen that the nonlinear models were able to better fit the data and the damage was more clearly detected. However, the model fit was far from perfect, resulting in several false identifications of damage, especially during the cold season. Also, the performance of one nonlinear method was not distinctively better than the other.

6.4 Summary

The mixture of factor analysers model was applied to eliminate non-linear environmental or operational effects from the monitoring data. The main advantage is that the environmental variables need not be measured. Compared to the linear case, the proposed method is more involved. The assessment of the number of factors is more important. Also, the estimation of the number of factor analysers is needed and clustering needs more attention. The results are better than when using linear models, provided the model parameters are correctly assessed. The method was also compared to nonlinear principal component analysis. As in the case of many other engineering problems, linear methods should be preferred as the first choice before applying any nonlinear models.

7 Mechanical Engineering Application: A Hydraulic Crane

Diverse applications can profit from structural health monitoring (SHM), including: civil infrastructures, aircraft, space structures, land and marine structures, offshore structures, and critical devices. In mechanical engineering, especially with rotating machinery, the term "condition monitoring" is traditionally used. However, there are also other mechanical engineering systems, for which SHM could be beneficial. Such systems include different cranes and forest harvesters. Monitoring of the manipulator structure of those systems is a relatively new application. Diagnostics of the hydraulic system is also important. Moreover, the new ergonomic regulations urge monitoring of vibrations experienced by the operator.

The monitoring of mobile hydraulic machines is motivated both by safety and economic interests. A structural failure can cause a loss of human lives or increased costs due to an interrupted process or higher maintenance efforts; therefore, damage detection at an early stage is important before a complete failure occurs.

Some characteristics of mobile hydraulic machines related to SHM are listed in the following. (1) The structure is neither a fixed structure nor a rotating machine. The manipulator consists of joints and links. The system behaviour is therefore highly nonlinear due to a variable geometry. (2) A human operator is needed, resulting in a non-stationary process, which depends on the operator. (3) The machines are used in harsh environments subject to weather, debris, impacts, contacts, and vibrations that affect both the structure and the hydraulics as well as the operator. (4) The anticipated damage scenarios of such systems are: fatigue cracks in the welds or joints, which are often hidden; local buckling of joints; and leakage in the hydraulic system. (5) An advantage compared to civil engineering structures is that excitation is readily available and controllable via the hydraulic system. (6) The extraction of modal parameters (natural frequencies, modeshapes, and damping) is often difficult due to high damping in the hydraulic system; therefore, other damage-sensitive features must be studied. (7) Natural variability is present in the system due to variable joint positions, loads, friction in joints, hydraulic oil viscosity, etc. In order to detect damage of a desired level, their effects must be compensated for.

In this section, vibration-based SHM is applied to the manipulator of a vehicle crane. Monitoring of the hydraulic system is beyond the scope of this study. It should be noted, however, that monitoring is applied to the integrated system, and the viscosity of the hydraulic oil has an effect on the dynamic properties of the structure. In order to reduce the amount of nonlinearities and operational variations, a solution is proposed, in which scheduled vibration measurements are performed around a fixed static equilibrium with (almost) the same static load and a controlled dynamic excitation. A reasonable amount of training data should nevertheless be acquired, because the operational conditions most probably have a large variability.

Several features have been studied for damage identification when the excitation is not known. Some of them are listed in the following, but the list is by no means exhaustive. Natural frequencies and modeshapes have been utilised by several researchers (Doebling et al., 1996); AR coefficients have been studied e.g. by Sohn et al. (2000a) and Kullaa (2005); power spectra were investigated by Kullaa (2004); transmissibilities were applied by Worden et al. (1999) and Sampaio et al. (2000); wavelets were used by Sun and Chang (2002); and modal filters by Deraemaeker et al. (2006).

In this section, the objective is to study two features used in damage detection: coefficients of the auto-regressive (AR) model and transmissibilities (TR). They are estimated from experimental response data and their performance to detect damage is compared. The best features would be sensitive to damage and would result in few false indications of damage. An experimental monitoring test is performed on a vehicle crane by measuring accelerations from eight positions of the structure. The same data records are used to extract both features. Normal operational variability is eliminated from the features before damage detection using factor analysis.

7.1 Features

The two features studied in this application are the coefficients of the autoregressive (AR) model and the transmissibilities (TR). Both features are related to the spectral properties and are therefore global. The AR model is parametric, while TR is a non-parametric quantity. Therefore, the AR model often results in relatively few features, whereas the number of features in TR can be large. A high number of features can be advantageous, because it is usually not known in advance which features change as the structure is deteriorated. A disadvantage is that the statistical reliability of damage detection decreases with a large number of features (curse of dimensionality) and more training data are needed. AR coefficients are identified from a single sensor, while TR involves a pair of sensors. Independent sensors are technically less demanding, because synchronisation between sensors is not needed. However, synchronised sensors provide additional information of the structure, e.g. modeshapes. Both of the proposed features would be appropriate also in a wireless sensor network, in which local computing in the node is preferred due to a high power consumption in data transmission.

AR coefficients. The autoregressive (AR) model of order n is defined as,

$$\begin{aligned} y(k) &= -a_1 y(k-1) - \ldots - a_n y(k-n) + e(k) \\ &= \varphi^T(k)\theta + e(k) \end{aligned} \tag{72}$$

$y(k)$ is the response at time $k\Delta t$, where Δt is the time increment; a_i are the unknown AR coefficients, and $e(k)$ is the error term. The AR coefficients in vector θ can be solved from the latter equation using the minimum least squares algorithm. In this study, the order n is chosen to be 10 (Ljung, 1999).

Transmissibilities. Assuming a single-point excitation, the transmissibility (TR) function is the transfer function between two responses,

$$\mathrm{TR}_{ij} = \frac{Y_i(\omega)}{Y_j(\omega)} \tag{73}$$

where $Y_i(\omega)$ and $Y_j(\omega)$ are the Fourier transforms of responses $y_i(t)$ and $y_j(t)$, respectively. In practice, TR is computed using the spectral densities,

$$\mathrm{TR}_{ij} = \frac{Y_i(\omega)}{Y_j(\omega)} = \frac{Y_i(\omega)Y_j^*(\omega)}{Y_j(\omega)Y_j^*(\omega)} = \frac{S_{ij}(\omega)}{S_{jj}(\omega)} \tag{74}$$

where $S_{ij}(\omega)$ is the cross-spectral density between $y_i(t)$ and $y_j(t)$, and $S_{jj}(\omega)$ is the power spectral density of $y_j(t)$. $Y_j^*(\omega)$ is the complex conjugate of $Y_j(\omega)$. TR is a complex function having information on the amplitude and phase.

Since the lengths of the measured responses are finite, TR has to be estimated. The estimation is done easily using the Fast Fourier Transform (FFT). Welch's method (Newland, 1994) is used, in which the time series is divided into overlapping segments (50% overlap), the Hamming window is applied to each segment, and an n-point FFT is applied to the windowed data. Finally, the results of each segment are averaged to form the spectral density estimate. Averaging is done in order to decrease the estimation error.

In this section, 32 or 64 spectral lines are estimated in the frequency range of interest. The absolute transmissibilities using the FFT length $n = 32$ resulted in estimation at $n/2 + 1 = 17$ frequencies for each sensor pair. Correspondingly, the FFT length $n = 64$ resulted in 33 features.

It should be noted that the transmissibility concept, if defined as in equation 73, must be used with caution. In the crane application, a single DOF excitation is applied with the lift cylinder. In that case, the transmissibility between any two responses is justified. In a more general case, there can be several independent excitations acting at different DOF. In that case, the validity of using equation 73 directly in SHM is questionable. A more detailed theoretical development for the transmissibility with a number of forces can be found e.g. in Maia et al. (2007).

7.2 Monitoring Experiments and Results

Vibration-based monitoring tests for a vehicle crane HIAB022-2L (Figure 36) were performed in the laboratory. The features were extracted from the response of the crane vibrating around a slightly varying static equilibrium. The crane was excited with a random excitation using its own hydraulic lift cylinder (Figure 36). The excitation force was not measured. Eight accelerometers, the positions and directions of which are shown in Figure 36, were used. A constant static mass of 300 kg was attached to the tip. The first 400 measurements were recorded from the undamaged structure. Fifteen different structural changes were then made using additional masses of five different sizes at three locations. The magnitudes of the additional masses for each damage scenario were 1.4, 3.4, 5.4, 7.4, and 9.4 kg, and their locations are shown in Figure 36. Each damage scenario was monitored with 50 measurements. The time period of each measurement was 60 seconds and the data were stored after filtering with a 100 Hz low-pass filter. These tests were performed during a relatively short time period. Much later, control measurements were made without additional masses to check if the system had changed.

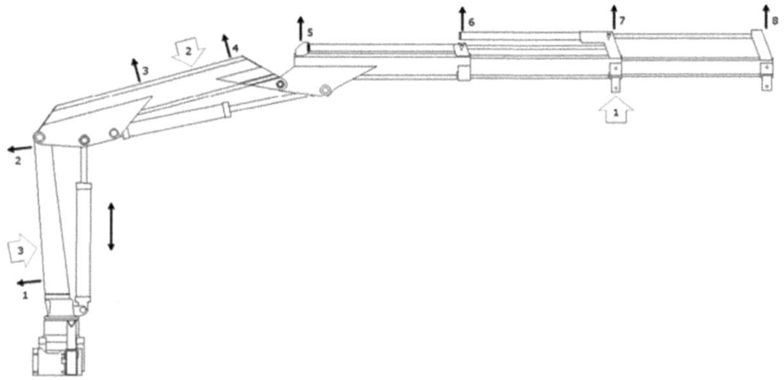

Figure 36. Vehicle crane HIAB-022-2L. The numbered broad arrows indicate the damage locations, the double arrow is the excitation, and the single arrows are the response positions.

Ten AR coefficients were extracted from each time record resulting in 80 features plotted in Figure 37 left including the undamaged structure (samples 1–400), damage position 1 (samples 401–650), and the control measurements (samples 651–777). Transmissibilities were estimated between sensor pairs 1–2, 3–4, 5–6, 5–7, and 5–8 (Figure 36). These pairs were chosen to see if it is easier to detect damage in a boom from the transmissibilities between the sensors on that boom, or if the transmissibilities give global information of the structure. The absolute transmissibilities using 32 spectral lines (85 features) of the undamaged structure and damage case 1 are shown in Figure 37 on the right.

From Figure 37 it can be seen that the variation of the features was large, especially with the transmissibilities. Also, occasional jumps in features can be clearly observed. They occurred due to applied lubrication to joints and larger motions driven with the crane between two measurements. These clearly affected the features.

Damage detection was performed using control charts (Montgomery, 1997). Every other sample between 1 and 400 was used as training data and

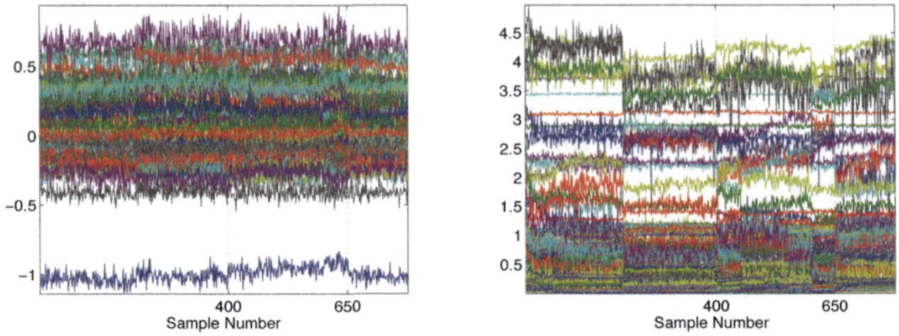

Figure 37. AR coefficients (left) and absolute transmissibilities (right). Damage case 1.

samples 1–300 were used as the in-control samples to compute the control limits for the control chart. Control charts for the features after factor analysis and subsequent principal component analysis are shown in Figure 38. It can be seen that damage was detected with either feature, but the control data signalled as well. This is an indication of a change in the system, a change that is not exactly known. It can also be due to a relatively small amount of training data or change in the operational conditions. Recall that the control measurements were performed much later than the actual monitoring experiments. It can be concluded that in practice the collection of the training data should be scheduled for several days. To see how the damage (added masses) can be observed with the proposed features, in the subsequent analyses the training data are extended to have also every second sample of the control data.

To see the effect of the factor analysis, AR coefficients for damage case 1 were analysed with and without factor analysis. The control charts are plotted in Figure 39. The advantageous effect of the factor analysis is obvious. The operational influences could be eliminated from the data, even without knowing the affecting quantities. Factor analysis is applied to all subsequent analyses.

The performance of the two features to detect the three damage cases is studied in the following. Control charts for damage cases 1, 2, and 3 are shown in Figures 40, 41, and 42, respectively. The analyses of the AR coefficients are shown in the left, and transmissibilities in the right. All damage cases were correctly detected with either feature and without false indications of damage. For damage case 1, the increase of the damage level can also be observed, whereas for damage cases 2 and 3 the damage level is

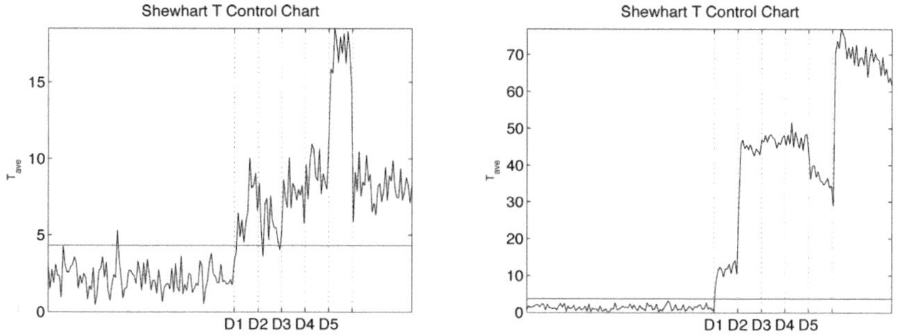

Figure 38. Control charts for AR coefficients (left) and absolute transmissibilities (right). Damage case 1.

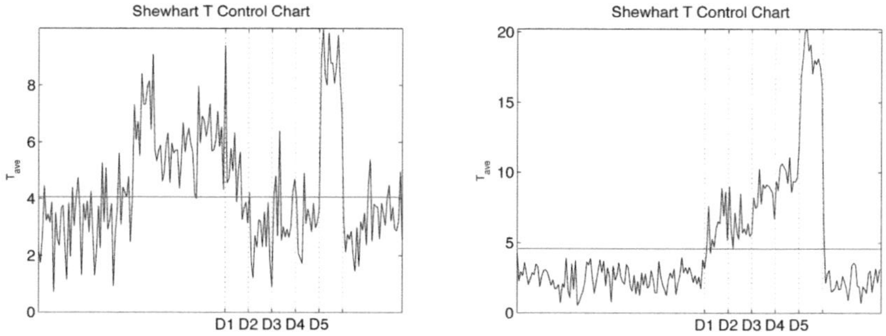

Figure 39. Control charts of the largest PC:s of the AR coefficients with damage case 1. Left: without FA; right: with FA.

not clearly seen from the control charts. Transmissibilities proved to be more sensitive to damage than the AR coefficients. Transmissibilities showed no intrinsic capability to localise damage, probably because local changes in the structure are propagated to transmissibilities around the structure. All damage cases were detected from transmissibilities both near damage and remote from damage.

The most sensitive results were obtained using complex transmissibilities (Figure 43). Transmissibilities having 32 and 64 spectral lines were estimated, resulting in 160 and 320 features, respectively. It should be noted that the training data consisted of only 264 samples, which was less than the dimensionality of the feature vector. Using the proposed approach with factor analysis, followed by the principal component analysis, good re-

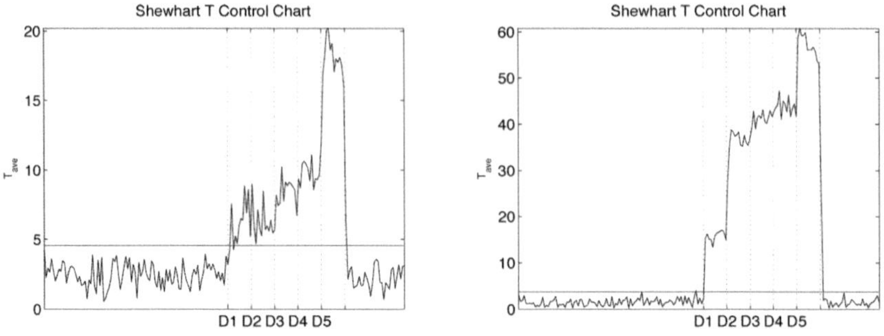

Figure 40. Damage case 1. Left: AR coefficients. Right: absolute transmissibilities.

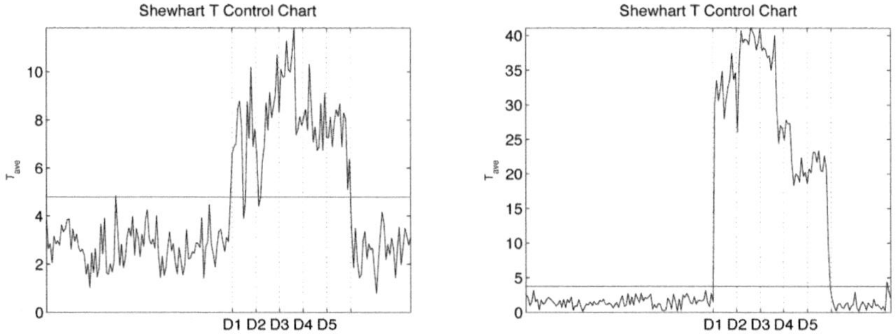

Figure 41. Damage case 2. Left: AR coefficients. Right: absolute transmissibilities.

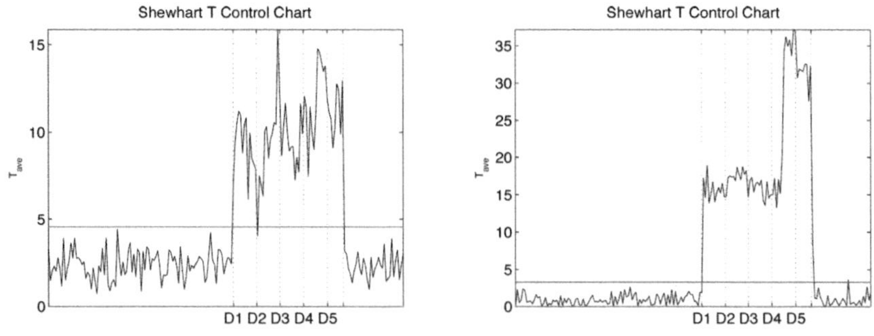

Figure 42. Damage case 3. Left: AR coefficients. Right: absolute transmissibilities.

sults were obtained. The most sensitive results were obtained with complex transmissibilities having 64 spectral lines. These features also showed no false indications of damage.

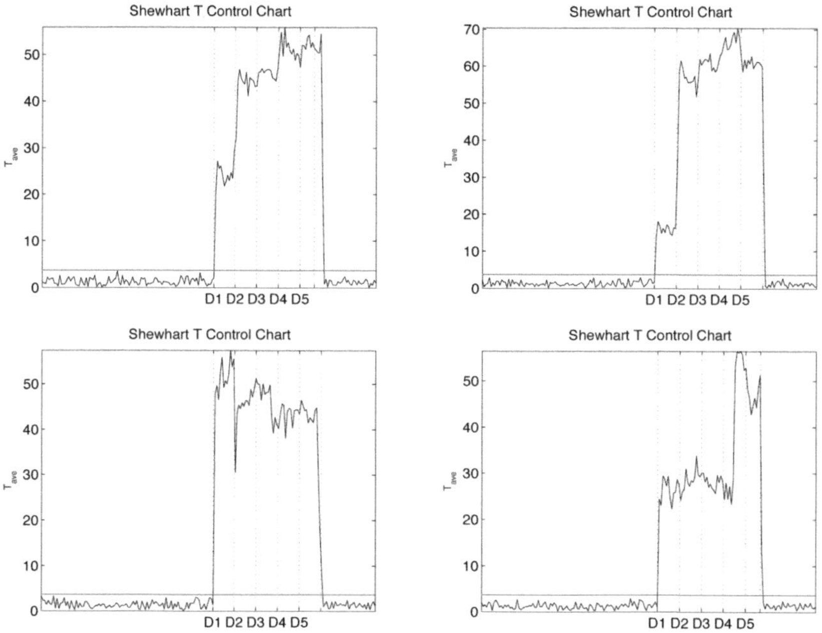

Figure 43. Control charts for complex transmissibilities. Top left: damage case 1, 32 spectral lines. Top right: damage case 1, 64 spectral lines. Bottom left: damage case 2, 64 spectral lines. Bottom right: damage case 3, 64 spectral lines.

7.3 Summary

The condition of a vehicle crane manipulator structure was monitored using multichannel measurements. Monitoring was solely based on the measurement data; no mathematical model of the structure was used. Damage detection was based on features extracted from the vibration measurements. Training data were first collected from the healthy structure. The performance of two different features to detect damage was studied: AR parameters and transmissibilities. The operational variations were first removed from the features using factor analysis. After this data normalisation procedure, the effectiveness of different features could be compared. Both features correctly detected all damage cases. Transmissibilities proved to be more sensitive to the introduced damage than the AR coefficients. The AR model may not correctly represent the dynamics of structures but more appropriate would be the ARMA model. However, its estimation is harder as it is a nonlinear problem. Complex transmissibilities with more features than the training samples ($p > n$) showed the most sensitive detection capability with no false indications of damage. It should be noted, however, that the training data must include all operational variations and consequently the collection of the training data should be scheduled for a long time period. The conclusions were drawn for artificial damage with added masses. A similar study should also be made for more realistic damage scenarios with stiffness degradation.

8 Conclusion

The most important functions needed in a vibration-based structural health monitoring system were introduced and some of them were studied in more detail. The main contribution was in damage detection and sensor validation under variable environmental or operational conditions using statistical tools applied to the features extracted from the measurement data. All techniques were data-based, and no mathematical model of the structure was needed.

Sensor validation was based on MMSE, while for damage detection of the structure under different environmental or operational conditions, factor analysis was applied. It should be noted that these two techniques are comparable with each other and either of them can be used for each function.

Control charts were used for damage detection, because they are simple, visual, and possible for automation. The importance of the visualisation cannot be overestimated; it gives the operator more confidence than just a simple message that the structure may be damaged. Occasional false alarms are also possible. Therefore, a good rule of thumb is to have at least two alarms in a sequence before taking action.

In many cases, the linear methods to eliminate the environmental or operational effects from the features work well. They are simple, fast, and automatic. Factor analysis requires one parameter, the number of factors, whereas MMSE needs no parameters. In contrast, nonlinear models are more complex, relatively slow, and need a lot of tuning due to a higher number of parameters. Nevertheless, in all methods, the training can be performed off-line when the data from the undamaged structure have been acquired. After training, the model is fixed and the performance of both the linear and nonlinear models is fast. However, it is probable that the operator needs to configure the SHM system, for example using different features. In that case, the models have to be re-trained, which is automatic with the linear techniques.

Linear models should always be applied first, and only after they fail, nonlinear methods should be utilised. In many cases, even if the physical phenomenon from the latent variables to the features is nonlinear, a linear correlation model is often sufficient. This effect takes place when there are several features to which the same nonlinear model applies, resulting in a linear correlation between them so that the nonlinear effect can be removed. nonlinear models are most useful in cases, where the number of features is small and linear correlation models are not capable of removing the influences of the environmental or operational conditions.

When designing an SHM system, the following configuration can be used as the first choice: Use MMSE to detect a sensor fault using both the Shewhart and S charts. After extracting the features from the time history measurements, select the training samples. Use factor analysis to remove the environmental or operational effects and compute the principal components of the standardised residuals using all data. Select 1–3 largest principal components, depending on the information contents of the components. Then use the multivariate Shewhart T control chart for damage detection.

Acknowledgements

The support of the Multidisciplinary Institute of Digitalisation and Energy (MIDE) research programme at Helsinki University of Technology is acknowledged.

Bibliography

S. Alampalli. Influence of in-service environment on modal parameters. In *Proceedings of the 16th International Modal Analysis Conference*, pages 111–116, Santa Barbara, California, February 2–5, 1998. Society for Experimental Mechanics, Bethel, Connecticut.

C.M Bishop. *Pattern recognition and machine learning.* Springer, New York, 2006.

P. Cornwell, C.R. Farrar, S.W. Doebling, , and H. Sohn. Structural testing series: Part 4. environmental variability of modal properties. *Experimental Techniques*, 23(6):45–48, 1999.

R.B. Crosier. Multivariate generalizations of cumulative sum quality-control schemes. *Technometrics*, 30(3):291–303, 1988.

A. Deraemaeker, E. Reynders, G. De Roeck, and J. Kullaa. Vibration based shm: comparison of the performance of modal features vs features extracted from spatial filters under changing environmental conditions. In P. Sas and M. De Munck, editors, *Proceedings of ISMA2006, International Conference on Noise and Vibration Engineering*, pages 849–863, Leuven, Belgium, September 18–20, 2006.

S.W. Doebling, C.R. Farrar, M.B. Prime, and D.W. Shevitz. Damage identification and health monitoring of structural and mechanical systems from changes in their vibration characteristics: A literature review. Technical Report Report LA-13070-MS, Los Alamos National Laboratory, 1996.

R. Dunia, S.J. Qin, T.F. Edgar, and T.J. McAvoy. Identification of faulty sensors using principal component analysis. *AiChe Journal*, 42(10):2797–2812, 1996.

C.R. Farrar, S.W. Doebling, P.J. Cornwell, and E.G. Straser. Variability of modal parameters measured on the alamosa canyon bridge. In *Proceedings of IMAC-XV: A Conference on Structural Dynamics*, pages 257–263, Orlando, Florida, USA, February, 1997. Society for Experimental Mechanics, Bethel, Connecticut.

K. Fukunaga. *Introduction to Statistical Pattern Recognition.* Academic Press, Boston, 2nd edition, 1990.

Z. Ghahramani and G.E. Hinton. The em-algorithm for mixtures of factor analyzers. Technical Report Technical Report CRG-TR-96-1, University of Toronto, 1996.

G. Kerschen, P. De Boe, J.-C. Golinval, and K. Worden. Sensor validation for on-line vibration monitoring. In W.J. Staszewski C. Boller, editor, *Proceedings of the Second European Workshop on Structural Health Monitoring*, pages 819–827, Munich, Germany, July 7–9, 2004. DEStech Publications.

M.A. Kramer. Nonlinear principal component analysis using autoassociative neural networks. *AiChe Journal*, 37(2):233–243, 1991.

J. Kullaa. Elimination of environmental influences from damage-sensitive features in a structural health monitoring system. In D.L. Balageas, editor, *First European Workshop on Structural Health Monitoring*, pages 742–749, Paris, July 10–12, 2002. Onera, DEStech Publications.

J. Kullaa. Is temperature measurement essential in structural health monitoring? In Fu-Kuo Chang, editor, *Proceedings of the 4th International Workshop on Structural Health Monitoring*, pages 717–724, Stanford, CA., September 15–17, 2003. Stanford University, DEStech Publications.

J. Kullaa. Structural health monitoring of a crane in variable configurations. In P. Sas and M. De Munck, editors, *Proceedings of ISMA2004, International Conference on Noise and Vibration Engineering*, pages 457–469, Leuven, Belgium, September 20–22, 2004.

J. Kullaa. Damage detection under a varying environment using the missing data concept. In Fu-Kuo Chang, editor, *5th International Workshop on Structural Health Monitoring*, pages 565–573, Stanford, CA., September 12–14, 2005. Stanford University, DEStech Publications.

J. Kullaa. Sensor fault identification and correction in structural health monitoring. In P. Sas and M. De Munck, editors, *Proceedings of ISMA2006, International Conference on Noise and Vibration Engineering*, pages 873–884, Leuven, Belgium, September 18–20, 2006.

L. Ljung. *System identification: theory for the user.* Prentice Hall, Upper Saddle River (NJ), 2nd edition, 1999.

C.A. Lowry, W.H. Woodall, C.W. Champ, and S.E. Rigdon. A multivariate exponentially weighted moving average control chart. *Technometrics*, 34 (1):46–53, 1992.

N.M.M. Maia, A.M.R. Ribeiro, M. Fontul, D. Montalvão, and R.P.C. Sampaio. Using the detection and relative damage quantification indicator (DRQ) with transmissibility. In L. Garibaldi, W. M. Ostachowicz, C. Surace, and K. Holford, editors, *Damage Assessment of Structures VII: Proceedings of the 7th International Conference on Damage Assessment of Structures (Damas 2007)*, pages 455–460, Torino, Italy, June 25–27, 2007. Trans Tech Publications.

G. Manson. Identifying damage sensitive, environment insensitive features for damage detection. In *Proceedings of the Third International Conference on Identification in Engineering Systems*, pages 187–197, Swansea, April 15–17, 2002. Institute of Physics Publishing.

K.V. Mardia, J.T. Kent, and J.M. Bibby. *Multivariate analysis.* Academic Press, London, 1979.

D.C. Montgomery. *Introduction to statistical quality control.* John Wiley & Sons, New York, 3rd edition, 1997.

D.E. Newland. *An introduction to random vibrations, spectral and wavelet analysis.* Longman, Harlow, 3rd edition, 1994.

P. Van Overschee and B. De Moor. *Subspace identification for linear systems: theory - implementation - applications.* Kluwer Academic Publishers, Dordrecht, Netherlands, 1996.

B. Peeters. *System identification and damage detection in civil engineering.* PhD thesis, KU Leuven, 2000.

B. Peeters and G. De Roeck. One year monitoring of the z24 bridge: Environmental influences versus damage effects. In *Proceedings of IMAC-XVIII: A Conference on Structural Dynamics*, pages 1570–1576, San Antonio, Texas, USA, February 7–10, 2000. Society for Experimental Mechanics, Bethel, Connecticut.

R.G. Rohrmann, M. Baessler, S. Said, W. Schmid, and W.F. Ruecker. Structural causes of temperature affected modal data of civil structures obtained by long time monitoring. In *Proceedings of IMAC-XVIII: A Conference on Structural Dynamics*, pages 1–7, San Antonio, Texas, USA, February 7–10, 2000. Society for Experimental Mechanics, Bethel, Connecticut.

R.P.C. Sampaio, N.M.M. Maia, J.M.M. Silva, and A.M.R. Ribeiro. On the use of transmissibility for damage detection and location. In *Proc. of the European COST F3 Conference on System Identification & Structural Health Monitoring*, pages 363–376, Madrid, June 7–9, 2000. Universidad Politcnica de Madrid.

S. Sharma. *Applied multivariate techniques*. John Wiley & Sons, New York, 1996.

H. Sohn, J.A. Czarnecki, and C.R. Farrar. Structural health monitoring using statistical process control. *Journal of Structural Engineering*, 126 (11):1356–1363, 2000a.

H. Sohn, C.R. Farrar, and N.F. Hunter. Data normalization issue for vibration-based structural health monitoring. In *Proceedings of IMAC-XVIII: A Conference on Structural Dynamics*, pages 432–437, Orlando, Florida, USA,, February, 2000b. Society for Experimental Mechanics, Bethel, Connecticut.

H. Sohn, C.R. Farrar, N.F. Hunter, and K. Worden. Structural health monitoring using statistical pattern recognition techniques. *Journal of dynamic systems, measurement, and control*, 123:706–711, 2001.

H. Sohn, K. Worden, and C.R. Farrar. Statistical damage classification under changing environmental and operational conditions. *Journal of Intelligent Material Systems and Structures*, 13(9):561–574, 2003.

H.W. Sorenson. *Parameter estimation. Principles and problems.* Marcel Dekker, New York, 1980.

Z. Sun and C.C. Chang. A wavelet packet based method for structural damage assessment. In *Proceedings of the Third World Conference on Structural Control, Volume 2*, pages 3–8, Como, Italy, April 7–12, 2002. Wiley.

S. Vanlanduit, E. Parloo, B. Cauberghe, P. Guillaume, and P. Verboven. A robust singular value decomposition for damage detection under changing operational conditions and structural uncertainties. *Journal of Sound and Vibration*, 284:1033–1050, 2004.

M.A. Wahab and G. De Roeck. Effect of temperature on dynamic system parameters of a highway bridge. *Structural Engineering International*, (4):266–270, 1997.

K. Worden, G. Manson, and N.R.J. Fieller. Damage detection using outlier analysis. *Journal of Sound and Vibration*, 229(3):647–667, 1999.

A.-M. Yan, G. Kerschen, P. De Boe, and J.-C. Golinval. Structural damage diagnosis under varying environmental conditions - part i: A linear analysis. *Mechanical Systems and Signal Processing*, 19:847–864, 2005a.

A.-M. Yan, G. Kerschen, P. De Boe, and J.-C. Golinval. Structural damage diagnosis under varying environmental conditions - part ii: Local pca for non-linear cases. *Mechanical Systems and Signal Processing*, 19:865–880, 2005b.

Structural Health Monitoring using Pattern Recognition

Keith Worden

Dynamics Research Group, Department of Mechanical Engineering,
University of Sheffield, Mappin Street, Sheffield S1 3JD, UK.

Abstract There are two main approaches to the diagnostic phase of Structural Health Monitoring (SHM): the first, is based on the solution of inverse problems, and the second, is based on *pattern recognition* or *machine learning*. The first approach usually depends on the construction of a model of the structure based on physical principles, while the second relies on building a model based on measured data. The complexity of many modern structures and their environments sometimes makes the second option an attractive proposition. While many engineers are familiar with the process of building physics-based models e.g. finite element models, familiarity with the principles of pattern recognition is less common. The objective of this chapter is to provide an introduction to the concepts of data-based modelling and pattern recognition in the context of the SHM problem.

1 Introduction

Hopefully, the previous chapters in this book have convinced the reader that Structural Health Monitoring (SHM) is a good idea. The objective of this chapter is to argue that one particular approach to SHM is a good idea. This approach is based on the discipline of *machine learning* or, more specifically *pattern recognition*. The idea is that one can *learn* relationships from data. In the context of SHM, this means that one can learn to assign a damage state or *class* to a given measurement vector from the structure or system of interest. The measurement vectors must be formed from measurements which are sensitive to the damage; in the normal terminology of pattern recognition, they are referred to as *features*. An example of a feature vector for SHM might simply be the first five natural frequencies, or it might be more sophisticated e.g. a set of wavelet coefficients. Once features have been established, the map between the features and the diagnosis can be constructed, and many algorithms are available for this purpose. The most popular algorithm in use today is the Artificial Neural Network (ANN) and this will be the basis for much of the discussion in this chapter. The use of pattern recognition offers the possibility of automating the SHM process i.e. removing the need for the intervention of human experts as far as possible. So that one does not lose this benefit, the rest of the SHM process should be automated as far as possible and this leads one to the idea of *intelligent fault detection.*

The layout of this chapter is as follows: Section 2 describes the basis for carrying out intelligent fault detection and discusses the role of data fusion. Section 3 introduces the idea of novelty detection, the lowest level of damage identification; a basic statistical algorithm is used for illustration. Section 4 describes how artificial neural networks have developed as a means of learning relationships in data; the historical development of the subject is briefly outlined. Section 5 returns to the idea of novelty detection and describes a more sophisticated approach based on neural network technology. Section 6 describes the theoretical basis for statistical pattern recognition and sketches the relationship with neural networks. Section 7 provides illustrations of the use of pattern recognition for damage detection and also for the more difficult problem of damage location. The chapter finishes with a section of discussion and conclusions.

2 Intelligent Fault Detection

2.1 Terminology

The purpose of this section will be to describe a framework for intelligent fault detection where the disciplines of pattern recognition or machine learning in general play a major part. Before proceeding, it is necessary to make a digression on the subject of taxonomy. In the past, engineers adopted a conservative design strategy whereby the structure or system was, by the considered application of design safety factors, required to function safely without problems for a prescribed lifetime. This *safe-life* approach has given way in many cases to *fail-safe* or *damage tolerant* philosophies. In the latter case, the structure is anticipated to sustain damage and still perform in a satisfactory manner. The following terms will be adopted in the current study in order to describe imperfections in systems in a consistent manner.

- A *defect* is inherent in the material and statistically all materials will contain some unknown amount of defects; this means that the structure can operate at its design condition if the constituent materials contain defects
- *Damage is* when the structure is no longer operating in its ideal condi-

tion but can still function satisfactorily, i.e. in a sub-optimal manner

- A *fault* is when the structure can no longer operate satisfactorily. If one defines the quality of a structure or system as its fitness for purpose or its ability to meet customer or user requirements, it suffices to define a fault as a change in the system that produces an unacceptable reduction in quality.

The above notation allows a hierarchical relationship to be developed; i.e., *defects* lead to *damage* and *damage* leads to *faults.* In the approach discussed here, it is necessary to introduce monitoring systems in order to obtain a damage tolerant structure, so that it can be decided when the structure is no longer operating in a satisfactory manner. This means that a fault has to have a strict definition, e.g. the stiffness of the structure has deteriorated beyond a certain level. In some cases a simple definition based on one parameter may not be sufficient. A good example of this is when a crack is propagating in a stable manner; there will be an increase in strain in the component and hence a reduction in component stiffness. Once the strain has increased above a certain level a decision may be made to take the component out of service. On inspection it may be found that the crack is growing in such a direction that the component will fail safe, so service could have been prolonged. In this situation, along with the strain monitor it would have been useful to monitor the direction of crack growth by some means. The choice of fault monitoring system clearly needs to take into account the material type and the operating environment. This approach ensures that the quality of the measurement will be optimised and hence an important issue will be the limitations of the sensor system. From the above it can be seen that the question - *what is a fault* - is not only based on the structures operating environment but also the type of monitoring system that is used.

2.2 Intelligence

Having established a precise terminology relating to sub-optimal behaviour of structures and systems, the discussion can proceed to matters of detection and how it can be achieved with intelligence. The first observation one might make is that fault detection is in a sense trivial, as a fault is defined as a change in the condition of the structure that produces an unacceptable reduction in quality. By implication, such a change will be evident. Thus, intelligent fault detection actually entails detecting the damage that will, if not corrected, lead to a fault.

Detection of damage is a facet of the broader problem of damage awareness or damage identification. The objective of a monitoring system must be to accumulate sufficient information about the damage for appropriate remedial action to be taken to restore the structure or system to high-quality operation or at least to ensure safety. Also, efficiency demands that only the necessary information should be returned by the monitor. With this in mind, it is helpful to think of the identification problem as a hierarchical structure, in the same way as one can think of the evolution of the fault as a hierarchical structure. This train of thought began with Rytter in his PhD thesis (Rytter (1993)). The original specification cited four levels:

DETECTION. The method gives a qualitative indication that damage might be present in the structure.

LOCALISATION. The method gives information about the probable position of the damage.

ASSESSMENT. The method gives an estimate of the extent of the damage.

PREDICTION. The method offers information about the safety of the structure e.g. estimates a residual life.

The vertical structure is clear; each level (essentially) requires that all lower-level information is available. Note that the damage identification scheme should if at all possible, be implemented on-line i.e. during operation of the structure; in this case prediction must also be understood as an estimate of the residual safe-life of the structure obtained during operation. For an aircraft in flight, for example, this is critical; if the diagnostic system signals serious damage but fails to indicate that there is time to land, the aircraft may be lost needlessly and at great expense when the crew bail out. Note that the primary concern is that the crew do bail out; issues of life safety far outweigh economic considerations.

Few would argue that the structure above summarises the main issues in SHM with one major exception; this is remedied by the introduction of a new level. At the risk of repetition, the new structure is:

DETECTION. The method gives a qualitative indication that damage might be present in the structure.

LOCALISATION. The method gives information about the probable position of the damage.

CLASSIFICATION. The method gives information about the type of damage.

ASSESSMENT. The method gives an estimate of the extent of the damage.

PREDICTION. the method offers information about the safety of the structure e.g. estimates a residual life.

Classification is important, if not vital, for effective identification at level 5 and possibly at level 4. Level 5 is distinguished from the others in that it cannot be accomplished without an understanding of the physics of the damage, i.e. characterisation. Level 1 is also distinguished in the sense that it can be accomplished with no prior knowledge of how the system will behave when damaged. In order to explain this, a digression on pattern recognition or machine learning is needed.

The approach to SHM discussed here is based on the idea of pattern recognition (PR). In the broadest sense, a PR algorithm is simply one that assigns to a sample of measured data a class label, usually from a finite set. In the case of damage identification, the measured data could be vibration modeshapes, full-field thermoelastic data, scattered wave profiles etc. The appropriate class labels would encode damage type, location etc. In order to carry out the higher levels of identification using PR, it will almost certainly be necessary to obtain examples of data corresponding to each class. That is, in order to establish that a given set of measurements from a composite panel shows the presence of a delamination, the algorithm must have prior knowledge of what data from a delaminated panel looks like as opposed to one with say, a resin-rich area. Each possible fault class should usually have a training set of measurement vectors that are associated uniquely with it. Many PR algorithms work by training a diagnostic; for example a neural network can learn by example - it is shown the measurement data and asked to produce the correct class label; if the result differs from the desired label, the network is corrected. Typically many presentations of data are required. This type of learning algorithm in which the diagnostic is trained by showing it the desired label for each data set is called *supervised learning.*

If supervised learning is required, there will be serious demands associated with it; data from every conceivable damage situation should be available. The two possible sources of such data are modelling and experiment. Modelling presents problems if the structure or system of interest is geometrically or materially complex, for example finite element analysis of structures requiring a fine mesh can be extremely time-consuming even if the material is well understood. Structures with composite or viscoelastic elements may not even have accurate constitutive models. The damage itself may be difficult to model; it may also make the structure dynamically nonlinear, i.e. an opening-closing fatigue crack, and this also presents a formidable problem. Unfortunately, the situation is no better for experiment. In order to accumulate enough training data, it would be necessary to make copies of the system of interest and damage it in all the ways that might occur naturally; for high-value structures like aircraft, this is simply not possible.

Fortunately there is an alternative to supervised learning - *unsupervised learning.* However, this mode of learning only applies to level 1 diagnostics, i.e. it can only be used for detection. The techniques are often referred to as novelty detection or anomaly detection methods (Bishop (1994); Tarassenko et al. (1997); Worden (1997)). The idea of novelty detection is that only training data from the normal operating condition of the structure or system is used to establish the diagnostic. A model of normal condition is created. Later, during monitoring, newly-acquired data are compared with the model. If there are any significant deviations the algorithm indicates novelty. The implication is that the system has departed from normal condition, i.e. acquired damage. The advantage of such an approach is clear; if the training data is generated from a model, only the unfaulted condition is required and this will simplify matters considerably. From an experimental point of view, there is no need to damage the structure of interest. Although novelty detection is only a level 1 approach, there are many situations where this suffices, i.e. safety-critical systems where any fault on the system would require it to be taken out of service.

It is an important qualifier that the novelty detectors should flag only significant deviations from normal operating condition. All real systems are subject to measurement noise and usually operate in a changing environment; the monitor must be able to distinguish between a statistical fluctuation in the data and a real deviation from normality. This means that of the various flavours of pattern recognition existing (Schalkoff (1992)), the most appropriate one is Statistical Pattern Recognition (SPR). Another important observation is that there may be variations in the normal condition that are not statistical, i.e. the characteristics of the structure may vary with changing environmental conditions, and this must be addressed (This is discussed in more detail in the chapter by Kullaa in this volume). In general, it is important that the algorithms used for damage identification should account properly for sources of uncertainty and variation in the data. The algorithms should also, as far as possible, return a confidence interval with their diagnosis.

The term *normal operating condition* requires some discussion. As implied earlier, the nature of engineering materials means that there will always be defects and sometimes even damage present in a structure to some extent. The normal operating condition therefore means a state of the system when there is some assurance, statistical or otherwise, that the system is fit-for-purpose. In some cases there may be macroscopic damage i.e. a fatigue crack; however, if it is known that the crack will not grow under the standard loadings on the system, the state qualifies as a normal operating condition. Novelty detection will then look for new cracks or unexpected growth of the old crack.

The discussion above is intended to show that there is often a trade-off between the level of a diagnostic system and the expense of training it adequately. Given this fact, the main requirement of an intelligent fault detection system is that it should return information at the apposite level for the context. It should measure the appropriate data and process this with the appropriate algorithm. It should take proper account of uncertainty in the data and return a confidence level in its diagnosis.

2.3 Data Processing and Fusion for Damage Identification

Once an operational evaluation stage has passed and a sensor network has been designed (and these are, by no means, trivial processes), the health monitoring system can begin to deliver data. The choice and implementation of algorithms to process the data and carry out the identification is arguably the most crucial ingredient of an intelligent fault detection strategy. Before even choosing the algorithm, it is necessary to choose between two complementary approaches to the problem:

- damage identification is an inverse problem;
- damage identification is a pattern recognition problem.

The first approach usually adopts a physics-based model of the structure and tries to relate changes in measured data from the structure to changes in the model, sometimes locally linearised models are used to simplify the analysis. The algorithms used are mainly based on linear algebra or optimisation theory, and an excellent survey of the dominant methods can be found in Doebling et al. (1996).

The second approach is based on the idea described above, whereby measured data from the system of interest are assigned a damage class by a pattern recognition algorithm. This is the approach that is chosen here for detailed discussion. There is no implied criticism of the inverse problem approach; the author is simply concentrating on an alternative and self-consistent framework. For a critical appraisal of inverse problem approaches to damage identification, the reader can consult Friswell and Penny (1997).

The data processing element of a monitoring system comprises all actions on the data upstream from the point of acquisition by the sensors. The ultimate product of the analysis is a decision as to the health of the system. The analysis has been neatly summed up by Lowe (2000) as the D2D (Data to Decision) process; the basic steps are summarised in Figure 1.

Beyond the sensing level which generates the raw data, the first stage is signal processing. This should more properly be called pre-processing. The purpose is to prepare the data for feature extraction, but more of that later. The pre-processing stage can encompass two tasks. The first of these is data cleansing. Examples of cleansing processes are: filtering to remove noise,

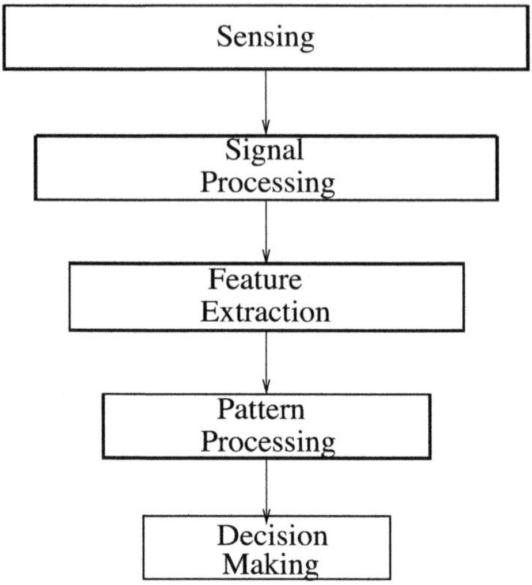

Figure 1. The D2D Process.

spike removal by median filtering, removal of outliers (care is needed here as the presence of outliers is one indication that the data is not from normal condition), and treatment of missing data values. The second (optional) pre-processing stage is a preliminary attempt to reduce the dimensions of the data vectors and further de-noise the signal. For example, given a random time-series with many points, it is often useful to convert the data to a spectrum by Fourier transformation. The number of points in the spectrum can be much lower than in the original time-history and noise can be averaged away. Another advantage of treating the time signal this way is that the data vector obtained should be independent of time. If the original time-series is random, it makes little sense to compare measurements at different starting times. The pre-processing is usually carried out on the basis of engineering judgement and experience. At this stage, the aim would be to reduce the dimension of the data set from possibly many thousands to perhaps a hundred.

The second stage is feature extraction. The term *feature* comes from the pattern recognition literature and is short for 'distinguishing feature'. The fundamental problem of pattern recognition is to assign a class label to a vector of measurements; this task is made simple if the data contains dominant features that distinguish it from data from other classes. In general, the components of the signal that distinguish the various damage classes will be hidden by features that characterise the normal operating condition of the structure, particularly when the damage is not yet severe. The task of feature extraction is to magnify the characteristics of the various damage classes and suppress the normal background. Suppose the raw data from the sensors is a time-series of accelerations from the outside of a gearbox casing. Further suppose that the time data has been pre-processed and converted into an averaged spectrum. Feature extraction in this situation could be extracting only the spectral lines at the meshing frequency and its harmonics as these lines are known to be sensitive to damage. So feature extraction can be carried out on the basis of engineering judgement also. Alternatively, statistical algorithms can be used to reduce the dimension like Principal Component Analysis (Bishop (1998)). The resulting low-dimensional data set is the feature vector or pattern vector that the pattern recognition algorithm will use to assign a class. The aim of this stage would be to generate a feature vector of dimension less than ten. A low-dimensional feature vector is a critical element in any pattern recognition problem as the number of data examples needed for training grows explosively with the dimension of the problem. Care must be taken at this stage that the information discarded in the dimension reduction is not relevant for diagnosing the damage. Feature extraction should only discard components of the data that are irrelevant for the purposes of identifying damage.

The next stage is pattern processing. This is the application of an algorithm which can decide the damage state on the basis of the given feature vector. An example would be a neural network that has been trained to return the damage type and severity when presented with say, condensed spectral information from a gearbox. Three types of algorithm can be distinguished depending on the desired diagnosis.

1. *Novelty detection.* In this case, the algorithm is required to simply indicate if the data comes from normal operating condition or not. This is a two-class problem which has the advantage that unsupervised learning can be used. Methods for novelty detection include: outlier analysis (Worden (1997)), kernel density methods (Tarassenko (1998)), autoassociative neural networks (Pomerleau (1993)), Kohonen networks (Taylor et al. (1999)), growing radial basis function networks (Roberts and Tarassenko (1994)) and methods based on Statistical Process Control (SPC) control charts (Sohn and Farrar (2000)).
2. *Classification.* In this case, the output of the algorithm is a discrete class label. In order to apply such an algorithm, the damage states must be quantised, i.e. for location, the structure should be divided into labelled substructures. In this case, the algorithm could only lo-

cate to within a sub-structure, so resolution of what is essentially a continuous parameter may not be good unless many labels are used. However, this type of algorithm is useful in the sense that the algorithms can be trained to give the probability of class membership; this gives an inbuilt confidence factor in the diagnosis. In the case where the desired diagnosis is from a discrete set, e.g. for diagnosing damage type, this class of algorithms is singled out. Examples of algorithms include: neural network classifiers trained with the *1 of M* rule, linear and quadratic discriminant analysis, kernel discriminant analysis and nearest neighbour classifiers. A comparison of some of these approaches on a damage classification problem is given in Worden and Manson (2000).

3. *Regression.* In this case the output of the algorithm is one or more continuous variables. For location purposes, the diagnosis might be the Cartesian coordinates of the fault, for severity assessment it could be the length of a fatigue crack. The regression problem is often nonlinear and is particularly suited to neural networks. As in the classification case, it is often possible to recover a confidence interval for a neural network prediction (Lowe and Zapart (1999)).

In all cases, the pattern processing is subject to an important limitation. There is a trade-off between the resolution of the diagnosis and the noise-rejection capabilities of the algorithm. Put simply, if the data is always noise-free, there will be very little fluctuation in the measurement from normal operating condition; in this case, small damages will cause detectable deviations. If there is much noise on the training data, it will be difficult to isolate deviations due to damage unless the damage is severe. One of the tasks of feature extraction is to eliminate as far as possible, fluctuations on the normal condition data. This optimisation for performance is a requisite feature of intelligent fault detection.

The final stage in the D2D chain is the decision. This is a matter of considering the outputs of the pattern recognition algorithm and deciding whether action needs to be taken, and what that action should be.

Having stressed the importance of adopting an intelligent approach to SHM, the next sections will outline in greater detail how pattern recognition and machine learning facilitate this process.

3 Novelty Detection

As discussed in the previous section, the property that sets apart damage detection from the higher levels of identification is that it can often be approached using unsupervised learning; the body of techniques concerned are called novelty detection methods. The philosophy of novelty detection is simply stated; if one is in possession of data guaranteed to be from the normal condition of a system or structure, one can construct a statistical (or other) model of that data. Any subsequent data from the system can be tested to see if they conform in some strict sense with the model of normality; noncomformity can then be said to infer damage.

The main advantage of novelty detection is substantial. At the risk of repetition, the major problem in machine learning approaches to SHM is the question of where the data corresponding to damage is obtained from. If the data is to be obtained from modelling, it is clear that for say, wave scattering from damage in composite structures, one will have a formidable modelling task which may be very difficult and may be expensive to run. Alternatively, unless the structures of interest are extremely inexpensive, it will not be possible to obtain the data from an experimental program. The nature of novelty detection means that one only ever needs to model or take measurements from the undamaged structure. The main problem with novelty detection is that it is sometimes accomplished by inferring the probability density function of the normal condition data in some way or another. If this exercise is attempted in its full generality, it is essentially the hardest of all machine learning problems.

The objective of this section is to illustrate the technique of novelty detection in a number of different contexts. First, it is shown how the simplest technique - one based on outlier analysis - is applied to a simulated system. In Section 5 a more general methodology based on auto-associative neural networks is discussed

This section is by no means intended as a comprehensive review of novelty detection methods, this has already been conducted by Markou and Singh (2003a,b). The reader is referred to these papers for a survey of the many alternative techniques available and also of the rest of the literature.

3.1 Gaussian-Distributed Normal Condition - Outlier Analysis

The method discussed in this section is tailored to the Gaussian distribution by its implicit assumption that the data can be characterised by its first two statistical moments. This is a situation where the probability density of the normal data is estimated in a simple parametric fashion; however, the assumption of Gaussianity is by no means always merited.

A discordant outlier in a data set is an observation that is surprisingly different from the rest of the data and therefore is believed to be generated by an alternate mechanism to the other data. The discordancy of the candidate outlier is a measure which may be compared against some objective criterion allowing the outlier to be judged as statistically likely or unlikely to have come from the assumed generating model. The standard reference on outlier analysis is Barnett and Lewis (1994).

Outlier detection in the case of univariate data is relatively straightforward in that any outliers must 'stick out' from one end or other of the data set. There are numerous discordancy tests but one of the most common, and the one whose extension to multivariate data will be employed later, is based on deviation statistics and given by,

$$z = \frac{|x_\zeta - \overline{x}|}{\sigma_x} \tag{1}$$

where x_ζ is the potential outlier and $\overline{x}$ and σ_x are the mean and standard deviation of the undamaged data sample respectively. The latter two values may be calculated with or without the potential outlier in the sample depending upon whether inclusive or exclusive measures are preferred. This discordancy value is then compared to some threshold value and the observation declared, or not, to be an outlier.

This discordancy measure is not restricted to a Gaussian normal condition and will work to some extent for data from any unimodal distribution; however if one is confident of Gaussianity, then a rigorous definition of confidence interval or threshold is available i.e. the 95% confidence level for an outlier is given by ± 1.96 for z.

The case of outliers in multivariate data is a little more complicated. A multivariate data set consisting of observations in p variables may be represented as points in a p-dimensional feature space. It becomes clear that detection of outliers in multivariate data is more difficult than the univariate situation due to the potential outlier having more 'room to hide'.

The discordancy test which is the multivariate equivalent of Equation (1) is the Mahalanobis squared-distance measure given by,

$$D_\zeta^2 = (\underline{x}_\zeta - \underline{\overline{x}})^T \Sigma^{-1} (\underline{x}_\zeta - \underline{\overline{x}}) \tag{2}$$

where $\underline{x}_\zeta$ is the potential outlier, $\underline{\overline{x}}$ is the mean of the sample observations and Σ, the sample covariance matrix.

As with the univariate discordancy test, the mean and covariance may be inclusive or exclusive measures. In many practical situations the outlier is not known beforehand and so the test would necessarily be conducted inclusively; in general for SHM however, the potential outlier is always known beforehand and so it is more sensible to calculate a value for the Mahalanobis squared-distance without this observation 'contaminating' the statistics of the normal data. Whichever method is used, the Mahalanobis squared-distance of the potential outlier is checked against a threshold value, and its status determined. For the multivariate case, the threshold value is dependent on both the number of observations and the number of dimensions of the problem being studied.

A Monte Carlo method (based on extreme-value statistics) can be used to arrive at the threshold value. The procedure for this is to construct a matrix with each element a randomly generated number from a zero mean and unity standard deviation normal distribution. The Mahalanobis squared-distances are calculated for all the elements, using (2) where the statistics are inclusive measures, and the largest value stored. This process is repeated for say, 1000 trials, whereupon the array containing all the largest Mahalanobis squared distances is then ordered in terms of magnitude. The critical values for the 5% and 1% tests of discordancy for a p-dimensional sample of n observations are then given by the Mahalanobis squared-distances in the array above which 5% and 1% of the trials occur. The calculation of the threshold again ties the method to the Gaussian distribution.

As an illustration, A data set from a computer simulation described in Worden et al. (2000) is considered, it is formed from the responses of the three-degree-of-freedom (3-DOF) lumped-parameter system shown in Figure 2. The equations of motion of this system are:

$$m\ddot{y}_1 + c\dot{y}_1 + k(2y_1 - y_2) = x_1(t)$$
$$m\ddot{y}_2 + c\dot{y}_2 + k(2y_2 - y_1 - y_3) = x_2(t)$$
$$m\ddot{y}_3 + c\dot{y}_3 + k(2y_3 - y_2) = x_3(t) \tag{3}$$

The values $m = 1$, $c = 20$ and $k = 10^4$ were used for the unfaulted condition.

The feature which was used for the detection process was the transmissibility function between the two top masses. It was computed by simulating the responses to a harmonic excitation on mass 3 for a frequency range between 0 and 50 Hz. In this case the magnitude of the transmissibility function was sampled at 50 regularly spaced points on the frequency range to give the pattern to be used as the unfaulted condition in the analysis (Figure 3).

The fault in this system was simulated by reducing the stiffness between the top two masses in Figure 2 by 1%, 10% and 50% of the original value and the three faulted patterns were calculated in the same manner as above with the stiffness altered in the equations of motion.

In order to construct a suitable mean vector and covariance matrix, for the normal condition data, the unfaulted pattern was copied 1000 times and each copy was subsequently corrupted with different Gaussian noise vectors of RMS 0.05. This procedure was repeated for the three damage

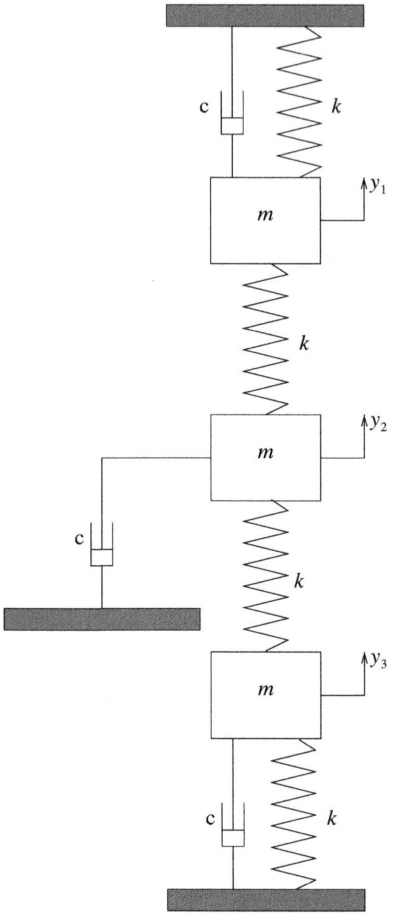

Figure 2. The three-degree-of-freedom simulated system.

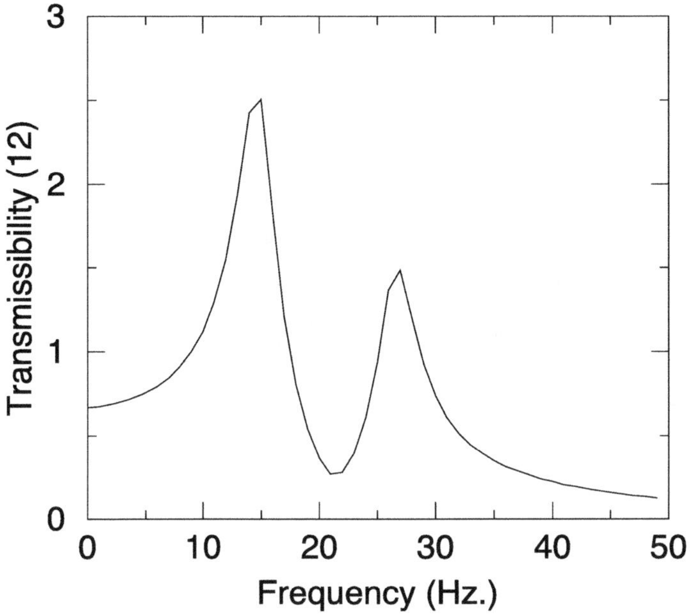

Figure 3. Transmissibility function for the unfaulted data.

cases which would form the testing patterns here. These three data sets were then concatenated on to the normal data to give a 4000 observation testing data set.

The exclusive Mahalanobis squared-distances for each of these 4000 observations were then calculated using Equation (2) and the results plotted as shown in Figure 4. The 99% threshold value for a 1000 observation, 50-dimensional problem was found to be 101 after 1000 trials. The plot shows that the unfaulted data set (first 1000 observations) were all correctly labelled as inliers, as expected, and that all the observations corresponding to the 10% and 50% stiffness reductions (third and fourth sets of 1000 observations respectively) were correctly diagnosed as outliers. Unfortunately, the method is unable to classify virtually any of the 1% reduction observations (second set of 1000 observations) as outliers. This is a general property of novelty detectors, as mentioned previously, there is a trade-off between the noise rejection capabilities and the sensitivity to damage, more details of this can be found in Worden et al. (2007).

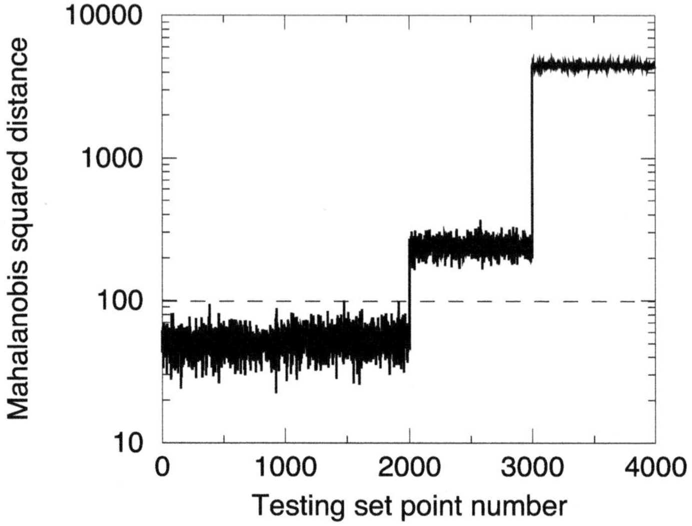

Figure 4. Mahalanobis squared-distances for the unfaulted and faulted data.

4 Neural Networks

4.1 Biological Neural Networks

Advanced as contemporary computers are, none have the capability of carrying out certain tasks - notably pattern recognition - as effectively as the human brain (or mammalian brain for that matter). The reason is that there are essential differences in the way in which the brain and standard serial machines compute. A conventional Von Neumann computer operates by passing instructions sequentially to a single processor. The processor is able to carry out moderately complex instructions very quickly. As an example, at one point many IBM compatible Personal Computers were based on the Intel 80486 microprocessor. This chip operated with a clock cycle of 66 MHz, and was capable of carrying out approximately 60 distinct operations. Averaging over long and short instructions, the chip was capable of performing about 25 Million Instructions per Second (MIPs). (There is little point in describing the performance of a more modern processor as it will without doubt be obsolete by the time this book is published.) State-of-the-art vector processors may make use of tens or hundreds of processors.

In contrast, neurons - the processing units of the brain - can essentially carry out only a single instruction. Further, the delay between instructions is of the order of milliseconds; the neuron operates at approximately 0.001 MIPs. The essential difference with an electronic computer is that the brain comprises a densely interconnected network of about 10^{10} processors operating in parallel.

It is clear that any superiority that the brain enjoys over electronic computers can only be due to its massively parallel nature; the individual processing units are considerably more limited. (In tasks where an algorithm is serial by nature, the brain cannot compete).

The construction of *Artificial Neural Networks* (ANNs) has been an active field of research since the mid-1940s. In the first case, it was hoped that theoretical and computational models would shed light on the properties of the brain. Secondly, it was hoped that a new paradigm for a computer would emerge which would prove more powerful than a Von Neumann serial computer when presented with certain tasks.

Before proceeding to a study of Artificial Neural Networks, it is useful to discuss the construction and behaviour of biological neurons in order to understand the properties which have been incorporated into model neurons.

The Biological Neuron As discussed above, the basic processing unit of the brain is the nerve cell or *neuron*; the structure and operation of the neuron is the subject of this section. In brief, the neuron acts by summing stimuli from connected neurons. If the total stimulus or *activation* exceeds a certain threshold, the neuron 'fires' i.e. it generates a stimulus which is passed on into the network. The essential components of the neuron are shown in the schematic Figure 5.

The cell body, which contains the cell nucleus, carries out those biochemical reactions which are necessary for sustained functioning of the neuron. Two main types of neuron are found in the cortex (the part of the brain associated with the higher reasoning capabilities), they are distinguished by the shape of the cell body. The predominant type have a pyramid-shaped body and are usually referred to as *pyramidal* neurons. Most of the remaining nerve cells have star-shaped bodies and are referred to as *stellate* neurons. The cell bodies are typically a few microns in diameter. The fine tendrils surrounding the cell body are the *dendrites*, they typically branch profusely in the neighbourhood of the cell and extend for a few hundred microns. The *nerve fibre* or *axon* is usually much longer than the dendrites, sometimes extending for up to a metre. The axon only branches at its extremity where it makes connections with other cells.

The dendrites and axon serve to conduct signals to and from the cell body. In general, input signals to the cell are conducted along the dendrites,

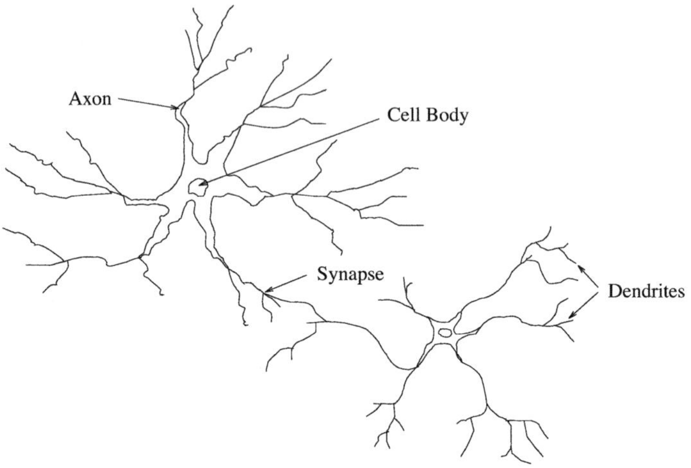

Figure 5. The biological neuron.

while the cell output is directed along the axon. Signals propagate along the fibres as electrical impulses. Connections between neurons, called *synapses*, are usually made between axons and dendrites although they can occur between dendrites, between axons and between an axon and a cell body.

Synapses operate as follows: the arrival of an electrical nerve impulse at the end of an axon say, causes the release of a chemical - a *neurotransmitter* into the synaptic gap (the region of the synapse, typically 0.01 microns). The neurotransmitter then binds itself to specific sites - *neuroreceptors* usually in the dendrites of the target neuron. There are distinct types of neurotransmitters: *excitatory* transmitters which trigger the generation of a new electrical impulse at the receptor site, and *inhibitory* transmitters which act to prevent the generation of new impulses.

The operation of the neuron is very simple. The cell body carries out a summation of all the incoming electrical impulses directed inwards along the dendrite. The elements of the summation are individually weighted by the strength of the connection or synapse. If the value of this summation - the *activation* of the neuron - exceeds a certain threshold, the neuron fires and directs an electrical impulse outwards via its axon. From synapses with the axon, the signal is communicated to other neurons. If the activation is less than the threshold, the neuron remains dormant.

A mathematical model of the neuron, exhibiting most of the essential features of the biological neuron was developed as early as 1943 by McCulloch and Pitts (1943). This model forms the subject of the Section 4.2; the remainder of this section is concerned with those properties of the brain which emerge as a result of its massively parallel nature.

Memory Information is actually stored in the brain in the network connectivity and the strengths of the connections or synapses between neurons. In this case, knowledge is stored as a distributed quantity throughout the entire network. The act of retrieving information from such a memory is rather different from that for an electronic computer. In order to access data on a PC say, the processor is informed of the relevant address in memory, and it retrieves data from that location. In a neural network, a stimulus is presented (i.e. a number of selected neurons receive an external input), and the required data are encoded in the subsequent pattern of neuronal activations. Potentially, recovery of the pattern is dependent on the entire distribution of connection weights or synaptic strengths.

One advantage of this type of memory retrieval system is that it has a much greater resistance to damage. If the surface of a PC hard disk is damaged, all data at the affected locations may be irreversibly corrupted. In a neural network, because the knowledge is encoded in a distributed fashion, local damage to a portion of the network may have little effect on the retrieval of a pattern when a stimulus is applied.

Learning According to the argument in the previous section, knowledge is encoded in the connection strengths between the neurons in the brain. The question arises of how a given distributed representation of data is obtained. On way is that the initial state of the brain at birth is gradually modified as a result of its interaction with the environment. This development is thought to occur as an evolution in the connection strengths between neurons as different patterns of stimuli and appropriate responses are activated in the brain as a result of signals from the sense organs.

The first explanation of such learning in terms of the evolution of synaptic connections was given by Hebb (1949):

> "When a cell A excites cell B by its axon and when in a repetitive and persistent manner it participates in the firing of B, a process of growth or of changing metabolism takes place in one or both cells such that the effectiveness of A in stimulating and impulsing cell B is increased with respect to all other cells which can have this effect."

If some similar mechanism could be established for computational models of neural networks, there would be the attractive possibility of 'programming' these systems simply by presenting them with a sequence of stimulus-response pairs so that the network can learn the appropriate relationship by reinforcing some of its internal connections.

4.2 The McCulloch-Pitts Neuron

Having found a description of a biological neural network, the first stage in deriving a computational model was to represent mathematically the behaviour of a single neuron. This step was carried out in 1943 by the neurophysiologist Warren McCulloch and the logician Walter Pitts (McCulloch and Pitts (1943)).

The McCulloch-Pitts model (MCP model) constitutes the simplest possible neural network model. Because of its simplicity it is possible without too much effort to obtain mathematically rigorous statements regarding its range of application; the major disadvantage of the model is that this range is very limited. The object of this section is to demonstrate which input-output systems or functions allow representation as a MCP model. In doing this, a number of techniques which are generally applicable to more complex network paradigms are encountered.

Boolean Functions For a fruitful discussion, limits must be placed upon the range of systems or functions which the MCP model will be asked to represent; the output of a nonlinear dynamical system, for example, can be represented as a nonlinear *functional* of the whole input history. This is much too general to allow a simple analysis. For this reason, the objects of study here are the class of Multi-Input Single-Output (MISO) systems which have a representation as a *function* of the instantaneous input values, i.e.

$$y = f(x_1, x_2, \dots, x_n) \tag{4}$$

y being the output and $x_1, \dots, x_n$ being the inputs. A further constraint is imposed that the variables y and $x_1, \dots, x_n$ are only allowed to take the values 0 and 1. Functions of this type are call *Boolean*. They arise naturally in symbolic logic where the value 1 is taken to indicate truth of a proposition while 0 indicates falsity (depending on which notation is in use, $1 = T = .true.$ and $0 = F = .false.$). In the following, curly brackets shall be used to represent those Boolean functions which are represented by logical propositions, e.g. the function,

$$f(x_1, x_2) = \{x_1 = x_2\} \tag{5}$$

Given the inputs to this function, the output is evaluated as follows,

$$\begin{array}{lllllll} f(0,0) & = & \{0=0\} & = & .true. & = & 1 \\ f(0,1) & = & \{0=1\} & = & .false. & = & 0 \\ f(1,0) & = & \{1=0\} & = & .false. & = & 0 \\ f(1,1) & = & \{1=1\} & = & .true. & = & 1 \end{array}$$

A Boolean function which is traditionally of great importance in neural network theory is the *exclusive-or* function $XOR(x_1, x_2)$ which is true if one, but not both, of its arguments is true. It is represented by the Boolean,

	0	1
0	0	1
1	1	0

Note that this function also has a representation as the proposition $\{x_1 \neq x_2\}$.

There is a very useful pictorial representation of the Boolean functions with two arguments. The possible combinations of input values can be represented as the vertices of the unit square in the Cartesian plane (Figure 6).

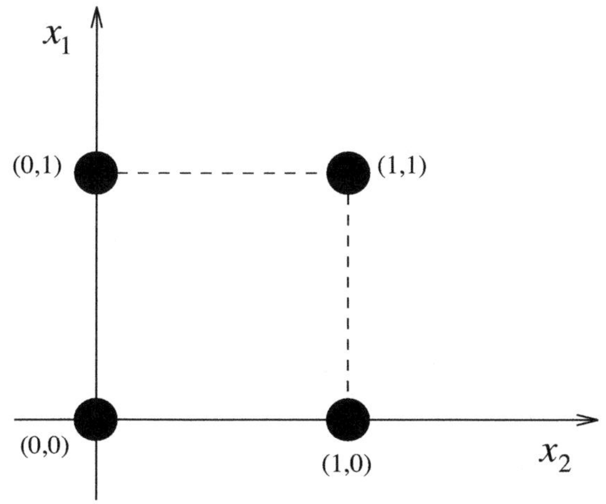

Figure 6. Pictorial representation of a two-input Boolean function.

Each Boolean function on this domain is now specified by assigning the value 0 or 1 to each point in the domain. If a point on which the function is true is represented by a white circle, and a point on which the function is false by a black circle, one obtains the promised pictorial representation. As an example, the XOR function has the representation shown in Figure 7. Unfortunately, the pictorial representation is only really useful for functions of two or three variables.

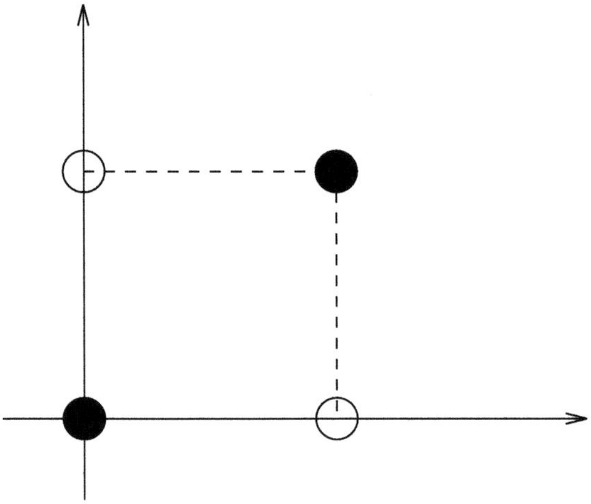

Figure 7. Pictorial representation of the XOR function.

The MCP Model Neuron In the MCP model, each input to a neuron is assumed to come from a connected neuron, the only information considered to be important is whether the connected neuron has fired or not (all neurons are assumed to fire with the same intensity). This allows a restriction of the possible input values to 0 and 1. On the basis of this information, the neuron will either fire or not fire, so the output values are restricted to be 0 or 1 also. This means that a given neuron can be identified with some Boolean function. The MCP model should therefore be able to represent an arbitrary Boolean function.

The MCP neuron can be illustrated as in Figure 8.

The input values $x_i \in \{0, 1\}$ are weighted by a factor w_i before they are passed to the body of the MCP neuron (this allows the specification of a strength for the connection). The weighted inputs are then summed and the MCP neuron fires if the weighted sum exceeds some predetermined threshold β. So the model fires if,

$$\sum_{i=1}^{n} w_i x_i > \beta \tag{6}$$

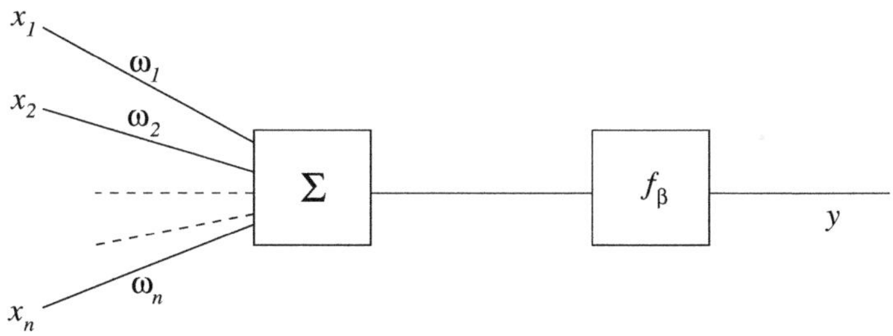

Figure 8. The McCullough-Pitts neuron.

and doesn't fire if,

$$\sum_{i=1}^{n} w_i x_i \leq \beta \tag{7}$$

Consequently, the MCP neuron has a representation as the proposition,

$$\{\sum_{i=1}^{n} w_i x_i > \beta\} \tag{8}$$

which is clearly a Boolean function. As a real neuron could correspond to an arbitrary Boolean function, there are two fundamental questions which can be asked:

1. Can a MCP model of the form (8) represent an *arbitrary* Boolean function $f(x_1, \ldots, x_n)$? i.e. do there exist values for $w_1, \ldots, w_n$ and β such that $f(x_1, \ldots, x_n) = \{\sum_{i=1}^{n} w_i x_i > \beta\}$?
2. If a MCP model exists, how can the weights and thresholds be determined? In keeping with the spirit of neural network studies one would like a *training* algorithm which would allow the MCP model to learn the correct parameters by presenting it with a *finite* number of input-output pairs; does such an algorithm exist?

The answer to Question 2 is yes; unfortunately the answer to the more fundamental Question 1 is no. The simplest way to determine the limitations of the class of MCP models is to consider the geometry of the situation. In n dimensions the equation,

$$\sum_{i=1}^{n} w_i z_i = \beta \tag{9}$$

represents a hyperplane which separates two regions of the n-dimensional input space. One region U consists of all those points $(z_1, \ldots, z_n)$ (where the z_i can take any real values), such that,

$$\sum_{i=1}^{n} w_i z_i > \beta \tag{10}$$

The other region L contains all those points such that,

$$\sum_{i=1}^{n} w_i z_i < \beta \tag{11}$$

This means that each MCP model (9) specifies a plane which divides the input space into two regions U and L (where L is now defined to include the plane itself). Further, by Equations (10) and (11) the MCP model takes the values 0 on L and 1 on U. This means that if one is to represent an arbitrary Boolean function f by a MCP model, there must exist a plane which splits off the points on which $f = 1$ from the points on which $f = 0$. Using the pictorial representation described earlier, such a plane should separate the white dots from the black dots. It is now obvious why there is no MCP model for the Boolean $\{x_1 = x_2\}$, no such plane exists (Figure 9).

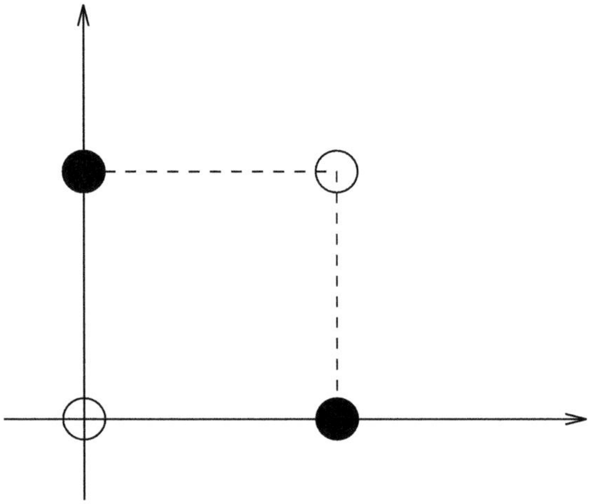

Figure 9. Decision boundary for problem Boolean.

The XOR function of Figure 7 is a further example. In fact, these are the only two-input Boolean functions which do not have a MCP model. There are actually 16 two-input Boolean functions, of which only two are not representable by a MCP model. This is not such a bad result. However, the percentage of functions which can be represented falls off rapidly with increasing number of inputs.

The solution to the problem is to use more than one MCP unit to represent a MISO system. For example, if two MCP units are used in the two-input case, one can partition the plane of the $XOR(x_1, x_2)$ function into four regions as below, and thereby solve the problem.

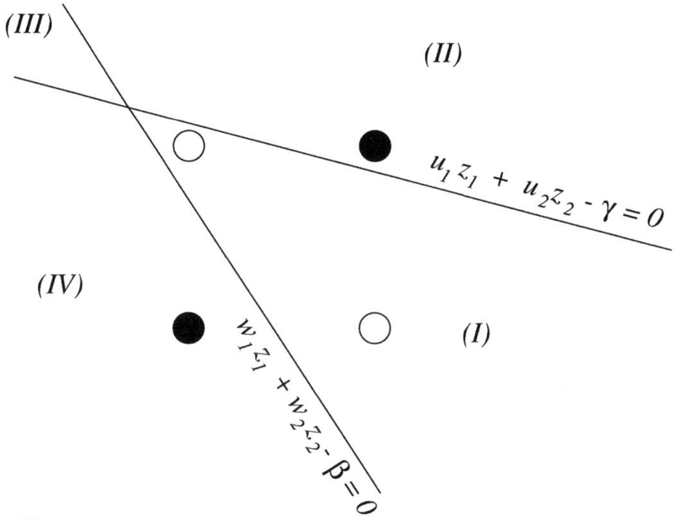

Figure 10. Partition of feature space for XOR function.

Consider the two lines in Figure 10. The parameters of the first line $w_1 z_1 + w_2 z_2 = \beta$ define a MCP model MCP_β, the parameters of the second $u_1 z_1 + u_2 z_2 = \gamma$ define a model MCP_γ. This configuration of lines separates the white dots (in region I) from the black dots as required. The points where the XOR function is 1 are in the region I where the outputs y_β and y_γ from MCP_β and MCP_γ are 1 and 0 respectively, all other pairs of outputs indicate regions where XOR is false. It is possible to define a Boolean function $f(y_\beta, y_\gamma)$ whose output is 1, if and only if $(y_\beta, y_\gamma) = (1, 0)$. The pictorial representation of this Boolean is shown in Figure 11.

It is clear from the figure that this function has a MCP model, say MCP_δ (with weights v_1, v_2 and threshold δ). Considering the network of MCP models shown in Figure 13, it is clear that the final output is 1 if, and only if, the input point (x_1, x_2) is in region I in Figure 12. Consequently, the network provides a representation of the XOR function. Note that the end result is a heterogeneous network structure in which there are three types of neurons:

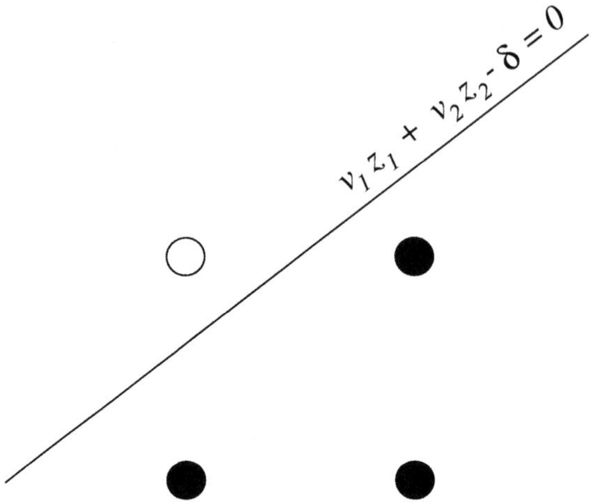

Figure 11. Decision function for multi-neuron solution to XOR problem.

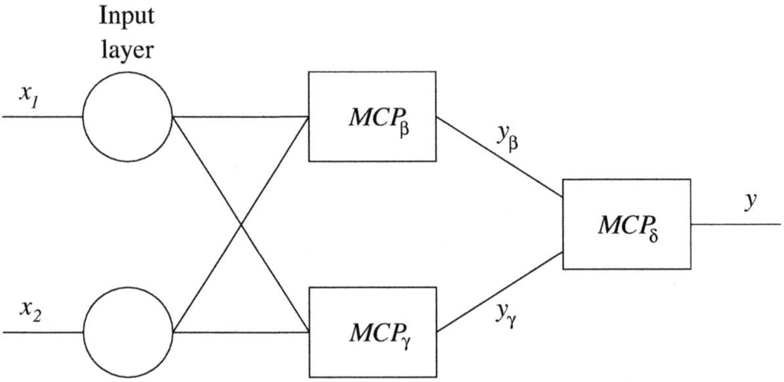

Figure 12. Network solution to XOR problem.

Input neurons Communicate directly with the outside world, but serve no computational purpose beyond distributing the input signals to all of the first layer of computing neurons.

Hidden neurons Do not communicate with the outside world; compute.

Output neurons Do Communicate with the outside world; compute.

Signals pass forward through the input layer and hidden layers and emerge from the output layer. Such a network is called a *feed-forward* network.

The constructions that have been presented suggest why the MCP model proves to be of interest, by passing to networks of MCP neurons, it can be shown fairly easily that any Boolean function can be represented by an appropriate network. Furthermore, a training algorithm exists for such networks which terminates in a *finite* time (Minsky and Papert (1988)).

4.3 Perceptrons

In the last section, it was shown how the failure of the MCP model to represent certain simple functions led to the construction of simple networks of MCP neurons which could overcome the problems. The first serious study of such networks was carried out by Rosenblatt and is documented in his 1962 book (Rosenblatt (1962)). Rosenblatt's *Perceptron* networks are composed of an input layer and two layers of MCP neurons as shown in Figure 13. The hidden layer is referred to as the *associative* layer while the output layer is termed the *decision* layer. Only the connections between the decision nodes and associative nodes are adjustable in strength; those between the input nodes and associative nodes have to be preset somehow before training takes place.

The neurons operate as threshold devices exactly as described in Section 4.2: if the weighted summation of inputs to a neuron exceeds the threshold, the neuron output is unity, otherwise it is zero.

It is immediately apparent that the perceptrons have applications in pattern recognition. For example, the input layer could be associated with a screen or retina as in Figure 14, such that an input of 0 corresponds to a white pixel and 1 to a black pixel. The network could then be trained to respond at the decision layer only if certain patterns appeared on the screen.

This pattern recognition problem is clearly inaccessible to an MCP model since there are no restrictions on the form of Boolean function which could arise. However, it is possible to show that *any* Boolean function can be represented by a perceptron network (Minsky and Papert (1988)). Unfortunately, it transpires that it can be necessary to use compute *all* possible

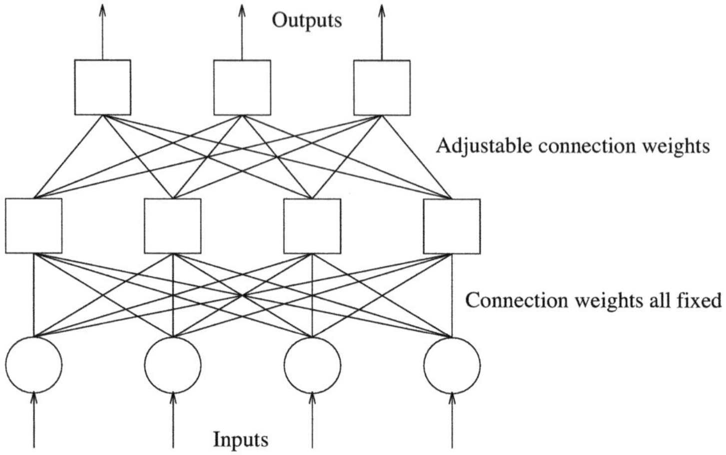

Figure 13. Structure of Rosenblatt perceptron.

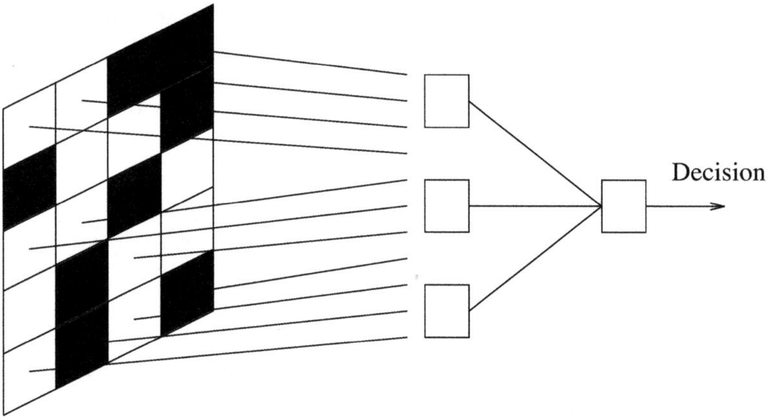

Figure 14. Perceptron image recognition.

products of inputs in forming the necessary networks. In this case, for N inputs, 2^N products and hence 2^N associative nodes (which effectively compute the products) are required. The XOR function above is an example of this type of network.

Having established that any Boolean function can be computed using a perceptron, the next problem is to establish a training algorithm which will lead to the correct connection weights. A successful learning rule was obtained by Rosenblatt (1962) called the *Delta rule*; this rule corrects the weights after comparing the network outputs for a given input with a set of desired outputs; the approach is therefore one of *supervised learning.*

Limitations of Perceptrons The results that have been presented indicate why perceptrons were initially received with enthusiasm; they can represent a Boolean function of arbitrary complexity and are provided with a training algorithm which is guaranteed to converge in finite time. The problem is that in representing a function with N arguments, the perceptron may need 2^N elements in the associative layer; the networks grow exponentially in complexity with the dimension of the problem.

A possible way of avoiding this problem was seen to be to restrict the number of connections between the input layer and associative layer, so that each associative node connects to a (hopefully) small subset of the inputs. A perceptron with this restriction is called a *diameter-limited* perceptron. The justification for such perceptrons is that the set of Booleans which require full connections might consist of a small set of uninteresting functions. Unfortunately, this has proved not to be the case, as shown in the book *Percetrons* (Minsky and Papert (1988)).

Another possible escape-route was the use of perceptrons with several hidden layers in the hope that the more complex organisation would avoid the exponential growth in the number of neurons. The problem here is that the adjustment of connection weights to a node by the Delta rule requires an estimate of the output error at that node. However, only the errors at the output layer are given, and at the time there was no means of assigning meaningful errors to the internal nodes. This was referred to as the *credit-assignment problem.* The problem remained unsolved until 1974 (Werbos (1974)). Unfortunately, Minsky and Papert's book resulted in the almost complete abandonment of neural network research until Hopfield's paper of 1982 brought about a resurgence of interest. As a result of this, Werbos' 1974 solution of the credit assignment problem was overlooked until after Rumelhart *et al* independently arrived at the solution in 1985 (Rumelhart et al. (1986)). The new paradigm the latter introduced - the Multi-Layer Perceptron (MLP) - is probably the most widely-used neural network to date.

4.4 Multi-Layer Perceptrons

The network is a natural generalisation of the perceptrons described in Section 4.3. The main references for this discussion are Bishop (1998) or Rumelhart and McClelland (1988). A detailed analysis of the network structure and learning algorithm is given in Bishop (1998), but a brief discussion is given here.

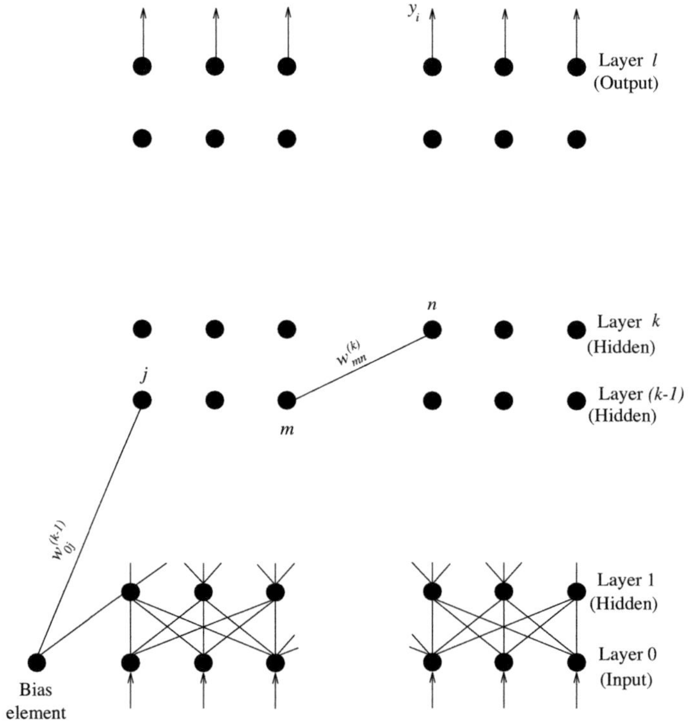

Figure 15. Multi-Layer Perceptron (MLP).

The MLP is a feedforward network with the neurons arranged in layers (Figure 15). Signal values pass into the *input layer* nodes, progress forward through the network *hidden layers*, and the result finally emerges from the *output layer*. Each node i is connected to each node j in the preceding and following layers through a connection of weight w_{ij}. Signals pass through the nodes as follows: in layer k a weighted sum is performed at each node i

of all the signals $x_j^{(k-1)}$ from the preceding layer $k-1$, giving the excitation $z_i^{(k)}$ of the node; this is then passed through a nonlinear *activation function* f to emerge as the output of the node $x_i^{(k)}$ to the next layer i.e.,

$$x_i^{(k)} = f(z_i^{(k)}) = f(\sum_j w_{ij}^{(k)} x_j^{(k-1)}) \tag{12}$$

Various choices for the function f are possible, the hyperbolic tangent function $f(x) = \tanh(x)$ is a good choice. A novel feature of this network is that the neuron outputs can take any value in the interval [-1,1]. There are also no explicit threshold values associated with the neurons. One node of the network, the *bias* node, is special in that it is connected to all other nodes in the hidden and output layers; the output of the bias node is held fixed throughout in order to allow constant offsets in the excitations z_i of each node.

The first stage of using a network is to establish the appropriate values for the connection weights w_{ij} i.e. the *training* phase. The type of training usually used is a form of *supervised* learning and makes use of a set of network inputs for which the desired network outputs are known. At each training step, a set of inputs is passed forward through the network yielding trial outputs which can be compared with the desired outputs. If the comparison error is considered small enough, the weights are not adjusted. If however a significant error is obtained, the error is passed *backwards* through the net and the training algorithm uses the error to adjust the connection weights so that the error is reduced. The learning algorithm used is usually referred to as the *backpropagation* algorithm, and can be summarised as follows. For each presentation of a training set, a measure J of the network error is evaluated where,

$$J(t) = \frac{1}{2} \sum_{i=1}^{n^{(l)}} (y_i(t) - \hat{y}_i(t))^2 \tag{13}$$

and $n^{(l)}$ is the number of output layer nodes. J is implicitly a function of the network parameters $J = J(\theta_1, \ldots, \theta_n)$ where the θ_i are the connection weights, ordered in some way. The integer t labels the presentation order of the training sets. After presentation of a training set, the standard steepest-descent algorithm requires an adjustment of the parameters according to,

$$\triangle\theta_i = -\eta \frac{\partial J}{\partial \theta_i} = -\eta \nabla_i J \tag{14}$$

where ∇_i is the gradient operator in the parameter space. The parameter η determines how large a step is made in the direction of steepest descent and therefore how quickly the optimum parameters are obtained. For this reason η is called the *learning coefficient.* Detailed analysis (Bishop (1998)) gives the update rule after the presentation of a training set,

$$w_{ij}^{(m)}(t) = w_{ij}^{(m)}(t-1) + \eta\delta_i^{(m)}(t)x_j^{(m-1)}(t) \tag{15}$$

where $\delta_i^{(m)}$ is the error in the output of the ith node in layer m. This error is not known *a priori* but must be constructed from the known errors $\delta_i^{(l)} = y_i - \hat{y}_i$ at the output layer l. This is the source of the name *backpropagation*, the weights must be adjusted layer by layer, moving backwards from the output layer.

There is little guidance in the literature as to what the learning coefficient η should be; if it is taken too small, convergence to the correct parameters may take an extremely long time. However, if η is made large, learning is much more rapid but the parameters may diverge or oscillate. One way around this problem is to introduce a *momentum* term into the update rule So that previous updates persist for a while, i.e.,

$$\triangle w_{ij}^{(m)}(t) = \eta\delta_i^{(m)}(t)x_j^{(m-1)}(t) + \alpha\triangle w_{ij}^{(m)}(t-1) \tag{16}$$

where α is termed the *momentum coefficient.* The effect of this additional term is to damp out high-frequency variations in the backpropagated error signal. This is the form of the algorithm used throughout the case studies later.

Existence of Solutions Before advocating the use of neural networks in representing functions and processes, it is important to establish what they are capable of. As described above, artificial neural networks were all but abandoned as a subject of study following Minsky and Papert's book (Minsky and Papert (1988)) which showed that perceptrons were incapable of modelling very simple logical functions. In fact, recent years have seen a number of rigorous results (Cybenko (1989), is a good example), which show that a MLP network is capable of approximating a given function with arbitrary accuracy, even if possessed of only a single hidden layer. Unfortunately, the proofs are not constructive and offer no guidelines as to the complexity of network required for a given function. A single hidden layer may be sufficient but might require many more neurons than if two hidden layers were used.

Uniqueness of Solutions This is the problem of local minima. The error function for a MLP network is an extremely complex object. Given a converged MLP network, there is no way of establishing if it has arrived at the global minimum. Some attempts to avoid the problem are centred around the association of a temperature with the learning schedule. Roughly speaking, at each training cycle the network may randomly be given enough 'energy' to escape from a local minimum. The probable energy is calculated from a network temperature function which decreases with time. Recall that molecules of a solid at high temperature escape the energy minimum which specifies their position in the lattice. An alternative approach is to seek network paradigms with less severe problems e.g. Radial-basis function networks (Bishop (1998)).

5 Novelty Detection Again

5.1 Non-Gaussian Normal Condition - Neural Networks

The outlier approach made an implicit assumption of the Gaussian nature of the normal condition data through the use of the Mahalanobis distance. In many cases, this will be an unwarranted assumption and more general techniques will be needed. One approach which is able to cope with non-Gaussian normal conditions is that based on the idea of an Auto-Associative Neural Network (AANN); the application of this network was first proposed in Pomerleau (1993) within a slightly different context.

The approach taken here is simply to train an (AANN) on the patterns. This simply means a feed-forward Multi-Layer Perceptron (MLP) network (Bishop (1998)), which is asked to reproduce at the output layer, those patterns which are presented at the input. This would be a trivial exercise except that the network structure has a 'bottleneck' i.e. the patterns are passed through hidden layers which have fewer nodes than the input layer (Figure 16). This forces the network to learn the significant features of the patterns; the activations of the smallest, central layer, correspond to a compressed representation of the input. Training proceeds by presenting the network with many versions of the pattern corresponding to normal condition corrupted by noise and requiring a copy at the output.

The novelty index $\nu(\underline{x})$ corresponding to a pattern vector $\underline{x}$ is then defined as the Euclidean distance between the pattern and the result of presenting it to the network $\hat{\underline{x}}$,

$$\nu(\underline{x}) = ||\underline{x} - \hat{\underline{x}}|| \tag{17}$$

It is clear how this works. If learning has been successful, then for all data in the training set so $\nu(\underline{x}) \approx 0$ if $\underline{x}$ represents the normal condition. If $\underline{x}$ corresponds to damage, $\nu(\underline{x})$ will be significantly non-zero. Note that there is no guarantee that ν will increase monotonically with the level of damage,

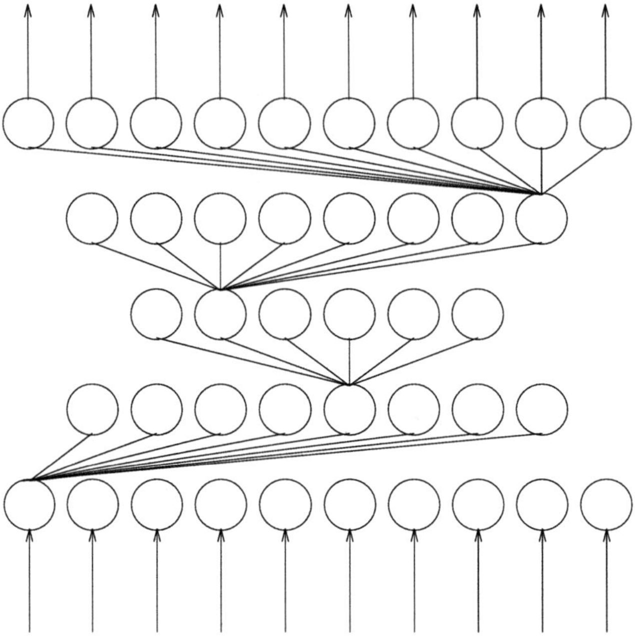

Figure 16. Auto-associative neural network.

this is why novelty detection only gives a level one diagnostic. Note that universal approximation property of the neural network (Bishop (1998)), means that the novelty detector can learn the properties of any normal condition distribution, it does not have to be Gaussian or even unimodal. In fact, the case study following will consider the extreme case of a normal condition set defined on two disconnected components.

As an illustration, the three degree-of-freedom system with concentrated masses described in Equation (3) and shown in Figure 2 was simulated again. In the first normal condition (NCI) the following values were used: $m = 1$, $c = 20$ and $k = 10^4$. In the second normal condition (NCII) the same parameters were used with the exception that the bottom mass was reduced by 50%. This problem with two normal conditions with different masses is intended to mimic the situation where an aeroplane drops a store. The fault in the system was simulated by decreasing the stiffness between the top two masses in Figure 3 by different degrees.

In both conditions, the response transmissibility function between the top two masses was calculated in terms of the transverse displacement for a harmonic excitation applied at the top mass.

In the simulations performed, the frequency of the harmonic exciting force was varied in the range from 0 to 50 Hz in order to encompass the three natural frequencies of the system. Figure 17 illustrates the transmissibility

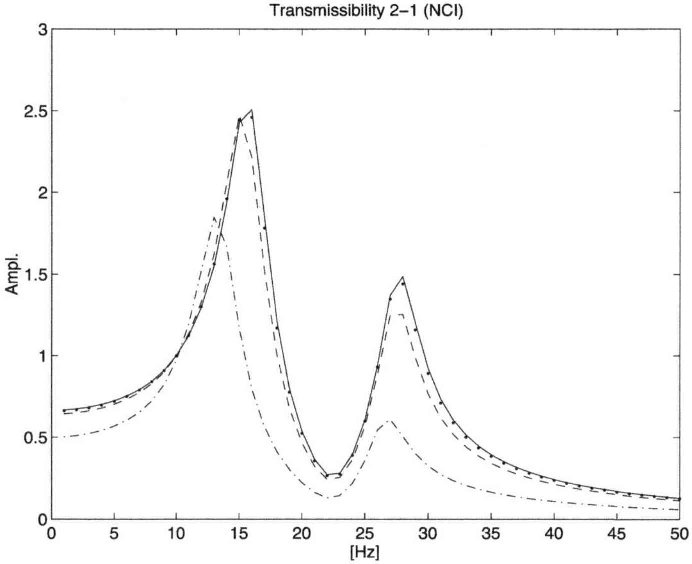

Figure 17. Transmissibilities for normal condition 1 (solid line) and corresponding damage cases (dashed lines)

function corresponding to the structure in NCI together with the functions relative to three damage conditions, denoted DCI, with 2%, 10% and 50% reductions in stiffness of the spring of interest respectively. Figure 18 shows equivalent functions corresponding to NCII together with damaged cases with the same reductions in stiffness, denoted DCII.

Note that the normal conditions NCI and NCII are significantly different from each other compared to the differences between a normal condition and its corresponding damage states. This means that an outlier analysis conducted on a given normal condition would strongly indicate damage for data from the other normal condition.

The training data for the neural network diagnostic system was obtained by making 500 copies of the transmissibility functions corresponding to the undamaged structure for each of the two normal conditions, and then pol-

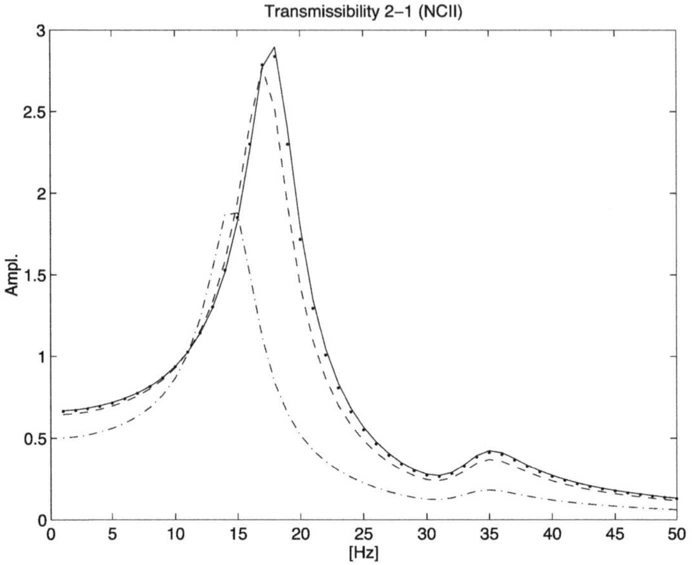

Figure 18. Transmissibilities for normal condition 2 (solid line) and corresponding damage cases (dashed lines)

luting each of these independently by adding Gaussian noise with an RMS value equal to 1% of the peak value of the transmissibility function. The testing sets were constructed by concatenating 500 noise-corrupted copies of the transmissibility functions relative to the different damage scenarios considered for both operating conditions.

For the task of pattern recognition, an AANN with 5 layers and node structure 50:40:30:40:50 was selected and trained for 100000 presentations of individual training patterns in random order. A slight modification of the novelty index in Equation (17) was used here, in which the two normal condition sets were normalised to the same level, details can be found in Surace and Worden (1997).

The results shown in Figure 19 correspond to a 10% stiffness reduction. As one can see, the normalised novelty index permits unambiguous identification of the presence of the fault and gives the required near-zero response on both normal conditions. As in the case of outlier analysis, a confidence threshold can be computed (Worden (1997)).

This example illustrates a case where it is necessary to design a novelty detector which does not fire when operational variations occur. In the field of SHM it is often even more important to be able to neglect environmental

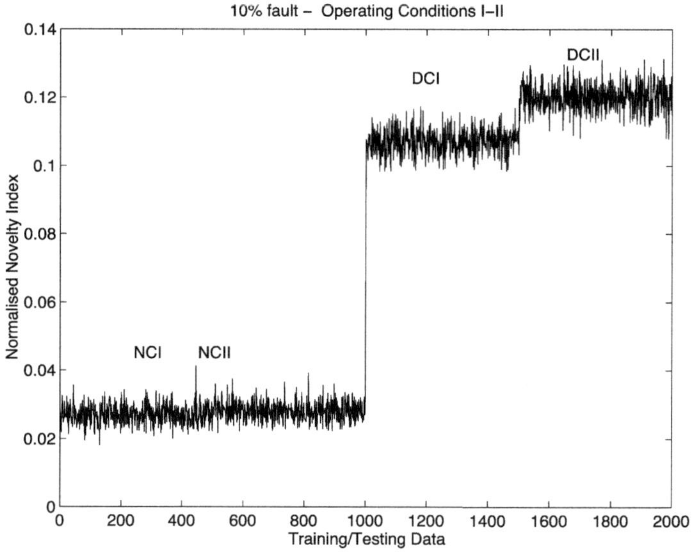

Figure 19. Novelty detection of the two 10% damage conditions.

changes e.g. changes due to variation in temperature. A useful survey of these issues can be found in Sohn (2007), and the matter is also discussed at length in the later chapter in this book by Kullaa.

6 Statistical Pattern Recognition

6.1 Introduction

There are currently three entirely separate theoretical frameworks for developing pattern classifiers. The most recent, based on the use of neural networks, is the main approach described here. The other two, the *statistical* and *syntactic* approaches, are more well established. In some respects, the three approches complement each other, so in attacking a particular problem the ideal situation would be to know something of all three. This is the approach taken in Schalkoff (1992), which provides the main source of reference for this lecture. However, as the syntactic approach is rather abstract and calls for a certain amount of mathematical sophistication, the discussion here is restricted to the statistical approach.

In brief, the problem of pattern recognition is to associate classes $C_i, i = 1, \ldots, N_c$ with measured data. Statistical methods allow two distinct approaches.

(SPR1) Given a measurement/feature vector $\underline{x}$, calculate the probability that this is associated with a given class C_i; this is the conditional probability $P(C_i|\underline{x})$. Repeat for all possible classes and choose that which gives the highest probability.

(SPR2) Form a measure of the error associated with choosing a particular class and then pick the class which minimises it.

The discussion below concentrates on *SPR1*. Statistical methods are necessary because in general, many different measurement vectors will correspond to noisy or distorted versions of the same basic pattern or *template* and thus require assignment to the same class.

It is assumed throughout that *training data* is available i.e. a sequence of measurement vectors $\underline{x}^{(k)}, k = 1, \ldots, N_t$ are known, together with the correct class for each vector $C_i^{(k)}$. This allows the construction of the *a priori* conditional probability density function (PDF) $p(\underline{x}|C_i)$ which specifies the probability that a measurement vector $\underline{x}$ can arise from a class C_i. An application of Bayes' theorem (see the Appendix to this chapter) yields the required $P(C_i|\underline{x})$ via,

$$P(C_i|\underline{x}) = \frac{p(\underline{x}|C_i)P(C_i)}{p(\underline{x})} \tag{18}$$

where $p(\underline{x})$ is the unconditional density function which can also be computed from the training set. $P(C_i)$ is the probability of finding an example from class C_i without considering any measurement information. The requirement of a training set makes this approach one of *supervised learning* in the terminology of neural networks.

To give an idea how this all leads to a decision rule, consider the following simple example from Schalkoff (1992). Suppose there are two classes C_1 and C_2 with equal *a priori* (i.e. before measurement) probabilities i.e $P(C_1) = P(C_2)$. Further suppose that there is a single distinguishing feature x and that the PDFs $p(x|C_1)$ and $p(x|C_2)$ have been established as Gaussians with the same variance σ but different means μ_1 and μ_2. So,

$$p(x|C_i) = \frac{1}{\sqrt{2\pi}\sigma} \exp\left\{-\frac{1}{2}\left(\frac{x-\mu_i}{\sigma}\right)^2\right\} \qquad i = 1, 2$$

The basic decision rule is,

$$\text{Choose } C_1 \text{ if } P(C_1|x) > P(C_2|x), \text{ otherwise choose } C_2$$

Bayes theorem shows that $P(C_1|x) > P(C_2|x)$ implies,

$$\frac{p(x|C_1)P(C_1)}{p(x)} > \frac{p(x|C_2)P(C_2)}{p(x)}$$

so using the fact that $P(C_1) = P(C_2)$ yields,

$$P(C_1|x) > P(C_2|x) \Longrightarrow p(x|C_1) > p(x|C_2)$$

and this allows a decision rule,

$$\text{Choose } C_1 \text{ if } p(x|C_1) > p(x|C_2), \text{ otherwise choose } C_2$$

which is based only on PDFs calculated from the training set. (In this case, determining the PDF simply requires estimates of μ_1, μ_2 and σ.) The decision boundary is a single number α, fixed by the condition,

$$p(\alpha|C_1) = p(\alpha|C_2)$$

which gives, on using the functional form for the Gaussian,,

$$\alpha = \frac{\mu_1 + \mu_2}{2}$$

So α is mid-way between μ_1 and μ_2, this agrees with intuition as the distributions have the same width (same σ). The situation is illustrated in Figure 20 (the densities shown are schematics and not intended to be representative of true Gaussians). The decision regions R_1 and R_2 are simply the half-lines meeting at α.

As the probability density functions for the two classes overlap, there will always be the possibility of choosing incorrectly on the basis of a given measurement x. It is possible to compute the probability of making such a wrong decision. By the total probability theorem it follows that,

$$\begin{aligned} P(\text{error}) &= P(\text{error}|C_1)P(C_1) + P(\text{error}|C_2)P(C_2) \\ &= P(x \in R_2|C_1)P(C_1) + P(x \in R_1|C_2)P(C_2) \\ &= P(x > \alpha|C_1)P(C_1) + P(x < \alpha|C_2)P(C_2) \end{aligned}$$

So, finally,

$$P(\text{error}) = P(C_1)\int_{\alpha}^{\infty} p(z|C_1)dz + P(C_2)\int_{-\infty}^{\alpha} p(z|C_2)dz \qquad (19)$$

And this can be computed from *a priori* information i.e. the training set.

In the general problem where many measurements/features are used to distinguish between many classes, the approach *SPR1* can be summarised as follows:

1. Establish a training set $\{\underline{x}^{(j)}, C_i^{(j)}\}, j = 1, \ldots, N_t$ for each class C_i.

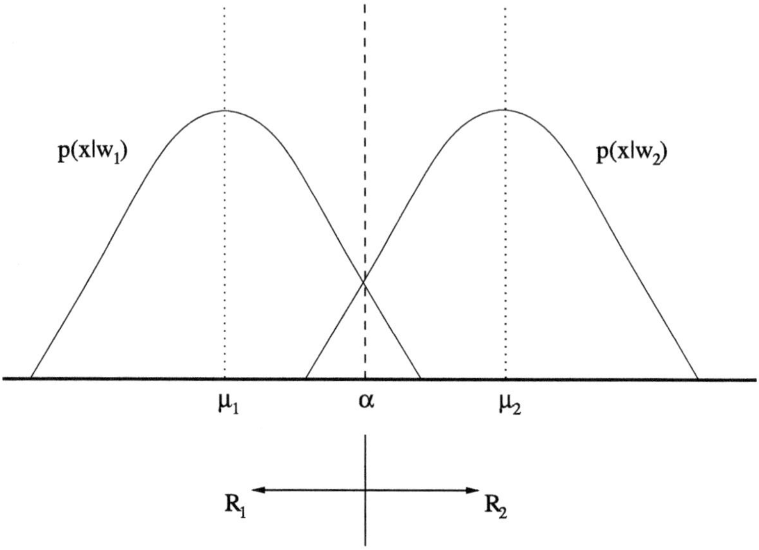

Figure 20. Solution to simple classification problem.

2. Compute the *a priori* information $p(\underline{x}|C_i)$, $P(C_i)$ and $P(C_i)$.
3. Given a new unclassified measurement $\underline{y}$, use Bayes' theorem to obtain the measurement conditioned probability $P(C_i|\underline{y}) = p(\underline{y}|C_i)P(C_i)/p(\underline{y})$ for each class C_i.
4. Choose C_i such that $P(C_i|\underline{y}) > P(C_j|\underline{y})$ for all $i \neq j$.

In point 3 above, the denominator of the expression, $p(\underline{y})$ is common to all of the measurement-conditioned probabilities, so step 4 above can be modified to,

$$\text{Choose } C_i \text{ if } p(\underline{y}|C_i)P(C_i) > p(\underline{y}|C_j)P(C_j) \text{ for all } i \neq j \tag{20}$$

and the measurement PDF need not be computed at step 2.

Note that step 2 is not trivial. There are two possibilities here, a parametric form can be assumed for the PDF with the training set used to determine the values of the parameters, or a non-parametric form can be obtained. In the latter case, the PDF can be constructed from a frequency histogram of the training data and stored as an array of values. It is most usual to use the former approach, and in the vast majority of cases, the PDF is assumed to be Gaussian, i.e.

$$p(\underline{x}|C_i) = \frac{1}{(2\pi)^{\frac{N}{2}}\sqrt{|\Sigma_i|}} \exp\left\{-\frac{1}{2}(\underline{x} - \underline{\mu}_i)^T \Sigma_i^{-1} (\underline{x} - \underline{\mu}_i)\right\} \tag{21}$$

for an N-component measurement/feature vector. Σ_i and $\underline{\mu}_i$ are respectively the covariance matrix and mean vector for the measurement vectors associated with class C_i. $|\Sigma_i|$ is the determinant of Σ_i. In this case, the problem of parameter estimation is reduced to obtaining Σ_i and $\underline{\mu}_i$ - standard statistical estimates.

6.2 Connection to Neural Networks

All of the above theory allows the construction of effective classifiers which can be used in SHM problems. However, in recent years it has become common practice to implement the classifier using a neural network structure; the MLP structure described in Section 4 is most common. At first, when such techniques were used, a certain amount of suspicion arose as a result of the black-box nature of the networks and the difficulty of determining precisely how they worked. Fortunately, researchers soon realised that the neural networks, if trained appropriately, will work according to the SPR principles discussed above. The main result, described in Bishop (1998) shows that neural networks trained according to certain principles will actually generate Bayesian posterior probabilities for the possible classes when presented with a data pattern.

The *appropriate* training strategy to make contact with statistical pattern recognition is the *1 of M strategy* (Bishop (1998)). This approach is quite simple; each pattern class is associated with a unique network output on presentation of a pattern during training, the network is required to produce a value of unity at the output corresponding to the desired class and zero at all other outputs. When the trained network is presented with a new pattern it will respond at each output with the posterior probability appropriate to that class. This means that such a network actually implements a Bayesian decision rule if each pattern vector is identified with the class associated with the highest output.

It is not quite as simple as this. Conditions are best if the following conditions are adhered to:

1. The training patterns for each class should be represented in the training data set in numbers proportional to their prior probabilities. In practice, most people adopt a *balanced* training set where possible, containing equal numbers corresponding to each class. It is possible to post-process the network outputs for a network trained on a balanced set in order to compensate for unequal priors (Tarassenko (1998)).
2. The network should ideally be trained to minimise the cross-entropy error function, although a standard least-squares function is adequate.

3. The nonlinear activation function for the output layer should ideally be the *softmax* function as described in Bishop (1998); however, this is not mandatory.

In practice, the procedure for training the neural network should make use of multiple data sets following the guidelines in Tarassenko (1998). The data should be divided into a training set, a validation set and a testing set. The training set is used simply to establish the network weights for a given network topology (number of layers, numbers of hidden units) and training parameters (learning rate etc.). However, the network topology itself should be determined in a principled manner. This can be accomplished by training networks with different numbers of hidden units and learning rates and minimising the error over an independent data set - the validation set. As the network has now been tuned to both the training set and validation set, a third independent data set - the testing set - should then presented to this optimised network in order to arrive at a final classification error. The number of input and output neurons is usually fixed by the nature of the data and do not require optimisation.

7 Experimental Illustrations

The work discussed in this section is concerned with a programme of experimental validation for a SHM methodology based on novelty detection and neural network classification. The work is reported in considerably more detail in Worden et al. (2008); Manson et al. (2003a,b). The philosophy of the programme of work was to develop methods which are robust enough to be successful on real aircraft structures. The programme spanned a period of around three years. The structure of interest was a Gnat trainer aircraft, or more specifically, the starboard wing of the aircraft as shown in Figure 21.

The first phase of the work was concerned with level one in the damage hierarchy - novelty detection.

7.1 Level One - Damage Detection

Damaged Inspection Panels As it was not permitted to damage the aircraft, damage was effectively introduced into an inspection panel. This was accomplished by making ten copies of the panel; one was left intact and the remaining nine received controlled damage. Figure 22 shows a schematic of the damage conditions. Damage states f1, f2 and f3 were holes of diameter 20mm, 38mm and 58mm respectively. States f4, f5 and f6 were saw-cuts across the panel width with f4 an edge cut of 50mm and f5

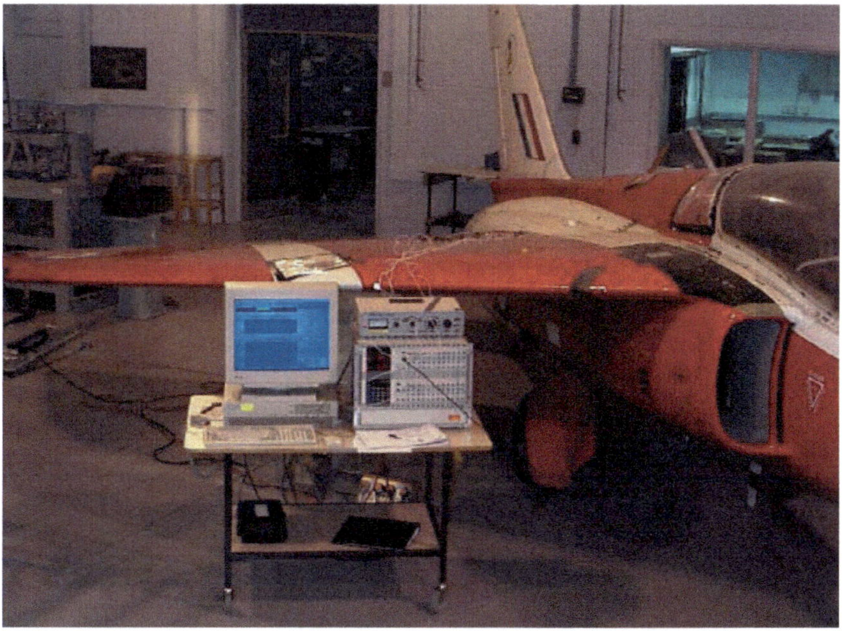

Figure 21. Gnat aircraft and acquisition system.

and f6 central cuts of extent 50mm and 100mm respectively. States f7, f8 and f9 were saw-cuts along the longer axis of the panel with f7 a 100mm edge cut and f8 and f9 central cuts 100mm and 200mm long respectively. The original point of introducing different damage types was to explore the possibility of classifying the different types and orientations. This could be, and sometimes is, added as an extra level in Rytters hierarchy (see Section 1).

Data Capture Transmissibilities were used as the base measurements from which novelty detection features would later be selected. The transmissibility between two points i and j is here defined as the ratio of the acceleration spectra measured at those points i.e. $T_{ij}(\omega) = A_i(\omega)/A_j(\omega) = Y_i(\omega)/Y_j(\omega)$, where $A_i(\omega)$ is an acceleration spectrum and $Y_i(\omega)$ is the corresponding displacement spectrum. These spectra were obtained by Fourier transforming acceleration time data obtained from piezoelectric accelerometers; appropriate windowing and averaging was employed. The reasons for the choice of transmissibility were based on their success in a previous

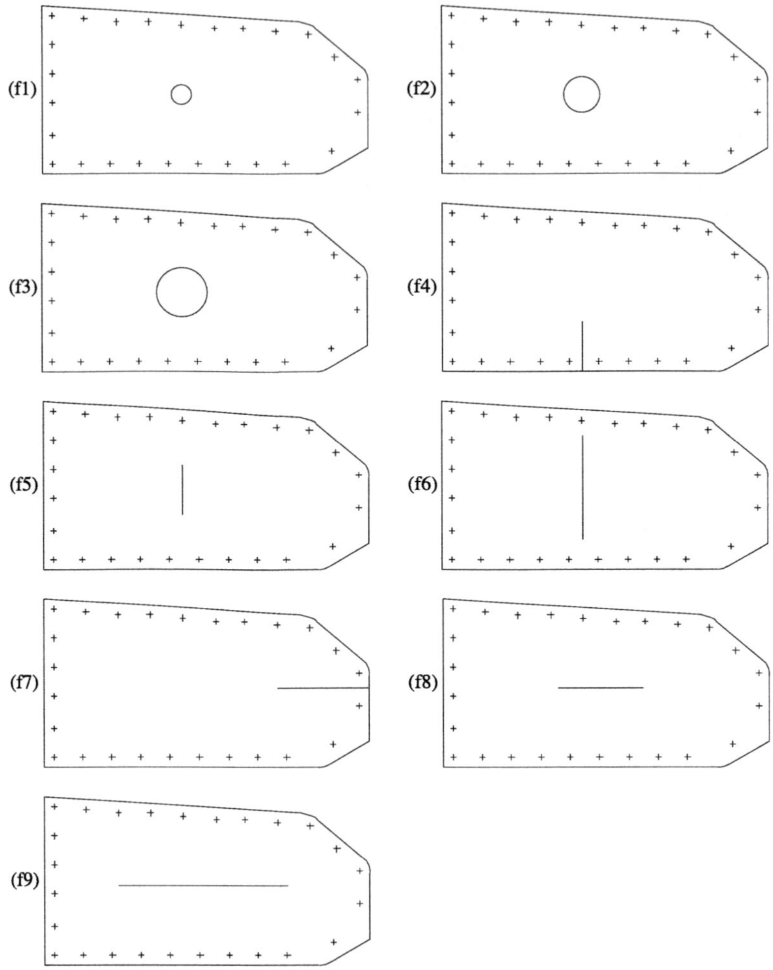

Figure 22. Different simulated damage states.

study on a laboratory structure (Worden et al. (2003)), and because they were thought to be susceptible to local changes in the region between the relevant sensors (accelerometers). The sensitivity to local changes is important for detecting small damage. Many other vibrational features are possible, like natural frequencies and modeshapes; however, many are global quantities which are not sensitive to small damage. Four sensors were used in all: one pair to establish the transmissibility across the panel in the length direction and one pair across the width (Figure 23).

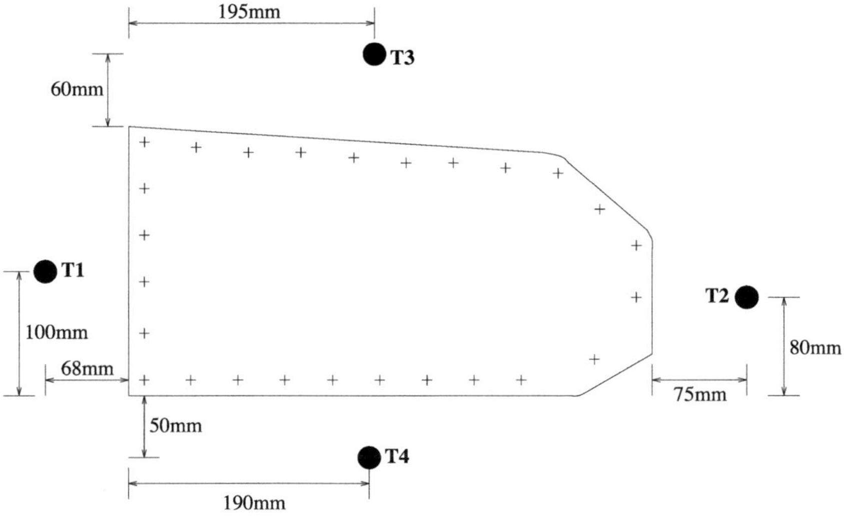

Figure 23. Panel and sensor layout.

The wing was excited with a white Gaussian excitation using an electrodynamic shaker attached directly below the inspection panel on the bottom surface of the wing. Transmissibilities were recorded in the 1-2 kHz range as this was found to be sensitive to the types of damage being investigated. In all cases 2048 spectral lines were recorded. Figure 24 shows two examples of the transmissibility (128 averages here for clarity) measured across the length of the panel area when the panel had been completely removed. (This shows the degree of variability in the measurements that is to be expected from environmental changes and instrumental drift. The degree of variability as a result of re-fixing the plate with the boundary screws was considerably higher; this is discussed later.)

For the undamaged panel a 128-average transmissibility and 110 one-shot transmissibilities (100 for the novelty detector training and 10 held

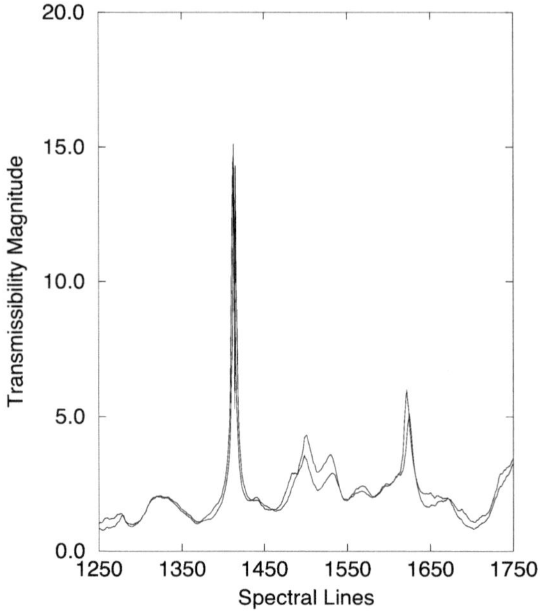

Figure 24. Examples of averaged transmissibility measurements.

back for the testing set) were obtained. Next, for each damaged panel and for the undamaged panel again (for testing purposes) a 128-average transmissibility and 10 one-shot transmissibilities were obtained. Finally, a set of measurements were recorded with the panel completely removed and, for repeatability purposes, a further four tests were carried out to obtain 128-average transmissibilities for the undamaged panel. The panel was removed and replaced between each of the latter tests. The test sequence, which identifies the purpose of each test is given in Table 1 (the labels for the fault conditions correspond to those in Figure 22).

Feature Selection In many situations there is a requirement for some pre-processing of the raw data signals before proceeding to the feature selection phase, this is represented by the Signal Processing box in Figure 1. In this study however, pre-processing of the transmissibilities was not deemed necessary.

It can be argued that the process of selecting or extracting good features is probably the most important and most difficult phase in the Data to Decision process (Section 1). Essentially, in the context of pattern recognition,

Test Label	Status	Number of Patterns	Number of Averages	Purpose
uff	Undamaged	1	128	Feature selection
uft	"	100	1	Training
uf	"	10	1	Testing
f1f	20mm diameter hole	1	128	Feature selection
f1	"	10	1	Testing
f2f	38mm diameter hole	1	128	Feature selection
f2	"	10	1	Testing
f3f	58mm diameter hole	1	128	Feature selection
f3	"	10	1	Testing
f4f	50mm width/edge cut	1	128	Feature selection
f4	"	10	1	Testing
f5f	50mm width/centre cut	1	128	Feature selection
f5	"	10	1	Testing
f6f	100mm width/centre cut	1	128	Feature selection
f6	"	10	1	Testing
f7f	100mm length/edge cut	1	128	Feature selection
f7	"	10	1	Testing
f8f	100mm length/internal cut	1	128	Feature selection
f8	"	10	1	Testing
f9f	200mm length/internal cut	1	128	Feature selection
f9	"	10	1	Testing
npf	Panel removed	1	128	Feature selection
np	"	10	1	Testing
uf2	Undamaged	10	1	Testing

Table 1. Test sequence for novelty detection experiment on Gnat.

what is meant by a feature is some set of values drawn from or calculated from the measured (or pre-processed) data. The choice of feature will depend upon the purpose of the novelty detector or classifier. For damage detection, one desires a feature that is capable of distinguishing between the undamaged and damaged states. It is obvious that a poor choice of feature will probably result in a poor novelty detector. Conversely, a good feature will often result in a successful novelty detector irrespective of the underlying method used to construct the detector.

In the case of the Gnat damage detection study here, the 128-average transmissiblities from all the undamaged and damaged cases were compared. It was found that there was a significant variability between the undamaged state transmissibilities due to the panel boundary conditions (23 screws). This raises the issue of robust features: a feature will clearly be of little value, even if it does distinguish between the damaged and undamaged states, if it results in a novelty detector which flags damage when there is merely a slight change in the boundary conditions. The issue of robustness against environmental variability will be discussed later and is highlighted elsewhere in this book (in the chapter by Kullaa).

The procedure for selecting potential features for the detection of one or more of the damage states was straightforward: each of the 128-average transmissibilities measured from the various damage conditions was compared to the five 128-average transmissibilities measured from the undamaged structure. This resulted in between one and three areas of interest being highlighted for each of the four main damage types (namely: no panel, holes, width-spanning cuts and length-spanning cuts) for each of the two transmissibilities. Figure 25 shows an example of one of these features, selected to detect length-spanning cuts in the panel. In total, ten areas of interest were highlighted from the transmissibilities across the panel length and eight from those recorded across the panel width. It is a simple matter to convert these areas of interest into feature patterns; the transmissibility function is simply sub-sampled over the required region to give an array of 50 sample points or a 50-dimensional pattern in multivariate statistics terminology. This means that there were 18 potential features in total, each of 50 dimensions.

One might argue that if data are available from the damaged and undamaged data, a better course of action would be to train a classifier. However, the intention here is to only to illustrate a detection method and unsupervised learning will suffice here. The damage state data is only used here to define novelty detectors which are likely to work well for illustrative purposes. One of the main problems with data-driven approaches, and this will be discussed in more detail later, is that data for the damage cases are rarely available. In the absence of damage data, one conceivable strategy is to define novelty detectors for a number of features and then observe them all to see if any signal damage.

Novelty Detection Once the features have been selected, the next step is to construct a novelty detector. There are many possible techniques for novelty detection; the method used for this study was outlier analysis as discussed in Section 3.1; this assumes that the training data can be represented by a multivariate Gaussian distribution. If this were not the case, a method capable of handling more complicated probability density functions such as the AANN of Section 5.1 could be used.

Novelty detectors were constructed for each of the 18 potential features discussed above. This was done using 500 artificially noise-contaminated unfaulted features. In the test programme, six separate tests were conducted under normal condition; however, although a 128-average transmissibility was obtained in each case, due to time constraints the 110 one-shot measurements were only obtained in one test (that labelled uf below). The test labelled uf2 below extracted 10 one-shot measurements. The use of five different normal condition data sets allowed robust features to be selected which were not subject to substantial variation as a result of the boundary conditions. Testing sets were constructed using the final 10 of the unfaulted

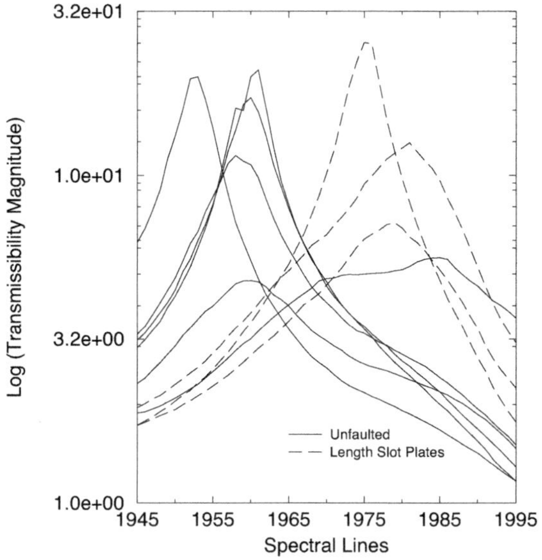

Figure 25. Novelty detection feature to detect length-spanning cuts.

patterns (denoted uf in Figure 26), followed by 10 of each fault condition f1 to f9. The testing set was completed with 10 patterns drawn from the panel removed condition (np) and 10 patterns with the unfaulted panel reattached (uf2) - see Table 1.

Four of the 18 features were capable of detecting some of the damage conditions whilst correctly classifying the 20 unfaulted patterns. The rest produced some false positives and were discarded. (In the event that no damage data were available to allow a judgement of performance and all novelty detectors were being monitored a problem arises as to which novelty detectors to trust. A voting scheme might help, but this is a difficult problem which requires further research.) Figures 26 and 27 show the novelty detector results for two of these features which, when combined, are able to detect all nine damage types and the panel removed condition while correctly returning below-threshold values for the unfaulted patterns.

All threshold values were calculated using the Monte Carlo method outlined in Section 3.1 based upon the critical value of 1% test of the discordancy.

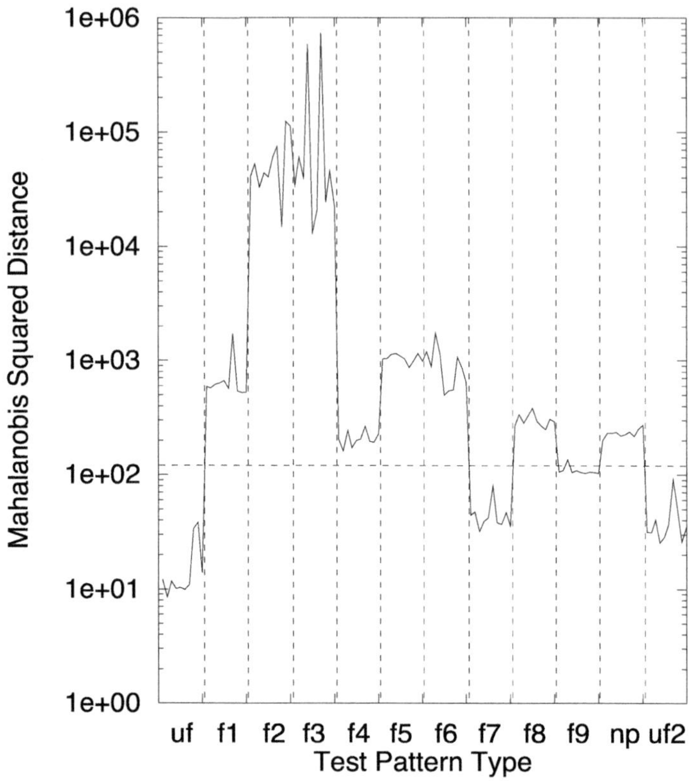

Figure 26. Outlier analysis results for feature from spectral lines 1800 to 1900 of $T_{12}(\omega)$.

7.2 Level Two - Damage Location

This phase of the work investigated the next level in Rytters damage hierarchy namely, damage location. Having detected that damage is present in the structure, there is generally a desire for further information regarding the location of the damage. This problem can be cast in a regression framework with the output being the coordinates of the damage. This was the framework used to locate impacts on a composite panel in Worden and Staszewski (2000). However, in the study here, due to the restrictions upon actually damaging the structure, the problem of damage location was changed into one of classification. As in the first phase of the study, the Gnat aircraft was the experimental structure and inspection panels on the

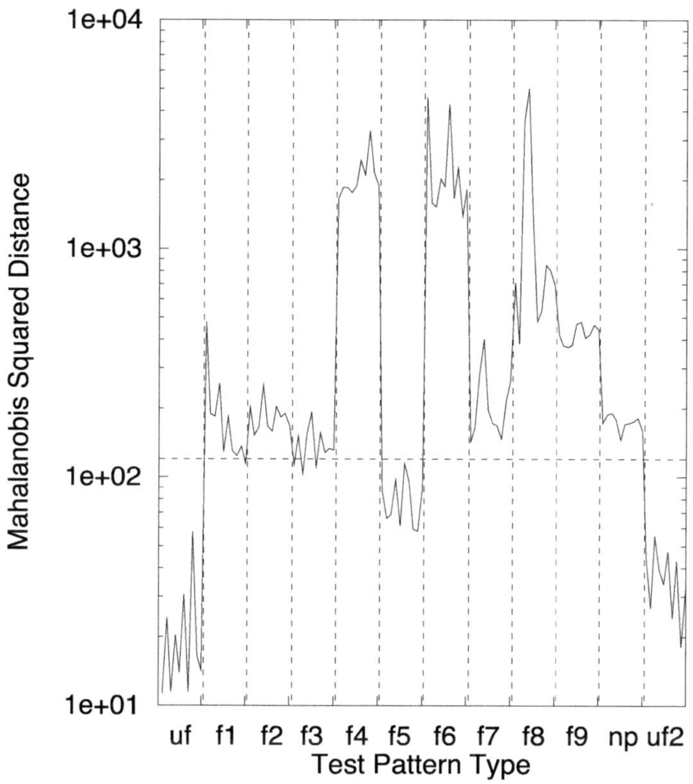

Figure 27. Outlier analysis results for feature from spectral lines 1900 to 2000 of $T_{34}(\omega)$.

starboard wing were used to introduce damage into the structure. The question on this occasion was concerned with identifying which of nine inspection panels had been removed. Although the casting of the problem in a classification framework was imposed by restrictions, it could be argued that this may be a more robust approach to the damage location problem. Consider the problem of damage in an aircraft wing: it may be sufficient to classify which skin panel is damaged rather than give a more precise damage location. It is likely that, by lowering expectations, a more robust damage locator will be the result.

Due to the success of using novelty detectors for the damage detection problem, it was decided to extend this approach to see whether it could be used for the Level Two problem. A network of sensors was used to establish a set of novelty detectors, the assumption being that each would be sensitive to different regions of the wing. Once the relevant features for each detector had been identified and extracted, a neural network was used to interpret the resulting set of novelty indices. (A further reason for using novelty indices for localisation features is that this substantially reduces the dimension of the feature vector.) As the panels were of different sizes, the analysis gave some insight into the sensitivity of the method.

Test Set-up and Data Capture As described above, damage was simulated by the sequential removal of nine inspection panels on the starboard wing: this also had the distinct advantage that each damage scenario was reversible and it would therefore be possible to monitor the repeatability of the measurements. Figure 28 shows a schematic of the wing and panels.

The area of the panels varied from about 0.008 m^2 to 0.08 m^2 with panels P3 and P6 the smallest. Transmissibilities were again used and were recorded in three groups, A, B and C as shown in Figure 29. Each group consisted of four sensors (a centrally placed reference transducer and three others). Only the transmissibilities directly across the plates were measured in this study. The excitation and recording of the transmissibilites were conducted in the same manner as during the first phase. One 16-average transmissibility (for feature selection) and 100 one-shot measurements (for training and testing) were recorded across each of the nine panels for the seven undamaged conditions (to increase robustness against variability caused by boundary conditions) and the 18 damaged conditions (two repetitions for the removal of each of the nine panels). The overall test sequence was:

1. Normal condition (all plates in place).
2. Plate P1 removed.
3. Plate P2 removed.
4. Plate P3 removed.
5. Normal condition.
6. Plate P1 removed.
7. Plate P2 removed.
8. Plate P3 removed.
9. Normal condition.
10. Plate P4 removed.
11. Plate P5 removed.
12. Plate P6 removed.
13. Normal condition.
14. Plate P4 removed.
15. Plate P5 removed.
16. Plate P6 removed.
17. Normal condition.

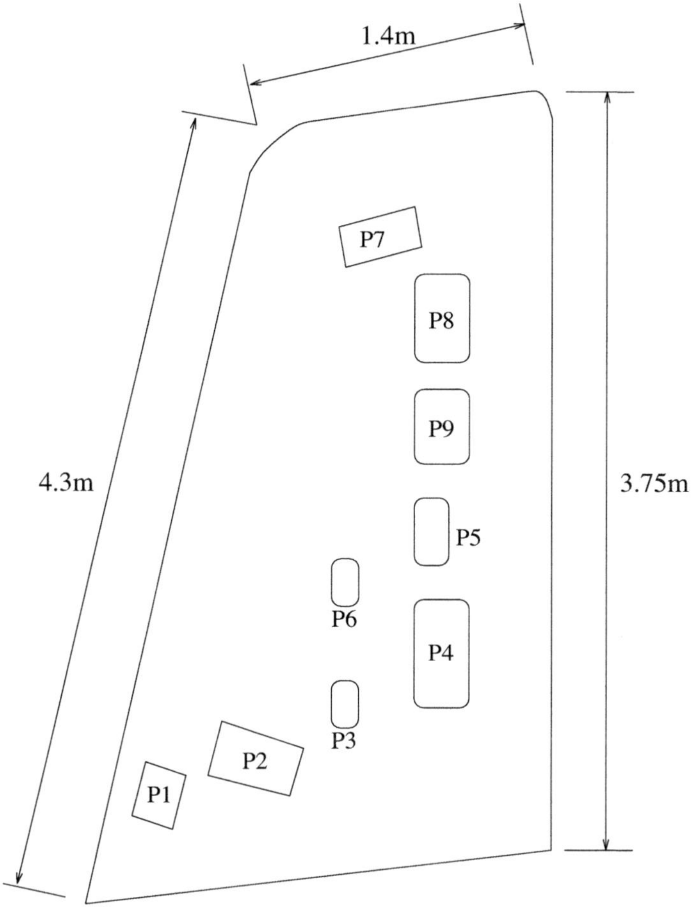

Figure 28. Schematic of the starboard wing inspection panels.

18. Plate P7 removed.
19. Plate P8 removed.
20. Plate P9 removed.
21. Normal condition.
22. Plate P7 removed.
23. Plate P8 removed.
24. Plate P9 removed.
25. Normal condition.

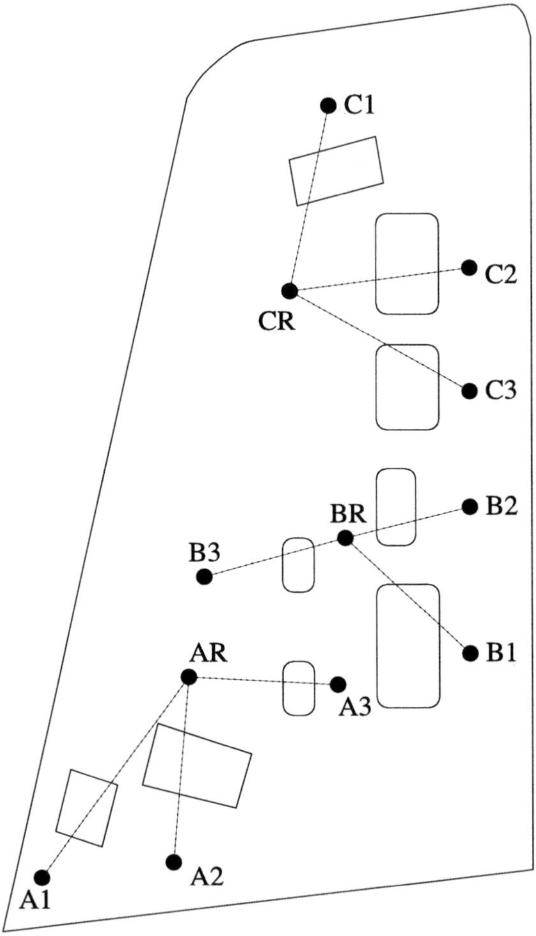

Figure 29. Schematic of the starboard wing transducer locations.

Feature Selection and Novelty Detection The feature selection process was essentially conducted in the same manner as previously with the only difference being a visual classification of potential features as weak, fair or strong. In order to simplify matters, only the group A transmissilibilities were considered to construct features for detecting the removal of the group A panels; similarly for groups B and C.

Candidate features were then evaluated using outlier analysis. The best features were chosen according to their ability to correctly identify the 200 (per panel) damage condition features as outliers while correctly classifying those features corresponding to the undamaged condition as inliers. Figure 30 shows the results of the outlier analysis for the feature that was designed to recognise removal of inspection panel 4. The data are divided into training, validation and testing sets in anticipation of presentation to the neural network classifier. As there are 200 patterns for each damage class, the total number of patterns is 1800. These were divided evenly between the training, validation and testing sets, so (with a little wastage) each set received 594 patterns, comprising 66 representatives of each damage class. The plot shows the discordancy values returned by the novelty detector over the whole set of damage states. The horizontal dashed lines in the figures are the thresholds for 99% confidence in identifying an outlier, they are calculated according to the Monte Carlo scheme described in Section 3.1. The novelty detector substantially fires only for the removal of panel 4, for which it has been trained. This was the case for most panels but there were exceptions (e.g. there were low sub-threshold discordancies for the smaller panels and some novelty detectors were sensitive to more than one damage type).

Network of Novelty Detectors for Damage Location The final stage of the analysis was to produce a damage location system. The algorithm chosen was a standard Multi-Layer Perceptron (MLP) neural network as described in detail in Bishop (1998). The neural network was presented with 9 novelty indices at the input layer and required to predict the damage class at the output layer.

Note that there are now two layers of feature extraction. At the first level, certain ranges of the transmissibilities were selected for sensitivity to the various damage classes. These were used to construct novelty detectors for the classes. At the second level of extraction, the 9 indices themselves are used as features for the damage localisation problem. This depends critically on the fact that the various damage detectors are local in some sense, i.e., they do not all fire over all damage classes. This was found to be true in this case.

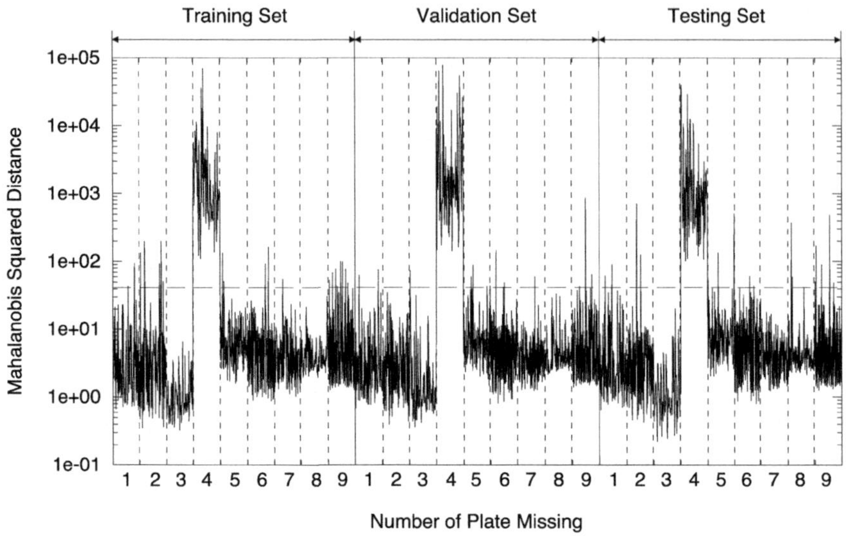

Figure 30. Outlier statistic for all damage states for the novelty detector trained to recognise panel 4 removal.

The procedure for training the neural network followed the guidelines in Tarassenko (1998). The data were divided into a training set, a validation set and a testing set. The training set was used to establish weights, whilst the network structure and training time etc. were optimised using the validation set. The testing set was then presented to this optimised network to arrive at a final classification error. For the network structure, the input layer necessarily had nine neurons, one for each novelty index, and the output layer had nine nodes, one for each class. A single hidden layer was assumed, as it is known that such networks are universal approximators (Cybenko (1989)).

In terms of pseudo-code, the training strategy was:

```
for number of hidden layer neurons = 1 to 50
    {
    for different random initial conditions = 1 to 10
        {
        train network on training data
        evaluate on validation data
        terminate training at minimum in validation set error
        }
    }
```

The training phase used Tarassenko's implementation of the *1 of M strategy* as discussed earlier. The best network had 10 hidden units and resulted in a testing classification error of 0.135 i.e. 86.5% of the patterns were classified correctly. The confusion matrix is given in Table 2. The main errors were associated with the two small panels P3 and P6 and the panels P8 and P9 whose novelty detectors sometimes fired when either of two panels was removed.

Prediction	1	2	3	4	5	6	7	8	9
True Class 1	62	1	0	0	2	0	0	1	0
True Class 2	0	61	0	0	5	0	0	0	0
True Class 3	0	1	52	0	7	4	0	2	0
True Class 4	1	0	3	60	0	1	0	1	0
True Class 5	2	1	0	0	60	3	0	0	0
True Class 6	2	0	6	0	8	52	0	0	0
True Class 7	1	0	4	0	1	1	58	1	0
True Class 8	0	0	0	0	1	1	0	62	2
True Class 9	2	1	1	0	0	0	0	15	47

Table 2. Confusion matrix from best neural network: testing set

Note that these results follow from a rather subjective approach to selecting features. Although a discussion of the method is beyond the scope of the current article, it is possible to select features by an optimisation procedure, Worden *et al.* used a genetic algorithm to select the 9 best features from the set of 44 candidates for the location problem above. When these features were used to train a neural network classifier as above, the confusion matrix of Table 3 resulted.

This corresponds to a classification rate of 98.1% or alternatively, only 11 misclassifications on the testing set of 594 patterns. This sort of performance

Prediction	1	2	3	4	5	6	7	8	9
True Class 1	65	0	0	0	0	0	0	0	1
True Class 2	0	65	0	1	0	0	0	0	0
True Class 3	1	0	62	0	0	1	0	1	1
True Class 4	0	0	0	66	0	0	0	0	0
True Class 5	0	0	0	0	66	0	0	0	0
True Class 6	0	3	0	0	0	62	0	1	0
True Class 7	0	0	0	0	0	0	66	0	0
True Class 8	1	0	0	0	0	0	0	65	0
True Class 9	0	0	0	0	0	0	0	0	66

Table 3. Confusion matrix from neural network trained with optimised features: testing set

is approaching a level where one might trust a neural network for automated SHM.

8 Discussion and Conclusions

There are many lessons to be learned from the examples in the previous sections. The overwhelming message is that an SHM system based on machine learning can only perform as well as the data that has been used to train the diagnostic, the adage *garbage in - garbage out* is particularly apt.

The most important question for SHM based on machine learning is already raised at the lowest level of detection; how does one acquire data corresponding to any damage states? The reason that this question is already pertinent at the detection level is that it is necessary to decide features that distinguish between the normal condition of the system or structure and the damaged conditions, and this is not possible without examples of the damage conditions. In the examples in Section 7 above, the features were selected regions of certain transmissibility functions. In the absence of examples from the damage cases it was not possible to assess if a given transmissibility peak was sensitive to a given type of damage, or in fact any type of damage. In the case of damage detection, this problem can potentially be overcome by training novelty detectors for each candidate feature and then monitoring all of them for threshold crossings on new data. This would be tedious but effective. In the case of a damage location system, data for each class of damage becomes essential. This can only be acquired in two ways, by modelling or from experiment. Both approaches have potential problems. If one considers modelling, one must hope that a low-cost model should suffice otherwise one simply invests all the effort that a model-driven approach like FE updating would require anyway. If one considers experiment, it will not generally be possible to accumulate data by damaging the structure in the most likely ways unless the structure is extremely cheap and mass-produced. It is obvious that a testing programme based on imposed damage could not be used on say, an aircraft wing. If one cannot impose real damage, one might be able to experimentally simulate the effects of damage.

The second major problem in damage identification was also raised in Section 7. Without careful feature selection, the variations in the measured data due to boundary condition changes in the structure swamped the changes due to damage. (This is also a problem for model-based approaches.) This is an observation that is particularly pertinent for Civil Engineering. If one wishes to carry out a program of automatic monitoring for an aircraft, it is conceivable that one might do it off-line in the reasonably well-controlled environment of a hangar. This is not possible for a bridge that is at the mercy of the elements. It is known that changes in the natural frequencies of a bridge as a result of daily temperature variation are likely to be larger than the changes from damage (Sohn (2007)). Bridges will also have a varying mass as a result of taking up moisture from rain etc. There are two possible solutions to this problem. The first is to accumulate normal condition data spanning all the possible environmental conditions. This is time-consuming and will generate such a large normal condition set that it is likely to be insensitive to certain types of damage. The second solution is to determine features that are insensitive to environmental changes but sensitive to damage. This of course raises the first problem discussed above, where is the damage-state data coming from?

A third problem relating to data-driven approaches is that the collection or generation of data for training the diagnostic is likely to be expensive, this means that the data sets acquired are likely to be sparse. This puts pressure on the feature selection activities as sparse data will usually require low-dimensional features if the diagnostic is ever going to generalise away from the training set. There are possible solutions to this, e.g. regularisation can be used in the training of neural networks in order to aid generalisation and this can be as simple as adding noise to the training data. Other possibilities are to use learning methods like Support Vector Machines (Cristianini and Shawe-Taylor (2000)) which are implicitly regularised and therefore better able to generalise on the basis of sparse data.

One issue which applies equally to data-driven and model-driven approaches is that they are more or less limited to Levels One to Three in

Rytters hierarchy. If one is to pursue damage prognosis, it is necessary to extrapolate rather than interpolate and this is a problem for machine learning solutions. If prognosis is going to be possible, it is likely to be very context specific and to rely critically on understanding the physics of the damage progression. In certain simple cases, it is already possible to make calculations. For example, for a crack in a metallic specimen with a simple enough geometry to allow the theoretical specification of a stress intensity, one can use the Paris-Erdogan law to predict the development of the crack given the loading history (or rather future). Even here there are problems. First of all, the loading future is uncertain and it may only be possible to specify bounds. Secondly, the constants of the Paris-Erdogan equation are strongly dependent on microstructure and would probably have to be treated as random variables in a given prediction. These observations are intended to show that prognosis is only likely to be possible in the framework of a statistical theory where the uncertainty in the calculation is monitored at all stages. Another major stumbling block in the application of prognosis is that most realistic situations will not be backed up by applicable theory, i.e. the laws of damage progression are not known for simple materials with complicated geometry or for complicated materials like composite laminates even with simple geometries.

The overall conclusion for this chapter is that if the conditions are favourable, machine learning algorithms can be applied to great effect on damage identification problems. In the light of the comments above, favourable conditions' largely means that data are available in order to train the machine learning diagnostics. Even if the conditions seem to exclude such a solution, one should bear in mind that even a model-driven approach will need appropriate data for model validation.

Acknowledgements

The author would like to thank a number of colleagues for contributing to the work presented here, they are: Dr Graeme Manson and Professor Wieslaw Staszewski of the University of Sheffield; Dr Cecilia Surace of the Politecnico di Torino, Italy and Dr Charles (Chuck) Farrar of the Los Alamos National Laboratory, USA. The author would also like to acknowledge his great debt to the late Dr David Allman, who essentially made most of the experimental work presented here possible.

Bibliography

V. Barnett and T. Lewis. *Outliers in Statistical Data.* John Wiley and Sons, Chichester, 1994.

C.M. Bishop. *Neural Networks for Pattern Recognition.* Oxford University Press, 1998.

C.M. Bishop. Novelty detection and neural network validation. *IEEE Proceedings - Vision and Image Signal Processing*, 141:217–222, 1994.

N. Cristianini and J. Shawe-Taylor. *An Introduction to Support Vector Machines and Other Kernel-Based Learning Methods.* Cambridge University Press, 2000.

G. Cybenko. Approximation by superpositions of sigmoidal functions. *Mathematics of Control, Signals and Systems*, 2:303–314, 1989.

S.W. Doebling, C.R. Farrar, M.B. Prime, and D. Shevitz. Damage identification and health monitoring of structural and mechanical systems from changes in their vibration characteristics. Technical Report LA-13070, Los Alamos National Laboratory, Los Alamos, New Mexico, USA, 1996.

M.I. Friswell and J.E.T. Penny. Is damage detection using vibration measurements practical? In *Proceedings of 2nd International Conference on Structural Damage Assessment using Advanced Signal Processing Procedures (DAMAS 97), Sheffield, UK*, pages 351–362, 1997.

D.O. Hebb. *The Organisation of Behaviour.* John Wiley, New York, 1949.

D. Lowe. Feature extraction, data visualisation, classification and fusion for damage assessment. In *Oral Presentation at EPSRC SIDANet Meeting, Derby, UK*, 2000.

D. Lowe and C. Zapart. Point-wise confidence interval estimation by neural networks: a comparative study based on automotive engine calibration. *Neural Computing and Applications*, 8:77–85, 1999.

G. Manson, K. Worden, and D.J. Allman. Experimental validation of a structural health monitoring methodology. part ii: novelty detection on an aircraft wing. *Journal of Sound and Vibration*, 259:345–363, 2003a.

G. Manson, K. Worden, and D.J. Allman. Experimental validation of a structural health monitoring methodology. part iii: damage location on an aircraft wing. *Journal of Sound and Vibration*, 259:356–285, 2003b.

M. Markou and S. Singh. Novelty detection a review. part i: statistical approaches. *Signal Processing*, 83:2481–2497, 2003a.

M. Markou and S. Singh. Novelty detection a review. part ii: neural network based approaches. *Signal Processing*, 83:2499–2521, 2003b.

W.S. McCulloch and W. Pitts. A logical calculus of the ideas imminent in nervous activity. *Bulletin of Mathematical Biophysics*, 5:115–133, 1943.

M.L. Minsky and S.A. Papert. *Perceptrons.* MIT Press, 1988.

D.A. Pomerleau. Input reconstruction reliability information. In *Advances in Neural Information Processing Systems, Volume 5*. Morgan-Kaufman, 1993.

S. Roberts and L. Tarassenko. A probabilistic resource allocating network for novelty detection. *Neural Computation*, 6:270–284, 1994.

F. Rosenblatt. *Principles of Neurodynamics*. Spartan, New York, 1962.

D.E. Rumelhart and J.L. McClelland. *Parallel Distributed Processing: Explorations in the Microstructure of Cognition*. MIT Press, 1988.

D.E. Rumelhart, G.E. Hinton, and R.J. Williams. Learning representations by back propagating errors. *Nature*, 323:533–536, 1986.

A. Rytter. *Vibration Based Inspection of Civil Engineering Structures*. Phd thesis, University of Aalborg, Denmark, Department of Building Technology, 1993.

R. Schalkoff. *Pattern Recognition: Statistical, Structural and Neural Approaches*. John Wiley and Sons, Singapore, 1992.

H. Sohn. Effects of environmental and operational variability on structural health monitoring. *Philosophical Transactions of the Royal Society A: Physical and Engineering Sciences*, 356:539–560, 2007.

H. Sohn and C.R. Farrar. Statistical process control and projection techniques for damage detection. In *Proceedings of European COST F3 Conference on System Identification and Structural Health Monitoring, Madrid, Spain*, pages 105–114, 2000.

C. Surace and K. Worden. Some aspects of novelty detection methods. In *Proceedings of 3rd International Conference on Modern Practice in Stress and Vibration Analysis, Dublin, Eire*, pages 89–94, 1997.

L. Tarassenko. *A Guide to Neural Computing Applications*. Arnold, 1998.

L. Tarassenko, P. Hayton, Z. Cerneaz, and M. Brady. Novelty detection for the identification of masses in mammograms. In *Proceedings of 4th International Conference on Neural Networks, Cambridge, UK*, volume 409, pages 442–447. IEE Publications, 1997.

O. Taylor, J. MacIntyre, C. Isbell, C. Kirkham, and A. Long. Adaptive fusion devices for condition monitoring: Local fusion systems of the neural-maine project. In *Proceedings of the 3rd International Conference on Damage Assessment of Structures (DAMAS 00), Dublin, Eire*, pages 205–216, 1999.

P.J. Werbos. *Beyond Regression: new tools for prediction and analysis in the behavioural sciences*. Phd thesis, Applied Mathematics, Harvard University, 1974.

K. Worden. Structural fault detection using a novelty measure. *Journal of Sound and Vibration*, 201:85–101, 1997.

K. Worden and G. Manson. Damage identification using multivariate statistics: kernel discriminant analysis. *Inverse Problems in Engineering*, 8: 25–46, 2000.

K. Worden and W.J. Staszewski. Impact location and quantification on a composite panel using neural networks and genetic algorithms. *Strain*, 36:61–70, 2000.

K. Worden, G. Manson, and N.R.J. Fieller. Damage detection using outlier analysis. *Journal of Sound and Vibration*, 229:647–667, 2000.

K. Worden, G. Manson, and D.J. Allman. Experimental validation of a structural health monitoring methodology. part i: novelty detection on a laboratory structure. *Journal of Sound and Vibration*, 259:323–343, 2003.

K. Worden, C.R. Farrar, G. Manson, and G. Park. The fundamental axioms of structural health monitoring. *Proceedings of the Royal Society Part A*, 463:1639–1664, 2007.

K. Worden, G. Manson, G. Hilson, and S.G. Pierce. Genetic optimisation of a neural damage locator. *Journal of Sound and Vibration*, 309:529–544, 2008.

9 A Little Probability Theory

In general, random variables will not all be mutually independent; the probability of an event A may well depend on the previous or simultaneous occurence of an event B. A is said to be *conditioned* on B. The need to incorporate this type of dependence into the theory results in the definition of the *conditional probability* $P(A|B)$, which is *the probability that A will occur given that B already has.* Note that the unconditional $P(A)$ can still be computed but will have a different value. A definition can be made via fairly intuitive reasoning.

Suppose that N experiments are conducted, one can define the conditional probability $P(A|B)$ as,

$$\frac{N(A \cap B)}{N(B)}$$

where $N(A \cap B)$ is the number of times A occurs when B occurs and $N(B)$ is the total number of times B occurs. This equation can be rewritten as,

$$\frac{N(A \cap B)/N}{N(B)/N} = \frac{P(A \cap B)}{P(B)}$$

so that,

$$P(A \cap B) = P(A|B)P(B)$$

where the RHS sort of decomposes the probability into two independent-looking events.

Interchanging variables in the defining equation gives,

$$P(B|A) = \frac{P(A \cap B)}{P(A)}$$

and combining the two equations produces *Baye's theorem*

$$P(A|B)P(B) = P(B|A)P(A) \tag{22}$$

or,

$$P(A|B) = \frac{P(B|A)P(A)}{P(B)}$$

Now suppose that A is actually conditioned on a continuous random variable X. What is $P(A|X = x)$? Consider first the probability that the value taken by X lies in a small interval $[x, x + \Delta x]$. From Bayes Theorem,

$$P(A|X \in [x, x + \Delta x])P(X \in [x, x + \Delta x]) = P(X \in [x, x + \Delta x]|A)P(A)$$

$$\Longrightarrow P(A|X \in [x, x + \Delta x]) \int_x^{x+\Delta x} p(z)dz = \int_x^{x+\Delta x} p(z|A)dz \; P(A)$$

where the conditional probability density function (PDF) is defined in the usual way. If Δx is small, the integrals are well approximated by assuming a rectangular area under the curves and

$$P(A|X \in [x, x + \Delta x])p(x)\Delta x \approx p(x|A)\Delta x P(A)$$

with the approximation getting better as Δx gets smaller. In fact, in the limit as $\Delta x \longrightarrow 0$

$$P(A|X = x)p(x) = p(x|A)P(A)$$

or,

$$P(A|x) = \frac{p(x|A)P(A)}{p(x)} \tag{23}$$

in an obvious shorthand. This last equation is the basis of the Bayesian approach to pattern recognition described in the main body of the text. $P(A)$ is called the *prior probability* of A i.e. before one sees the data or evidence x. The term $p(x|A)$ is called the *likelihood* and $P(A|x)$ is called the *posterior probability* i.e. after one has seen the evidence.

Elastic Waves for Damage Detection in Structures

Wieslaw Ostachowicz* and Pawel Kudela
* Faculty of Navigation,
Gdynia Maritime University, Gdynia, Poland
‡ Department of Mechanics of Intelligent Structures,
Polish Academy of Sciences, Gdansk, Poland

Abstract The main issues regarding damage detection in elements of structures are discussed in the particular case that the detection is conducted by the use of methods based on the phenomenon of elastic wave propagation. The emphasis is placed on modelling the phenomenon of elastic wave propagation in composite elements of structures, along with issues of wave interactions with damage and problems of damage location.

1 Introduction to SHM Methods Based on the Phenomenon of Elastic Wave Propagation

The scope of *Structural Health Monitoring* (SHM) includes constant monitoring of the structure's material condition (in real-time), for the elements of the structure as well as for the whole structure during its useful lifetime. The condition of the structures material is required to remain within the limits specified by the standards of the design process. Those standards, regarding the material, ought to take into consideration changes caused by standard exploitation wear during the operation process, changes caused by environmental conditions, in which the structure is being used, and coincidental situations influencing the condition of the material. Owing to the fact that the monitoring process is being conducted continuously during operation, there will be a record of the complete history of utilisation. Such information may be used for future condition prognosis as well as prediction of faults and the structure's safe utilisation time.

According to a number of publications, SHM is being defined as a new approach to non-destructive inspection of a structure (also called Non-Destructive Testing/Evaluation - NDT/E). The innovation of this approach is based on the continuous monitoring of the material's condition during the exploitation process of the device. It requires certain structural and utilisation solutions distinctive for this new discipline.

SHM is combined of such disciplines as sensorics, IT, electronics (especially microprocessor technology), mechanics and materials engineering. The effect of synergy is achieved by combining former disciplines, and it enables raised safety levels of the structure's utilisation along with lowering of the maintenance expenses.

Systems executing SHM processes ought to be structure integrated; this allows insertion of changes into the structure in such a way that the probability of a failure is minor. It also enables minimisation of the failure risk through management of the structure's utilisation and treating it as part of a bigger system. The first layer of a SHM system is the monitoring layer specified by the type of physical phenomenon that is being monitored by the sensors. It is dependent on the damage type to be detected and the type of physical phenomenon that is being used by the sensors in order to generate the signals (mostly electrical) containing features and processable information regarding damage. Several (perhaps up to a few dozen) connected sensors can work together in a system measuring environmental factors influencing the condition and process of the exploitation of the structure. Data gathered from all the sensors along with historical data from previous structures allow diagnostic synthesis of information (*signal fusion*) regarding the condition of the structure. Once the above-mentioned information is linked with all the data from the general system of knowledge about the phenomenon of damage and structural wear, it is possible to gain prognosis of condition and data defining the scope of any necessary repair. It is now common for such purposes that simulation systems are used; such systems enable extremely quick generation of results, similar to those obtained from the chain of sensors based on familiar damage models (virtual exploitation of the structure).

The motivation for applying such systems is:

- ability to avoid failures with catastrophic consequences;
- ability to optimise the utilisation process (minimisation of emergency stoppage time);
- gaining essential information for designers regarding structural modification;
- ability to minimise maintenance costs and to raise the efficiency of a device thanks to the use of a methodology of repair according to condition, as well as avoiding disassembly, and replacement of non-damaged and non-used elements;
- ability to avoid operator mistakes regarding evaluation of the condition of the structure.

The systems mentioned are being used for air force and aviation, military equipment, construction industry infrastructure, and machines crucial for industry (e.g. power industry, chemistry, etc.). An extremely important factor influencing the common use of SHM systems is the economical factor. The justification may be found in several papers and it lies in comparing maintenance costs with the efficiency of the structure. For the structure without SHM systems installed, costs rise along with the utilisation time and at the same time efficiency drops. Installation of SHM systems potentially enables one to fix maintenance costs with equally fixed efficiency of the structure. However, one condition of applying SHM systems must be met; this condition limits the general ability of putting them into practice i.e. the cost of the system itself ought to be lower than the positive economical effect connected with its application.

The necessity of reducing the cost of SHM systems is nowadays connected with the application of intelligent materials and structures; they enable integration of the structure and the built-in sensors into one system. In order to be effective, such actions must be taken during the stage of designing of the structure.

From the beginning of the 80's, a tendency towards intelligent structure applications are were observed, especially in aviation and construction industry. Their characteristic feature is adaptation of those structures to the exploitation conditions. In intelligent structures, this adaptation takes place autonomically. Within the range of intelligent structures, distinctions can be made as follows: structures sensitive to utilisation conditions, structures controlled within the range of their properties, and auto-adaptative structures that adjust their properties to their utilisation needs. In practice, homogeneous materials commonly used in structures are being replaced by composite materials or other multi-materials (materials composed of layers of various physical properties). Within the range of materials and intelligent structures one can distinguish structures of adjustable geometry (shape), structures with adjustable vibration behaviour and structures with adjustable condition. In particular, the last type of structure is constantly in use with SHM techniques. Most often, intelligence is expressed through structure-integrated sensors made from intelligent materials (*embedded sensors*) or executive modules (*embedded actuators*), for which the task is to identify defects and alleviate the effects of failure. Operation of such modules depends on generating deformations of the structure in such a way as to decrease stresses in these areas of concentration. Nowadays, the search for phenomena, and methods of their measurement, which enable continuous monitoring of structural condition through monitoring of the condition of its material, continues.

Research on the development of SHM systems is very often inspired by discoveries in the fields of biology and living organisms (*biomimetics*). Very similar research is being conducted in the fields of medicine and SHM method development. SHM systems find application not only in the lifetime of the structure but also in the time of its production, transport and installation. They also enable proper management of the structure's wear, through suitable choice of missions that must be executed, and service actions required for safe meeting of given criteria. Methods that could be used as the most economically effective and durable, in every stage of the product usage, are still being searched for.

To put it briefly, the foregoing survey regarding SHM allow one to say that it is a new interdisciplinary area, gathering such sciences as mechanics, materials engineering, electronics, computer science, physics, optics and many others. This area has applications in the utilisation of structures in aviation, construction industry, motor industry and power industry including those connected with nuclear power. The spectrum of applications is constantly expanding. Among the many methods used in SHM systems, one can distinguish between active or passive methods. In the passive methods, one can observe signals generated by the structure's inbuilt sensors and on this basis, the condition can be evaluated. Active methods depend on forcing disturbances by the use of properly inbuilt executive cores that cause structural responses, measured by sensors. On the basis of response signals registered by the sensors, the condition is evaluated. Some of the widespread SHM methods are those based on structural vibration measurement. Among them: symptomatic methods, in which the symptoms of damage are certain estimators from the signals of structural response, or model-based methods, where the symptoms of damage are parameter changes or changes to the structure of a model.

Many methods used in practice to build SHM systems are being adopted from widely-known and applied disciplines e.g. NDT. Classical NDT methods can be executed continuously, such as: measurement of acoustic emission, Lamb waves, temperature, or mechanical impedance or direct monitoring of the displacement field with the use of visual methods; all these are examples of applying NDT techniques in SHM systems. In those kinds of solution, two sets of methods may be distinguished: methods where the sensors are integrated with the structure and non-contacting methods. The latter have wider application owing to miniaturisation and the costs of SHM system installation.

One of such methods, that can be applied passively as well as actively, and where the measurements are executed in a non-contacting way, is the method based on examination of thermo-flexibility phenomena accompanying damage, especially its formation and propagation. Currently, interest in this method is increasing due to the opportunity of non-contact measurements of thermal phenomena within structures and intensive development in temperature measurement. Another method is the method of surface Lamb wave excitation within the structure. This method uses a grid of sensors/actuators. Registration and processing of transmitted waves as well as reflected waves is conducted. Evaluation of the condition of the structure takes place on the basis of wave profile deformations due to damage in the interrogated area of the structure.

A different method is one based on parameters of modal models of the monitored structure. One very effective method is the modal filter method and the statistical evaluation of detected changes in the model. More and more common, is the application of scanning pictures for evaluation of structural deformation. In this way, one may monitor static as well as dynamic changes. Depending on measurement requirements, methods using laser beam are applied (strain methods, holographic methods, and interferometry methods characterised by nanometer sensitivity of measurements of displacement fields) or non-coherent light methods (fotogrametric methods, the picture correlation method, moiré pattern techniques and pattern projection techniques). The most commonly-used technique of picture processing is correlation of images of non-deformed and deformed structures. Contemporary techniques of image measurement and recording enable monitoring of even minor changes in the condition of the structure owing to applications of the so-called phase methods of analysis of pattern images or subpixel techniques in fotogrametric methods or methods of picture correlation. Contemporary quick cameras enable recording of dynamic changes of deformation with frequencies up to 32000 Hz. Pulse laser sources of light enable transmission of highly sensitive interferencial methods from the lab, directly to the studied structure. In new structures, where utilisation safety is of the greatest importance, sensors in the form of intelligent materials, piezoelectrics and optical fibres (*fibre Bragg gratings*), are currently embedded; they become an inseparable part of the structure and continuously gather information about the structure's fatigue limit and condition of the material. This enables prediction of properties, estimation of time of safe utilisation and evaluation of planned repair ranges. As presented in the review of current knowledge, investigations conducted using SHM techniques are more and more precise and enable evaluation of condition within the range of local changes, especially, in the degradation of structural material. This enables more and more accurate predictions of the condition of the structure during its lifetime.

2 Modeling of Structural Stiffness Loss Due to Damage

Fatigue cracking and delamination are particularly dangerous and, at the same time, the most common kinds of damage in elements of machines and structures. It is of great importance for safe operation to ensure that elements of machines and structures are free of any fatigue cracks and delaminations and in the case of their presence, to determine their extent. Since existing non-destructive methods for detection of fatigue cracks and delaminations fail in many practical cases, vibration methods in diagnosis of such damage have been continuing for nearly twenty years. These methods are based on diagnostic relations between the size and location of failures and changes in some dynamic characteristics of constructional elements. In order to establish such relations and to identify changes of the dynamic characteristics, efficient models that facilitate the assessment of the influence of fatigue cracks and delaminations must be established. A review of the existing models used for analysis of the influence of fatigue cracks and delaminations on changes in dynamic characteristics of constructional elements is presented in this section.

2.1 Discrete models

In general, discrete models of fatigue damage are not restricted geometrically. Such restrictions are one of the biggest disadvantages of the continuous or discrete–continuous models. In order to create a discrete model of a constructional element with a fatigue crack, the Finite Element Method (FEM) is most usually applied. Although other methods like the boundary element method, graph method, transition matrix method and the analogue method are also used, these methods are not as popular and commonly used as the finite element method.

The simplest method applied to model constructional elements with fatigue damage is based on the use of classical finite elements. In this case a fatigue crack in the finite element is modelled by reduction of elastic coefficients of the element (Cawley and Adams, 1979), by reduction of its Young's modulus (Yuen, 1985), and by reduction of the cross–sectional area of the element at the crack position (Bachschmid et al., 1984). The main disadvantage of these approaches is the fact that the reduced parameters describing a fatigue crack are chosen arbitrarily. Generally, their values are not directly related to the actual size of a crack and due to that fact, a precise study of the influence of the crack depth on changes in dynamic characteristics cannot be made. The singular character of the stress and strain fields around the crack tip is also neglected in these methods.

In the last decade, new FEM-based models have been formulated. Some authors have assumed that the failure appears inside a special finite element (Ostachowicz and Krawczuk, 2001). A model of a truss finite element with an open one–sided transverse crack has been developed by Krawczuk (1992). Models of beam finite elements with fatigue cracks of different types can be found in the work of Haisty and Springer (1988), Gounaris and Dimarogonas (1988), Chen and Chen (1988). Krawczuk and Ostachowicz (1993b) investigated a mathematical, FEM-based model of a beam with a crack, loaded at the end with a constant tensile axial force; the authors assumed that the crack does not propagate and remains open during the beam's vibrations. Assumption of a complete opening of the crack in this case was correct because the beam was subjected to the action of a constant axial force.

Ostachowicz and Krawczuk (1992) also developed a model of a rotor shaft of constant cross–section with a crack. The shaft was modelled by finite elements; the crack was considered to be open. The stiffness matrix for the element with the crack was formulated. The model took into consideration the torsional–bending interaction in the rotor vibration.

The curved-beam finite element with a transverse, one-edged, nonpropagating, open crack has been investigated by Krawczuk and Ostachowicz (1997). The authors presented an analysis of the effect of the crack position and location on the changes of the in–plane natural frequencies and mode shapes of the clamped–clamped arch. The authors assumed that the crack only changes the stiffness of the element, with the mass of the element remaining unchanged. The investigated model of the cracked element was restricted to curved beams with rectangular cross–section.

A cracked-beam finite element that is based on elasto–plastic fracture mechanics has been formulated by Krawczuk et al. (2000, 2001). Crack tip plasticity at the cracked cross-section was included in the model of the local flexibility. The inertia and stiffness matrices took into account the effect of flexural bending deformation due to the crack presence; they were formulated in closed form.

Apart from one–dimensional models, special models of two or three–dimensional constructional elements with fatigue cracks have been also investigated. The cracks occurring in a plate can be modelled by the finite element method in various ways. Plate finite elements with fatigue cracks have been used by Qian et al. (1991), Krawczuk (1993), and Krawczuk and Ostachowicz (1994), while a solid finite element with a fatigue crack has been developed by Krawczuk and Ostachowicz (1993a), and a shell element by Krawczuk (1994).

Krawczuk (1993) presented a method of creating the stiffness matrix of a finite plate element with a non-propagating, internal open crack. The method is similar to the one described by Qian et al. (1991) but contrary to their approach, the stiffness matrix of the cracked element was given in closed-form. The additional flexibility matrix was calculated by taking into account the additional elastic stress energy due to the occurrence of the crack in the plate. The method is restricted to cracks whose length is smaller than the dimensions of the element; It is assumed that the crack changes only the stiffness of the element and the mass of the element remains unchanged.

A method of creating the stiffness matrix of a hexahedral eight–node finite element, with a single, nonpropagating, transverse, single-edge crack at half of its length, has been investigated by Krawczuk and Ostachowicz (1993a). The crack was modeled by adding an additional flexibility matrix to the non-cracked element. The terms of the additional matrix have been calculated by the use of an approximate model of the stress intensity factor.

Many researchers have studied damage models in composite structures extensively. Krawczuk et al. (1997) proposed the formulation of a finite composite beam element with an open crack. The damaged part of the beam was modelled by a special finite element with a crack, while the undamaged part was substituted by three-node beam element. The crack is placed in the middle of the element and remains open. The element has three nodes; each of them has two degrees of freedom: transverse displacements and rotations. In the paper (Krawczuk et al., 1997), only the case of flat bending was considered.

Krawczuk et al. (1997) have investigated a model of a layered, delaminated composite beam. The beam was modelled by beam finite elements with three nodes and three degrees of freedom per node. In the delaminated region, additional boundary conditions were applied. It was assumed that the delamination was open (i.e. the contact forces between lower and upper parts are neglected). The delaminated region was modelled by three finite elements connected at the delamination crack tip where additional boundary conditions were applied. Each element had three nodes with three degrees of freedom: axial displacements, transverse displacements, and the independent rotations. In addition to general conditions of beam theory, it was assumed that the extensional and bending stiffness were uncoupled.

A model of a finite delaminated plate element has been developed by Żak et al. (2000); Zak et al. (2001). The delamination was modelled by three plate finite elements and to connect them, additional boundary conditions were applied at the delamination front. Each finite element had eight nodes with five degrees of freedom per node. Later papers (Krawczuk and Ostachowicz, 2002; Ostachowicz et al., 2002, 2003; Zak et al., 2003) present results for the identification of the location of failures in both isotropic and composite structures by means of a genetic algorithm search technique based on changes in natural frequencies. The location and size of failures are determined by minimisation of an error function which expresses the difference between calculated and measured natural frequencies.

Laboratory experiments have been conducted to ensure the reality of analytical and numerical models; the results obtained are promising, particularly because they have confirmed investigated models. Results of experimental tests have been presented in (Krawczuk et al., 2000, 1997; Żak et al., 2000; Ostachowicz et al., 2003; Zak et al., 2003).

3 Lamb Waves

Elastic waves that propagate in solid media bounded by two free and parallel surfaces are known in the literature as Lamb waves or guided waves. Lamb waves are named after Horace Lamb in honour of his fundamental contributions in this area of research. Lamb developed a mathematical theory that describes this kind of elastic waves, but interestingly he never managed to generate this type of wave in a real structure. Lamb waves propagate both as symmetric (S0, S1, S2, ...) and antisymmetric (A0, A1, A2, ...) modes and the number of these modes depends on the product of the excitation frequency and the element thickness. For example, up to about 2 MHz·mm, only the two fundamental Lamb wave modes S0 and A0 will propagate in an Aluminum alloy plate.

Table 1. Characteristic phase velocities and wave lengths in a 1mm thick aluminium plate.

Frequency	Phase velocity [mm/μs]			Wavelength [mm]		
[kHz]	SH0	A0	S0	SH0	A0	S0
100	3182	964	5496	31.82	9.64	54.96
150	3182	1161	5495	21.21	7.74	36.63
200	3182	1318	5494	15.91	6.59	27.47
250	3182	1450	5492	12.73	5.80	21.97
300	3182	1564	5490	10.61	5.21	18.30

A characteristic feature of this type of wave motion is elliptical particles motion in contrast to Rayleigh (surface) waves, where the wave motion is circular.

The solution of the Lamb wave equations must be obtained numerically. As a result, dispersion relations for various Lamb wave modes are obtained, i.e. the dependence of the wave number on the frequencythickness product.

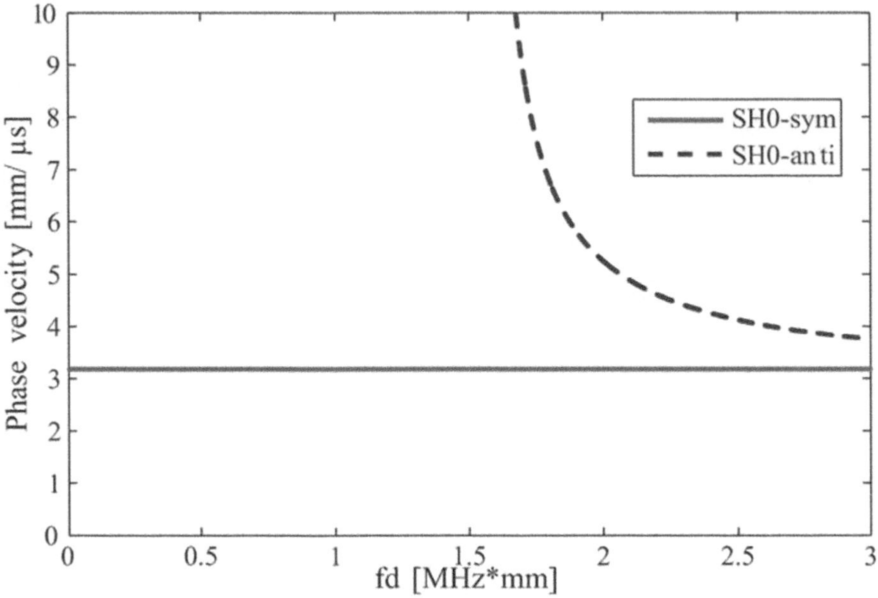

Figure 1. Phase velocity dispersion curves for an aluminium plate - shear horizontal waves.

It is interesting to look at the phase and group velocities of Lamb wave modes as well as shear modes corresponding to shear deformation. Typical results obtained for the group and phase velocity dispersion curves for the shear wave modes and the Lamb wave modes are presented in Figs. 1–4. These curves have been calculated analytically and obtained for an aluminium plate with the material properties as follows: Young's modulus 72.7 GPa, Poisson's ratio 0.33, mass density 2700 kg/m^3.

It can be noticed that in the frequency range up to around 2 MHz, only the fundamental modes can propagate in the plate. Some characteristic wave velocities and wave lengths in the case of the Aluminium plate under consideration are given in Table 1. The wave lengths are calculated from a simple equation,

$$\Lambda = \frac{c_{min}}{f} \tag{1}$$

where c_{min} is the minimal phase velocity and f denotes the carrier frequency of a wave packet. It can be noticed that the A0 mode has much shorter wavelengths in the lower frequency range than the SH0 and S0 modes; this means that A0 mode is well-suited to the detection of damage of rather small

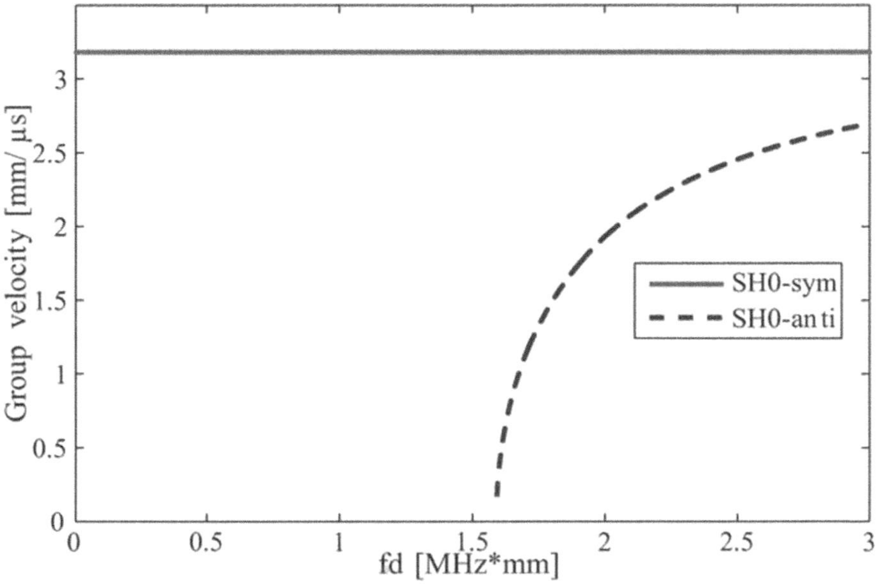

Figure 2. Group velocity dispersion curves for an aluminium plate - shear horizontal waves.

dimensions. In the case of the S0 and SH0 modes much higher frequency must be used in order to achieve comparable damage sensitivity. In contrast, the S0 mode is much less dispersive than A0 mode, while at the same time SH0 mode is almost *nondispersive* over all frequencies. (The term nondispersive means that a wave packet propagates in a structure without any observable distortion in shape.

Lamb waves propagating in a bounded solid media can be modelled assuming an appropriate displacement field, which in a accurate manner modes paths of particle motion through the thickness of the media. The displacement field can be approximated by,

$$
\begin{aligned}
u(x,y,z) &= \underline{u_0(x,y)} + \varphi_x(x,y)\cdot z + \underline{\psi_x(x,y)\cdot z^2} + \chi_x(x,y)\cdot z^3 \ldots \\
v(x,y,z) &= \underline{v_0(x,y)} + \varphi_y(x,y)\cdot z + \underline{\psi_y(x,y)\cdot z^2} + \chi_y(x,y)\cdot z^3 \ldots \\
w(x,y,z) &= w_0(x,y) + \underline{\varphi_z(x,y)\cdot z} + \psi_z(x,y)\cdot z^2 + \underline{\chi_z(x,y)\cdot z^3} \ldots
\end{aligned}
\qquad (2)
$$

where u_0, v_0 and w_0 represent the displacement components of the points located on a certain midplane surface, while φ_x and φ_y physically denote the rotations of appropriate solid sections about the x and y axes respectively. It is worth noting that the odd-order terms with respect to z in

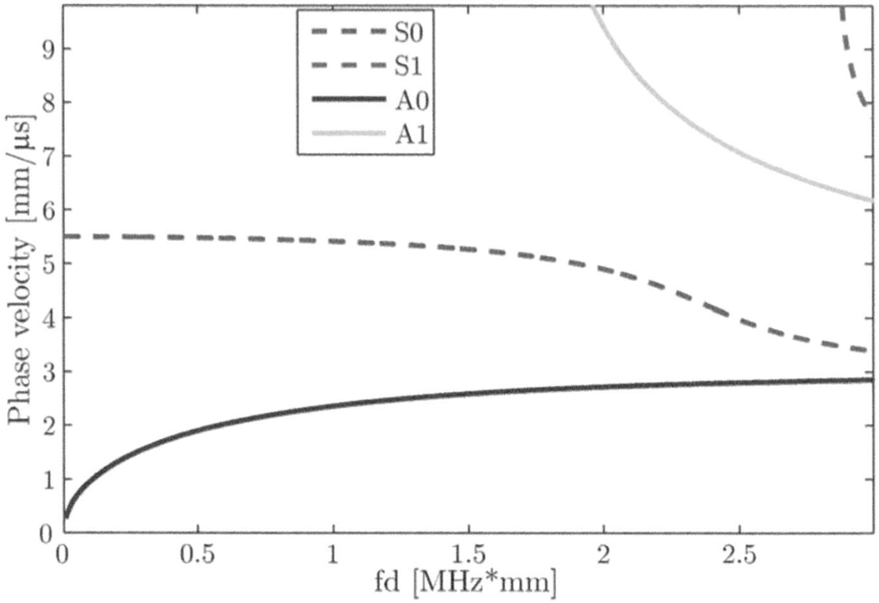

Figure 3. Phase velocity dispersion curves for an aluminium plate - symmetric and antisymmetric modes.

the x and y displacements together with the even-order terms in w with respect to z describe the antisymmetric wave modes and the other terms depict symmetric wave modes. Structural Health Monitoring (SHM) systems are usually based on the use of the fundamental modes of Lamb waves (S0 and A0), because in those cases it is usually much easier and convenient to analyse the received signals. An adequate approximation of the A0 mode requires at least the linear terms with respect to z in the u and v displacements and a constant term in the w displacement; this is consistent with the assumption of *First-Order Shear Deformation Theory* for plates. However, in order to capture the dispersion effect of the S0 mode, some additional terms must be included in the displacement field. It should be emphasised that application of the first-order shear deformation theory for plates in the case of spectral finite elements results in a diagonal form of the mass matrix for isotropic materials or symmetric laminates. In contrast, application of higher-order theories leads to generation of nonzero offdiagonal elements in the mass matrix; this means that the equations of motion are solved with much lower efficiency.

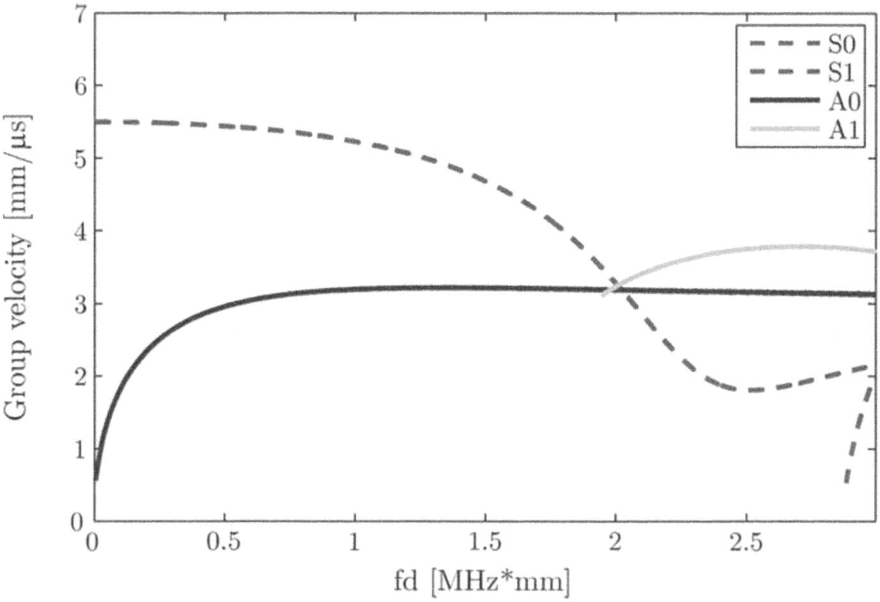

Figure 4. Group velocity dispersion curves for an aluminium plate - symmetric and antisymmetric modes.

4 Modelling of Elastic Waves

Wave propagation in structural elements has been studied over a considerable period of time. Although mathematical frameworks are well established and developed, wave propagation in real scale engineering structures still remains an open area of research. For simulation of stress wave propagation, the best way is to give an exact solution. However, even in some simple cases, such as elastic media, if local inhomogeneities (joints, inclusions, holes, etc.) are included, it is difficult to obtain exact solutions. For a specific geometry and finite periodic or semi-infinite boundary conditions, many solution techniques have been proposed and reported so far an excellent overview of theses techniques is given in (Bond, 1990).

In the case of SHM systems, piezoelectric actuators generate impulse wave signals and usually these are various modes of Lamb waves. The main problems in the analysis of high frequency (50 to 350 kHz) elastic wave propagation in structures with high velocities (1 to 6 km/s), are related to spatial discretisation. In order to obtain an accurate solution of the equation of motion, and to capture the effect of wave scattering at boundaries and structural discontinuities, a huge number of degrees of freedom (dof) is necessary. Conventional modal methods, when extended to high-frequency regimes, become computationally inefficient since many higher modes that should participate in the motion are misrepresented. For this reason, application of some approximation method is necessary.

Among many methods used for modelling and studying the phenomena of propagation of elastic waves, many numerical methods can be distinguished including: the finite difference method (FDM) (Strickwerda, 1989), the finite element method(FEM) (Yamawaki and Saito, 1992; Koshiba et al., 1984; Verdict et al., 1992; Alleyne and Cawley, 1992) and the boundary element method (BEM) (Cho and Rose, 1996). Unfortunately, the first two methods are not only time–consuming, but also require large computational memory even in the case of simple two–dimensional (2D) wave propagation problems. Moreover, they suffer from *numerical dispersion* which leads to improper wave velocity or false waves, which do not exist in the exact solution. In contrast, the boundary element method (BEM) is less time–consuming but application of the method to complex media with inhomogeneities is problematic.

Other methods are the finite strip element method (FSEM) and the semi–numerical method (SNM)(Cheung, 1976; Liu and Xi, 2002; Liu et al., 1990), which require much less memory storage space for necessary data due to a lower level of discretisation and application of the exact solution in one direction. SNM is very effective for the computation of forced wave motion in the frequency domain and can be used for much higher frequencies than the methods based on FEM. As with the BEM, the FSEM uses a Green's function but in a different manner. On the other hand, variable size of strip stiffness matrices and modification of spline functions at the boundary nodes are inconvenient in implementation.

A method that incorporates the advantages of FEM (discretisation) and the FDM (time integration schemes) is the unstructured grid method (UGM) (Liu et al., 2004, 2005). This method is based on the dynamic equilibrium equations of computational cells formed around auxiliary triangular grids. The solution is obtained by the calculation of nodal displacements and central point stresses of spatial grids alternately. A different approach has been proposed by Schechter et al. (1994) and extended by Yim and Sohn (2000). In the mass–spring–lattice–model (MSLM), inertia and stiffness properties are calculated using lumped parameters. More recent developments in this area include the new Local Interaction Simulation Approach (LISA) (Delsanto et al., 1992, 1994, 1997). This method simulates wave propagation heuristically, i.e. directly from physical phenomena and properties. It should be noted that the LISA approach suffers from inaccuracy in some cases (Ruffino and Delsanto, 1999).

More recently, various spectral methods have been proposed for the analysis of elastic wave propagation in complex media: the fast Fourier transform–based spectral finite element method (FFT–based SFEM) (Doyle, 1997) and the spectral element method (SEM) (Patera, 1984) also called the pseudospectral method. It should be stressed that despite the terminology, which may be misleading, these methods are completely different.

The FFT-based SFEM proposed by Doyle (1997) is very similar to the technique of the FEM as far as the assembly and the solution of the equation of motion is considered. The formulation of this method starts from exact solutions of the governing partial differential equations in the frequency domain. Excitation signals are transformed into a number of frequency components using the FFT. Next, as a part of a large frequency loop, the dynamic stiffness matrix is generated, transformed, and a solution is found for each unit impulse at each frequency. This yields directly the frequency response function (FRF) of the analysed problem. The calculated frequency domain responses are then transformed back to the time domain using the inverse fast Fourier transformation (IFFT).

The FFT–based SFEM proposed by Doyle is computationally efficient but the inverse Fourier transform is very difficult to do in an exact analytical manner. For this reason, many approximate and asymptotic schemes have usually been resorted to (Amaratunga and Williams, 1995). Such approaches reduce the problems associated with "wrap around"[1] due to the assumed periodicity of solutions in the FFT–based SFEM and thus may result in a decreased number of points in the time window for the same problem (Mitra and Gopalakrishnan, 2006). Further, FFT-based SFEM cannot be used for finite-length undamped structures. For such cases, a semi-infinite element (*throw-off* element (Doyle, 1997)) is normally used to allow some leakage of response, which in turn amounts to adding artificial damping through the release of trapped energy.

Consequently, the FFT-based SFEM is well suited to simple 1D problems (Palacz and Krawczuk, 2002; Krawczuk et al., 2003; Mahapatra and Gopalakrishnan, 2003), but becomes difficult to use for complex geometries. A comparative study of the FFT-based SEM with the LISA approach can be found in Lee et al. (2004). Despite problems with the periodic nature of the FFT, recent work in this area shows some application of the FFT-based SFEM to wave propagation phenomena in anisotropic plates and inhomogeneous layered media (Chakraborty and Gopalakrishnan, 2004, 2005,

[1]The "wrap around" effect means that the remaining part of the response beyond the chosen time window will start appearing first, which totally distorts the signal.

2006a,b).

The SEM, as proposed by Patera (1984), is much more versatile for the investigation of the propagation of elastic waves in structures of complex geometry. This method originates from the use of spectral series for the solution of partial differential equations (Boyd, 1989). The idea of the SEM is very similar to the FEM except for the specific approximation functions it uses. Elemental interpolation nodes are located at points corresponding to zeros of an appropriate family of orthogonal polynomials (Legendre or Chebyshev). A set of local shape functions consisting of Lagrange polynomials, which are spanned on these points, is built and used. As a consequence of this, as well as the use of the Gauss-Lobatto-Legendre integration rule, a diagonal form of the mass matrix is obtained. In this way, the cost of the time domain integration is much less expensive than in the case of the classic FE approach. Moreover, the numerical errors decrease faster than any power of $1/p$ (so-called *spectral convergence*), where p is the order of the applied polynomial (Pozrikidis, 2005). The main fields of application of SEM nowadays include fluid dynamics (Canuto et al., 1991), heat transfer (Spall, 1995), acoustics (Dauksher and Emery, 1997; Seriani, 1997), seismology (Komatitsch and Vilotte, 1998; Seriani, 1999), etc.

The application of SEM for problems of propagating waves in anisotropic crystals has been shown by Komatitsch et al. (2000).

The first attempt to use SEM for problems of propagation of elastic waves in 2D structural elements with cracks has been made by Żak et al. (2006). A 36-node spectral membrane element with two degrees of freedom per node has been developed. The crack has been modelled by simple splitting of the nodes between appropriate spectral elements. This approach has been extended to isotropic and composite plates (Żak et al., 2006; Kudela et al., 2007b,a). Also, the SEM found applications for the problems of wave propagation in anisotropic and inhomogeneous uncracked and cracked beams (Sridhar et al., 2006) as well as for the problems of cracked composite rods based on the three–mode theory of rods (Kudela and Ostachowicz, 2007).

A 3D spectral element has been developed and used for SHM purposes by (Kim et al., 2008).

It seems that the SEM is a most versatile and promising tool for wave propagation modelling and is becoming more and more popular in this field.

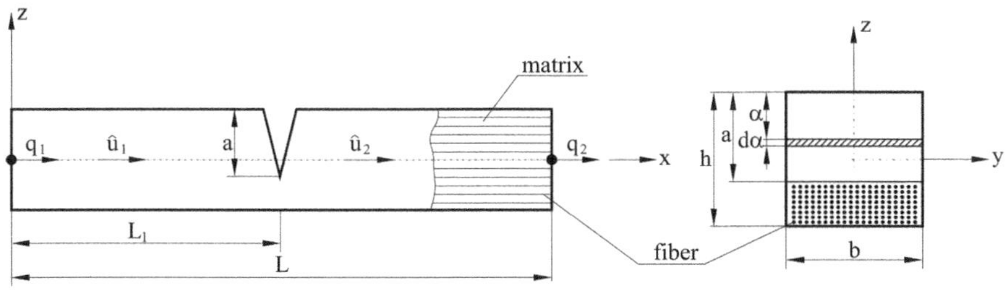

Figure 5. Composite spectral rod element.

4.1 The FFT–Based Spectral Finite Element Method – Cracked Rod

A multilayer composite spectral rod element is presented in Fig. 5. The crack is located at a distance of L_1 from the left hand end. The element has two nodes and one degree of freedom per node (longitudinal displacement). Nodal spectral displacements for the left and right parts of the rod are assumed as follows,

$$\hat{u}_1(x) = A_1 e^{-ik_n x} + B_1 e^{-ik_n(L_1-x)}, \quad x \in (0, L_1) \tag{3}$$

$$\hat{u}_2(x) = A_2 e^{-ik_n(x+L_1)} + B_2 e^{-ik_n[L-(L_1+x)]}, \quad x \in (0, L-L_1) \tag{4}$$

where the wave number is obtained from the equation,

$$k_n = \omega_n \sqrt{\frac{\mu}{D}} \tag{5}$$

The mass density per unit length can be expressed as,

$$\mu = \rho\, b\, h, \quad \mu = I_0 \tag{6}$$

for the isotropic and anisotropic cases, respectively. The material stiffness matrix has the forms,

$$D = E\, b\, h, \quad D = A_{11}\, b \tag{7}$$

for the isotropic and anisotropic cases, respectively. The mass density per unit length from Eq. (6) in the case of a composite rod can be expressed as,

$$I_0 = b \sum_{k=1}^{N} \rho_k \left(h_{k-1} - h_k\right) \tag{8}$$

where N is the number of composite layers, ρ_k is the mass density of the k-th layer and b is the rod width. The material stiffness matrix from Eq. (7) in the case of a composite rod can be expressed as (Vinson and Sierakowski, 1989),

$$\begin{aligned} D^e &= b\sum_{k=1}^{N}(\overline{Q}_{11})_k\,(h_{k-1} - h_k) \\ \overline{Q}_{11} &= E_{11}\sin^4\theta + 2(\nu_{12}E_{22} + 2G_{12})\sin^2\theta\cos^2\theta + E_{22}\sin^4\theta \end{aligned} \tag{9}$$

where E_{11} is the Young's modulus along the reinforcing fibres, E_{22} is the Young's modulus in the direction perpendicular to the direction of the fibres, ν_{12} is the Poisson ratio, G_{12} is the shear modulus and θ is the angle between the material axis parallel to the reinforcing fibres and the x axis.

The coefficients A_1, B_1, A_2 and B_2 can be calculated as functions of the nodal spectral displacements using the following boundary conditions,

- at the left end of the element,

$$\hat{u}_1(0) = q_1 \tag{10}$$

- at the crack location (total change of displacements and compatibility of shear forces),

$$\hat{u}_2(0) - \hat{u}_1(L_1) = \hat{\theta}\frac{\partial \hat{u}_1(L_1)}{\partial x} \tag{11}$$

$$\frac{\partial \hat{u}_1(L_1)}{\partial x} = \frac{\partial \hat{u}_2(0)}{\partial x} \tag{12}$$

- at the right end of the element:

$$\hat{u}_2(L - L_1) = q_2 \tag{13}$$

where $\hat{\theta} = Ebhc$ and c is the flexibility at the crack location (see section 4.3). Taking into account the formulae describing nodal spectral displacements for the left and right parts of the element, the boundary conditions can be written in a matrix form as,

$$\begin{bmatrix} A_1 \\ B_1 \\ A_2 \\ B_2 \end{bmatrix} = \mathbf{W}^{-1} \begin{bmatrix} q_1 \\ 0 \\ 0 \\ q_2 \end{bmatrix} \tag{14}$$

$$\mathbf{W} = \begin{bmatrix} 1 & e^{-ik_nL_1} & 0 & 0 \\ (ik_n\hat{\theta} - 1)e^{-ik_nL_1} & -1 - ik_n\hat{\theta} & e^{-ik_nL_1} & e^{-ik_n(L-L_1)} \\ -ik_n e^{-ik_nL_1} & ik_n & ik_n e^{-ik_nL_1} & -ik_n e^{-ik_n(L-L_1)} \\ 0 & 0 & e^{-ik_nL} & 1 \end{bmatrix} \tag{15}$$

The nodal spectral forces can be determined by differentiating the spectral displacements with respect to x, and then can be expressed as follows,

$$\hat{F}_1 = D\frac{\partial \hat{u}_1(0)}{\partial x} \tag{16}$$

$$\hat{F}_2 = D\frac{\partial \hat{u}_2(L - L_1)}{\partial x} \tag{17}$$

The relation between nodal displacements and forces can be shown as,

$$\begin{bmatrix} \hat{F}_1 \\ \hat{F}_2 \end{bmatrix} = \mathbf{K_d} \begin{bmatrix} \hat{q}_1 \\ \hat{q}_2 \end{bmatrix} \tag{18}$$

where the dynamic stiffness matrix is given by,

$$\mathbf{K_d} = D\begin{bmatrix} ik_n & -ik_n e^{-ik_nL_1} & 0 & 0 \\ 0 & 0 & -ik_n e^{-ik_nL} & ik_n \end{bmatrix}\mathbf{W}^{-1} \tag{19}$$

Unlike conventional finite elements, a special case is derived here where the rod is very long and application of any load at any location causes no secondary disturbances other than incident waves departing from that location. This simulates a condition, wherein the boundaries are at such a distance that the effect of reflected waves becomes negligible due to attenuation throughout their long travel, and do not reach the location under consideration within the time of observation. In other words, the *throw–off* element is a non–resonant single–node element that acts as a conduit to allow the propagation of the trapped energy out of the system. The nodal spectral displacement for the throw–off element is assumed in the following form,

$$\hat{u}(x) = C_1 e^{-ik_nx} + D_1 e^{-ik_n(L_1-x)}, \quad x \in (0, L_1) \tag{20}$$

After using transformations similar to Eqs. (10 – 19) the frequency dependent stiffness matrix $\mathbf{K_t}$ for the *throw–off* rod element can be defined as,

$$\mathbf{K_t} = Dik_n \tag{21}$$

4.2 The Time Domain Spectral Element Method – Cracked Rod

Spectral rod finite elements with crack, are formed by the connection of two classic spectral finite elements with nodes separated by the use of a spring. Spectral elements based on the elementary and three-mode theory are presented in Figure 6.

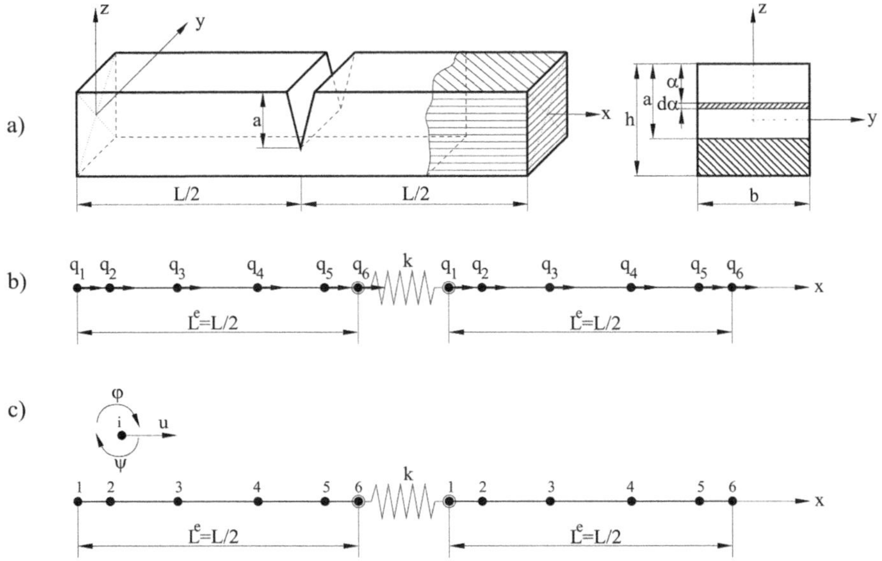

Figure 6. Schematic diagrams of models a) model of the composite rod fragment with crack, b) spectral finite element for the elementary theory, c) spectral finite element for the three–mode theory.

The stiffness of the spring modelling the size of a transverse, non-growing crack is calculated according to the laws of fracture mechanics. The effect of the crack is achieved at the stage of global stiffness matrix assembly.

The rod spectral finite element has been derived according to elementary theory; the element consists of 6 nodes. With each node, one degree of freedom is given, i.e. longitudinal displacement. Nodes are placed unequally. Local nodal coordinates $\xi_i \in [-1, 1]$, $i \in 1, \ldots, 6$ are obtained as roots of the equation,

$$\left(1 - \xi^2\right) P_5'(\xi) = 0 \tag{22}$$

where $P_5'(\xi)$ denotes the first derivative of Legendre's 5th-order polynomial. The obtained coordinates correspond to Gauss-Lobatto-Legendre integration points. On such selected points, Lagrange's approximation is defined which gives a set of shape functions. The same set of shape functions can be used for displacement field approximation inside the element as well as for element geometry approximation. A spectral rod element applied to the modelling of a composite rod with crack derived according to the elementary theory of rods is presented in Figure 6a. The element contains 6 nodes; each node has one degree of freedom – longitudinal displacement. In the elementary theory of the rod, displacements take the following form,

$$u(x, z) = u_0(x) \tag{23}$$

where u_0 is the average axial displacement. The strain field may be expressed by the equation,

$$\varepsilon_x(x, z) = \frac{\partial u_0}{\partial x} \tag{24}$$

Assuming an approximation of the displacement field within the element,

$$u^e(\xi) = \mathbf{N}^e \mathbf{q}^e = \sum_{i=1}^{6} N_i^e(\xi)\, q^e(\xi_i) \tag{25}$$

where $\mathbf{N}^e$ are shape functions, and $\mathbf{q}^e$ are nodal degrees of freedom within the element, and substituting into equation (24), one obtains the strain approximation,

$$\varepsilon_x^e(\xi) = \mathbf{B}^e \mathbf{q}^e = \sum_{i=1}^{6} B_i^e(\xi)\, q^e(\xi_i) \tag{26}$$

where $\mathbf{B}^e$ is the matrix connecting strains with nodal displacements calculated as,

$$\mathbf{B}^e = \frac{\partial}{\partial x}\mathbf{N}^e(\xi), \quad \frac{\partial}{\partial x} = J^{-1}\frac{\partial}{\partial \xi}, \quad J = \frac{\partial x}{\partial \xi} \tag{27}$$

Matrices of mass and stiffness are calculated numerically with the use of the Gauss-Lobatto-Legendre integration rule,

$$\mathbf{m}^e = \int_{\Omega_e} \left[\mathbf{N}^e(x)\right]^T I_0^e\, \mathbf{N}^e(x)\, \mathrm{d}x \approx \sum_{i=1}^{6} w_i \left[\mathbf{N}^e(\xi_i)\right]^T I_0^e\, \mathbf{N}^e(\xi_i)\, \det(J^e) \tag{28}$$

$$\mathbf{k}^e = \int_{\Omega_e} \left[\mathbf{B}^e(x)\right]^T D^e\, \mathbf{B}^e(x)\, \mathrm{d}x \approx \sum_{i=1}^{6} w_i \left[\mathbf{B}^e(\xi_i)\right]^T D^e\, \mathbf{B}^e(\xi_i)\, \det(J^e) \tag{29}$$

Quadrature weights $w_i > 0$, that are independent of the element are estimated from the formula,

$$w_i = \frac{2}{n(n-1)\left[P_{n-1}(\xi_i)\right]^2}, \qquad i \in 1, \ldots, n \tag{30}$$

where $n = 6$ for the 6–node element.

Time Domain Integration For the sake of the accuracy of solution, discretisation of the area with the use of the spectral finite element method ought to provide at least 5 nodes per wave length[2]. Moreover, the mesh of the elements should be, if possible, homogeneous over the entire area. After spatial discretisation with spectral elements, to achieve a solution, only the following system of differential equations in the time domain is needed,

$$\mathbf{M}\ddot{\mathbf{Q}} + \mathbf{K}\mathbf{Q} = \mathbf{F} \tag{31}$$

where $\mathbf{M}$ is the global inertia (mass) matrix, $\mathbf{K}$ is the global stiffness matrix, $\mathbf{Q}$ is the vector of global degrees of freedom, and $\mathbf{F}$ is the vector of time dependent forces. Damping is omitted, as it is possible to consider wave attenuation based on experimental measurements.

Discretisation in the time domain of the system of differential equations of the second degree (31), may be conducted with the classic Newmark's scheme or the central difference scheme. The given methods are conditionally stable methods of direct integration, where the equation of the motion is integrated step by step; this means that the equation of motion (31) ought to be fulfilled only at chosen moments in time. For the stability of the solution (to avoid accumulation of integration errors and rounding errors) the integration step Δt must be adequately small. The calculation of an adequate integration step in the simulation of wave propagation within composite elements is difficult owing to the great number of parameters of the problem (minimum and maximum speed, carrier wave frequency, time of analysis, and number of nodes per wavelength). In practice, in the application of the central difference method, the number of integration steps should be chosen individually for the considered problem. It may be assumed that the integration step Δt is proportional to p^{-2}, where p denotes the order of the approximating polynomial in the spectral element. This means that a high degree of polynomial leads to a significant cost of calculation. In the case of too small a number of integration steps, the algorithm is unstable and this manifests itself in a violent increase of the displacements with each time step.

It should be emphasised that in the method of spectral elements, spatial discretisation is very accurate, owing to the fact that it is based on high–order polynomials. On the contrary, in the case of time domain discretisation with the use of a central difference scheme only second–degree accuracy is obtained, which means that global accuracy is reduced. For this

[2]For 5th-order approximating polynomials

reason, application of a higher–order scheme in the time domain would be interesting.

In the method of central differences, one assumes changeability of the acceleration vectors in time in the form of:

$$\ddot{\mathbf{Q}} \simeq \frac{1}{\Delta t^2}\left(\mathbf{u}_{t+\Delta t} - 2\,\mathbf{u}_t + \mathbf{u}_{t-\Delta t}\right) \tag{32}$$

Substituting the difference formula (32) into equation (31) and marking the displacement vector at time t as $\mathbf{u}_t = \mathbf{Q}$, one obtains,

$$\frac{1}{\Delta t^2}\left(\mathbf{u}_{t+\Delta t} - 2\,\mathbf{u}_t + \mathbf{u}_{t-\Delta t}\right)\mathbf{M} + \mathbf{K}\mathbf{u}_t = \mathbf{F}_t \tag{33}$$

From equation (33) one calculates the sought displacement condition at the time step $t + \Delta t$, meaning $\mathbf{u}_{t+\Delta t}$. This is obtained based on the solution at time t. For this reason, this method is numbered among *explicit* methods. The great advantage of this manner of solving equation (31) is the fact that the matrix of stiffness does not have to be inverted.

One should draw attention to the fact that calculation of the results at the current time step, using results obtained at the previous time step, requires assuming a certain starting procedure. One assumes that the vectors $\mathbf{Q}_0$, $\dot{\mathbf{Q}}_0$, $\ddot{\mathbf{Q}}_0$ are known at the initial time, namely at the time $t = 0$. This way, using the difference formula for the second derivative (32) and a difference formula for the first derivative,

$$\dot{\mathbf{Q}} \simeq \frac{\mathbf{u}_{t+\Delta t} - \mathbf{u}_{t-\Delta t}}{2\Delta t} \tag{34}$$

one can calculate the displacement vector $\mathbf{u}_{t-\Delta t}$ at a fictional moment, which will precede the beginning of the motion,

$$\mathbf{u}_{t-\Delta t} = \mathbf{Q}_0 - \Delta t\dot{\mathbf{Q}}_0 + \frac{\Delta t^2}{2}\ddot{\mathbf{Q}}_0 \tag{35}$$

Frontal method Despite using various formats to store sparse matrices, which provides saving of the computer's RAM, wave issues are so complex that the computer's memory resources are usually not sufficient. It is possible, in some cases, to use the frontal method, in which algebraic equations of a matrix are calculated at the level of the finite element without the necessity to formulate global matrices.

(Kudela, 2008) proposed an integration method for the wave equations, where assembly of the global stiffness matrix does not take place. The algorithm of the method is presented below.

After ordering, equation (33) takes the form,

$$\underbrace{\left(\frac{1}{\Delta t^2}\mathbf{M}\right)}_{\mathbf{M}_0}\mathbf{u}_{t+\Delta t}=\mathbf{F}_t-\underbrace{\mathbf{K}\mathbf{u}_t}_{\hat{\mathbf{F}}}+\underbrace{\left(\frac{1}{\Delta t^2}\mathbf{M}\right)}_{\mathbf{M}_0}\mathbf{u}_t-\underbrace{\left(\frac{2}{\Delta t^2}\mathbf{M}\right)}_{\mathbf{M}_2}\mathbf{u}_{t-\Delta t} \quad (36)$$

In the equation (36), the vector $\hat{\mathbf{F}}$ may be calculated in such a manner that the use of the global stiffness matrix $\mathbf{K}$ is not necessary. The proposed explicit time integration algorithm has the following steps:

- Loop over elements e
  - For each element calculate the characteristic elemental matrices $\mathbf{k}^e$, $\mathbf{m}^e$
  - Assemble each diagonal mass matrix of the element e into the global vector $\mathbf{M}=\overset{n_{el}}{\underset{e=1}{\mathrm{A}}}\operatorname{diag}(\mathbf{m}^e)$, where $\overset{n_{el}}{\underset{e=1}{\mathrm{A}}}$ denotes the assembly operator
  - Successive elemental stiffness matrices $\mathbf{k}^e$ can be stored in a binary file
- End of loop over elements e
- Define constants $a_0=1/\Delta t^2$, $a_2=2a_0$
- Calculate the auxiliary vectors $\mathbf{M}_0=a_0\,\mathbf{M}$, $\mathbf{M}_2=a_2\,\mathbf{M}$, $\tilde{\mathbf{M}}^\alpha=1/\mathbf{M}_0^\alpha$
- The displacement vector $\mathbf{u}_{t-\Delta t}$ is calculated from Eq. (35)
- Apply the initial conditions at the time instant $t=t_0$
- Loop over time instants t
  - Set up a pointer to the elemental stiffness matrix $\mathbf{k}^e$ at the beginning of the file
  - Loop over elements e
    - Read the stiffness matrix $\mathbf{k}^e$ from the file and move the pointer
    - Perform multiplication $\hat{\mathbf{f}}=\mathbf{k}^e\mathbf{u}_t^I$, where I denotes the vector with numbers of degrees of freedom corresponding to element e
    - Assemble vector $\hat{\mathbf{F}}=\overset{n_{el}}{\underset{e=1}{\mathrm{A}}}\left(\hat{\mathbf{f}}^e\right)$
  - End of loop over elements e
  - Calculate effective vector $\tilde{\mathbf{R}}=\mathbf{F}_t-\hat{\mathbf{F}}+\mathbf{M}_0^\alpha\mathbf{u}_t^\alpha-\mathbf{M}_2^\alpha\mathbf{u}_{t-\Delta t}^\alpha$, where indices α denote that multiplication is performed element by element without summation
  - The solution of the equation of motion at the time instant $t+\Delta t$ is achieved by multiplication of element by element: $\mathbf{u}_{t+\Delta t}^\alpha=\tilde{\mathbf{M}}^\alpha\,\tilde{\mathbf{R}}^\alpha$

– $\mathbf{u}_{t-\Delta t} = \mathbf{u}_t$, $\mathbf{u}_t = \mathbf{u}_{t+\Delta t}$, $t = t + \Delta t$

- End of loop over time instants t

The proposed integration algorithm of the wave equation provides optimal use of computer memory resources - the global stiffness matrix is not assembled; this enables the solving of problems with a great number of degrees of freedom on an ordinary PC. The proposed algorithm is extremely efficient since it uses the diagonal form of the mass matrix so that inversion of the matrix is completely eliminated.

4.3 Flexibility at the crack location

The flexibility at the crack location for a spectral rod element can be calculated using the Castigliano theorem (Przemieniecki, 1968):

$$c_{ij} = \frac{\partial^2 U}{\partial S_i \partial S_j} \quad (\text{for } i = j = 1) \tag{37}$$

where U denotes the elastic strain energy of the element caused by the presence of the crack and the S_i are the independent nodal forces acting on the element. The following relation can express the elastic strain energy due to the crack,

$$U = \frac{1}{E} \int_A K_I^2 \, \mathrm{d}A \tag{38}$$

where A denotes the area of the crack and K_I is a stress intensity factor corresponding to the first mode of the crack formation (Tada et al., 1973). The stress intensity factor can be expressed as follows,

$$K_I = \frac{S_1}{bh} \sqrt{\pi \alpha}\, f\left(\frac{\alpha}{h}\right) \tag{39}$$

where α, b and h denote the crack depth, height of the rod and width of the rod at the crack location respectively (see Fig. 6), and f is the correction function in the form (Tada et al., 1973),

$$f\left(\frac{\alpha}{h}\right) = \sqrt{\frac{\tan(\pi\alpha/2h)}{\pi\alpha/2h}} \, \frac{0.752 + 2.02(\alpha/h) + 0.37[1 - \sin(\pi\alpha/2h)]^3}{\cos(\pi\alpha/2h)} \tag{40}$$

After some simple transformations, the flexibility of the elastic element, which is used for modeling the cracked cross section of the rod, can be rewritten as,

$$c = \frac{2\pi}{\overline{E}b} \int_0^{\overline{a}} \overline{\alpha}\, f^2(\overline{\alpha})\, \mathrm{d}\overline{a} \tag{41}$$

where $\overline{E}$ denotes Young's modulus (averaged with respect to any layers of composite).

It should be noted that application of the technique of static node condensation as presented in section 4.1 in the case of SEM, causes the mass matrix to lose diagonality. In order to maintain the diagonal form of the mass matrix it is necessary to split nodes between two spectral elements and add to the global stiffness matrix a special matrix. This special matrix is a consequence of the inverse form of the flexibility at the crack location and can be expressed as follows:

$$\mathbf{K_s} = \begin{bmatrix} 1/c & -1/c \\ -1/c & 1/c \end{bmatrix} \tag{42}$$

In such a case, the mass matrix of the elements located next to the crack is identical with that of classical spectral elements.

4.4 Comparative example

Numerical calculations were conducted for an unconstrained rod with dimensions: length 2 m, height 0.02 m and width 0.02 m. The following properties of materials were assumed: Young's modulus 210 GPa and mass density 7860 kg/m^3. Excitation, in the form of an impulsive force with amplitude 100 N was applied to the node on the left end of a rod. A forcing signal in the form of a sine with five cycles, modulated with a Hanning window, was applied.

The aim of the numerical example is to compare results obtained by the use of the spectral element method with the spectral element method based on FFT (Palacz and Krawczuk, 2002). In this example, the applied forcing signal had a carrier frequency of 100 kHz. However, for signal amplitudes to be compared, the excitation amplitude for the second method is twice as large. Such a procedure is necessary because in the method based on the FFT, on the left end of the rod an element of the *throw-off* type is added. This causes the actuated wave to propagate simultaneously in two directions: to the left it is led *ad infinitum* by the *throw-off* element, and to the right it propagates because of the element with the crack. The crack was inserted exactly in the centre of the rod. It was assumed that the depth of the crack was 15% of the height of the rod.

Figures 7–9 present comparisons of the signals obtained by the use of both methods. For legibility, the distance covered by the wave was placed on the horizontal axis. Distance was obtained by calibration of the time axis with the theoretical velocity $v = \sqrt{E/\rho} = 5168.9$ m/s. In addition, the beginning of the excitation was shifted adequately to half of the impulse time.

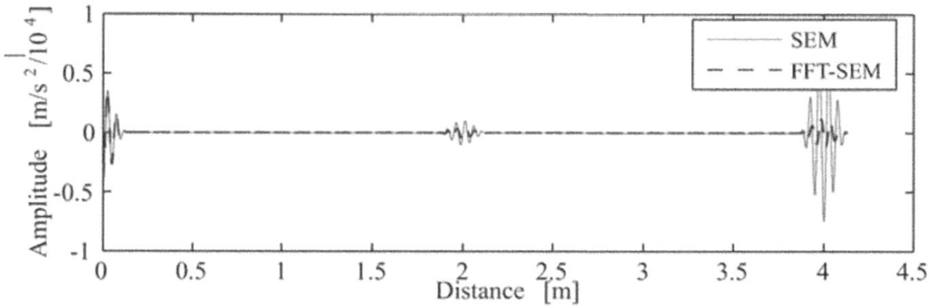

Figure 7. Comparison of signals on the left end of the rod, obtained with the method of spectral elements (SEM) and the method of spectral elements based on the FFT (FFT-SEM).

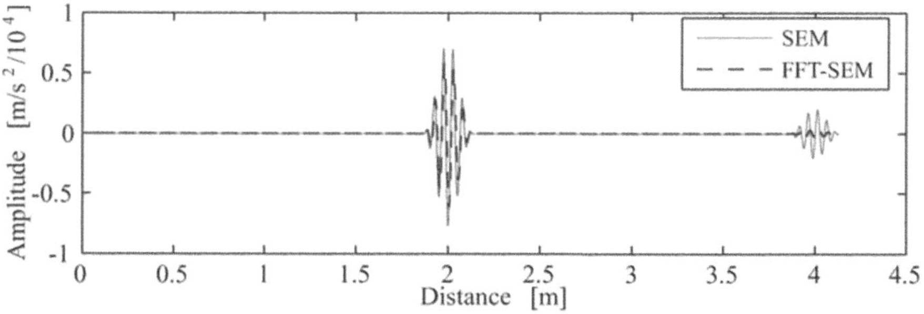

Figure 8. Comparison of signals on the right end of the rod, obtained with the method of spectral elements (SEM) and the method of spectral elements based on the FFT (FFT-SEM).

In Figure 7 one can observe that for both methods, the wave velocity corresponds to the theoretical velocity - the centre of the impulse occurs almost precisely at the distance of 2 m, i.e. once the wave covers the distance from the left end of the crack and back to the left end. The next centre of the impulse occurs at the distance equal to 4 m and it corresponds to the reflection from the right end of the rod. In both methods the shape of the signal remains the same, whereas amplitudes differ.

Taking into account the fact that issues of wave propagation in a medium without damping are discussed, the signal amplitude value should not undergo changes. This is the case only when the method of spectral elements is applied, as is presented in Figures 7-8, where the impulse amplitudes at the distance of 4 m (Figure 7) and 2 m (Figure 8) are identical. For the

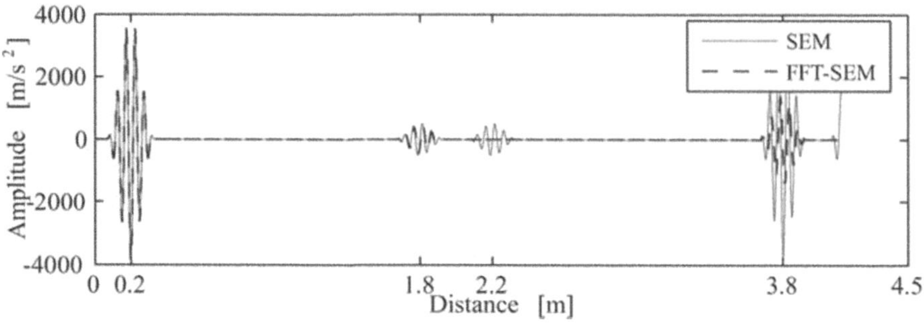

Figure 9. Comparison of signals in the point located 20 cm from the left end of the rod, obtained with the method of spectral elements (SEM) and the method of spectral elements based on the FFT (FFT-SEM).

spectral element method based on FFT, the signal amplitude decreases as a result of energy transfer from the system through the *throw-off* element.

Influence of the *throw-off* element on the signal received at the a location of 20 cm from the left end of the rod is presented in Fig. 9. The wave generated at the left end of the rod, while propagating, travels through discussed location (impulse at the distance of 0.2 m), it reflects from the crack and returns to this location (impulse at the distance of 1.8 m), next it reaches the left end of the rod, where the wave is transferred ad infinitum without reflection from the edge (no impulse at the distance of 2.2 m).

The comparative analysis allows one to state that the method of spectral elements shows a certain advantage over the FFT-based spectral elements method in the fidelity of modelling the phenomenon of wave propagation. When only the first reflection (from the crack) is analyzed, the considered methods lead to compatible results. The only condition is, that in the FFT-based spectral element method, the analyzed location will not agree with the degree of freedom where the *throw-off* was added. The advantage of the FFT-based spectral element method is that if one searches for solutions only at several locations, the time of calculation is shorter. Continuing the theoretical deliberation, the numerical examples presented will be based on the spectral element method.

Attention should be brought to the fact that the formalism presented in paragraphs 4.3-4.1 refers to the elementary theory of rods. However, expanding this formalism to other theories of rods, beams or plates is straightforward. From the point of view of SHM, the elementary theory of rods is not sufficient at higher frequency ranges because it does not take into account the dispersive character of the waves. The above aspect is clearly presented in Fig.10, where comparison of signals received in a composite rod with a crack modelled using elementary and three-mode theory was presented. Signals received by the use of the model based on elementary

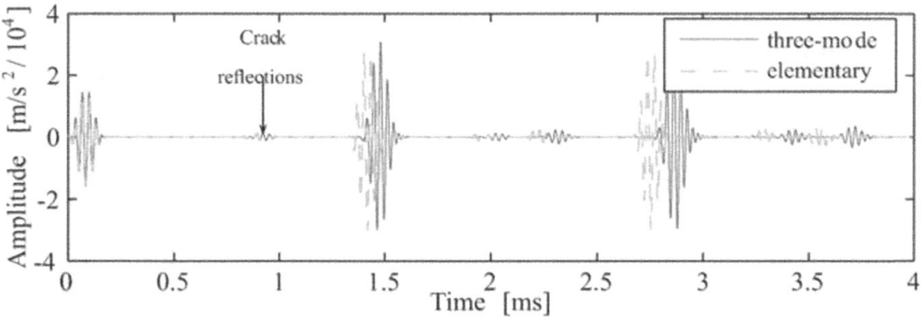

(a) Signal with carrier frequency 30 kHz.

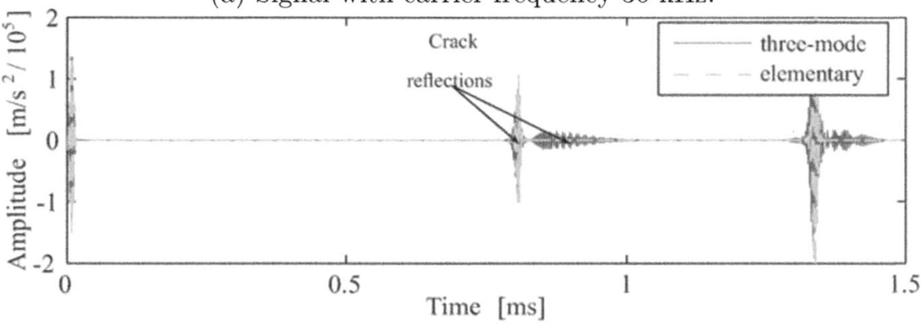

(b) Signal with carrier frequency 300 kHz.

Figure 10. Comparison of the signals received on the left end of the rod with crack at the depth of 15% of section height for the tree-mode and elementary theory.

theory, significantly differ from signals received with the use of the model based on three-mode theory, in regard to both wave velocity and shape of signal packet.

4.5 Influence of crack on wave propagation

Influence of crack location and crack depth on wave propagation Figure 11 presents the absolute values of signals obtained at the left end of the rod for various locations of a crack with 15% of the cross-section height. Location of the crack close to the left end of the rod causes multiple reflections and superpositions of waves. With the location of the crack at a shorter distance from the half–length of the rod (Figure 11), one can observe

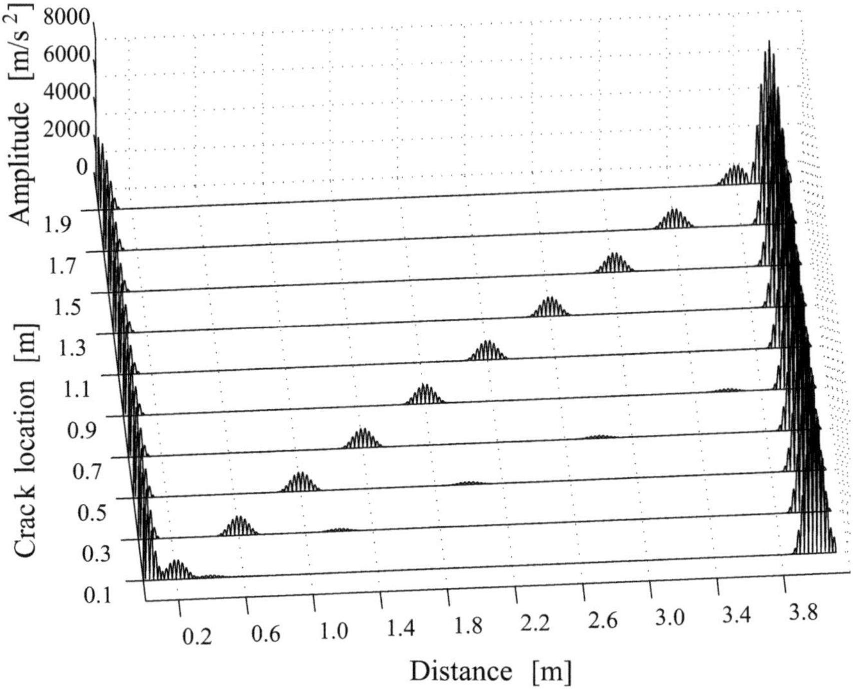

Figure 11. Influence of crack location at the depth of 15% of cross–section height, on wave propagation.

two reflections from the crack and a reflection from the right end of the rod. The example presented shows that the location of the crack may be easily identified based on the time of flight and the velocity of the wave packet.

In the next example, the effect of depth of the crack was studied. The crack was located at a distance of 1.2 m from the left end of the rod. With an increase in crack size, the amplitude of the signal reflected from the crack increases, and the amplitude of the signal passing through the crack decreases, as presented in Figure 12. For a crack depth of 5% of the cross-section height, the wave impulse reflected from the crack also occurs, although this is not very visible in Figure 12

Influence of signal frequency on amplitude of the reflected wave
It follows from Figure 13 that along with an increase of forcing signal frequency, there is increased amplitude of the signal reflected from the crack

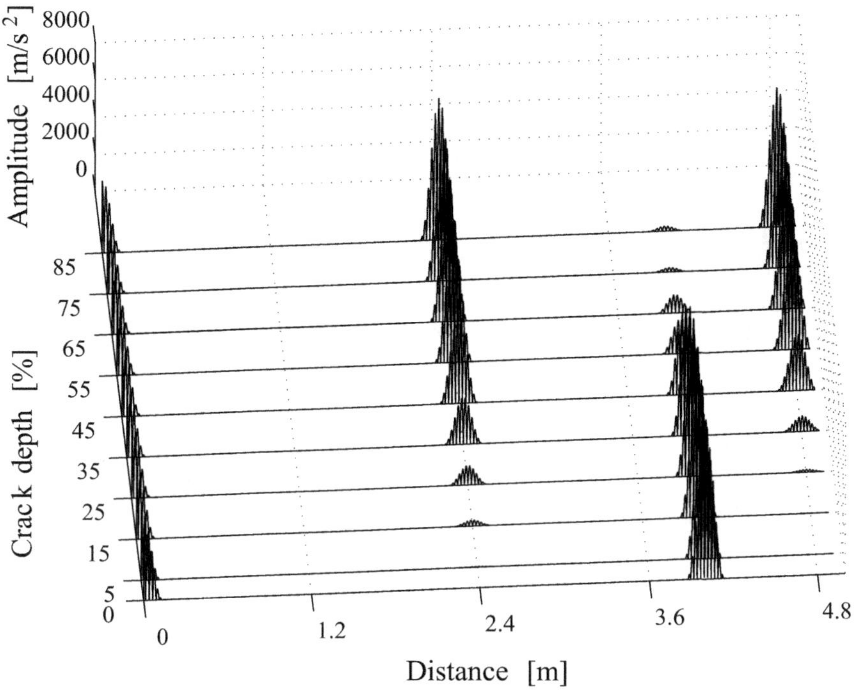

Figure 12. Influence of the crack depth, situated at the distance of 1.2 m from the left end of the rod, on wave propagation.

in relation to the amplitude of the signal received at the place of excitation. This means that sensitivity of damage detection methods based on changes in elastic wave propagation will be higher for higher frequency ranges. However, one must take into account the fact that, in reality one deals with elastic waves whose velocity depends on frequency, and therefore the signal undergoes dispersion. In order to take into consideration the above effect, one must apply Love's theory, Mindlin–Hermann's theory or three-mode theories, depending on the frequency range (Krawczuk et al., 2004).

5 Damage Identification in 1D Structures

Damage identification problems may be treated as inverse problems, but it is also possible to do damage identification based on knowledge in the field of wave propagation. In the former case, a high-fidelity numerical model of the analysed structure is necessary. Moreover, reference signals (signals

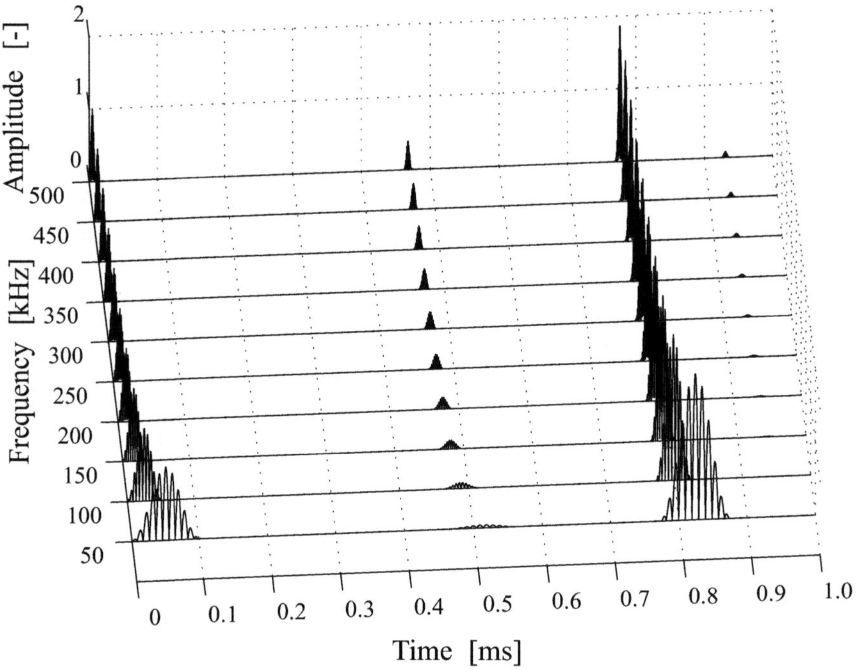

Figure 13. Signal frequency impact, on propagation of the wave reflected from a crack with the size of 15% of section height.

received for non-damaged structures) theoretically are not necessary. But in practice higher signal-to-noise ratio is obtained using differential signals. In the latter case, identification of damage may take place by processing the signals registered at sensors embedded in real objects. Monitoring of the structure in real-time is also possible because calculation time is very short in comparison with the former case. The use of reference signals means that better damage identification results can be obtained. Aspects of the application of both techniques, are illustrated by the example of the rod in Figure 14.

It was assumed that the geometry of the rod, material properties, and signal parameters were the same as in the comparative analysis of section 4.4. The rod was divided into 255 spectral elements. Damage of a size of 15% of the section height, at a distance of $L_c = 60.16$ cm from the left end of the rod, was inserted into the model. Simulation of experimental measurements was conducted, assuming an arbitrary distribution of two

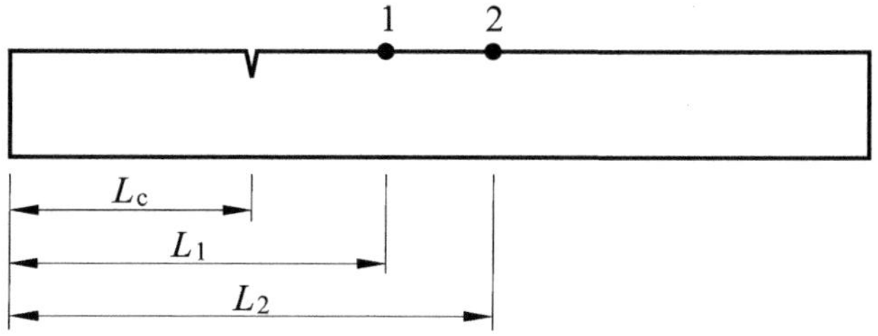

Figure 14. Schematic diagram of the rod with crack and sensors.

"PZT transducers" at distances of $L_1 = 89.84$ cm and $L_2 = 110.16$ cm. It was also assumed that each of the transducers could also work as a wave actuator and as a sensor at the same time. In this way, 4 signals were obtained. These were next contaminated by random noise with a quantity not exceeding 3% of the maximum amplitude of the signal. A similar simulation was conducted for the non–damaged rod.

Detection Because signals coming from real measurements contain noise, it should be minimised by filtering and the level of noise ought to be estimated. Giving an estimate of noise level is difficult in automation. The level of the noise may be estimated by making several measurements from the same sensors and conducting a statistical analysis. After performing an operation of subtraction on two signals registered in various operating conditions one will obtain the noise. If within the structure there is enough damage that there is the difference between signals, one will observe exceedance of the noise level (Fig. 15a). Unfortunately in this way, damages for which the amplitude of the reflected wave is lower than the noise level, will not be detected (Figure 15b). To overcome these obstacles, two methods of extracting features connected with damage from the signals could be proposed. The objective of these methods is to to get a smooth function with improved signal-to-noise ratio.

Method I The first method is simple weighted summation of signal amplitudes in a moving window (a type of cross-correlation) according to the equation,

$$e(t_j) = \sum_{i=1}^{N_w} \left[F(t_i)\, S\left(t_j + (i-1)\Delta t\right)\right]^2 \tag{43}$$

where S is the processed signal, F is a weight function, which may be a window modulating the signal (e.g. a Hanning window or a Gaussian window), Δt denotes the time step, N_w is the number of points in the moving window (in practice $N_w = t_e/\Delta t$, where t_e stands for the excitation time).

Method II The second method is a similar weighted summation, but corresponding to the power spectrum,

$$e(t_j) = |\mathcal{F}(f_c)| \approx |\text{DFT}\left[F(t_i)\, S\left(t_j + (i-1)\Delta t\right)\right]|, \quad i = 1, \ldots, N_w \quad (44)$$

where $\mathcal{F}$ denotes the linearly interpolated amplitude corresponding to the carrier frequency of the excited signal f_c, DFT denotes the discrete Fourier transform. To clarify, e is named as the *intensity function of reflected waves* because this function gives peaks corresponding to the location of reflected wave packets.

Application of the above–mentioned techniques regarding the signal presented in Figure 15b and alteration of the time scale causes amplification of the function e at the location of wave reflections (after transformation the signal is shorter by the width of moving window), which is presented in Figure 16. Method II introduces signal operations in the time domain as well as in frequency; this enables signal filtration to be more favourable than in the case of method I. In Figure 16 one can observe, that in method II the relation of the maximum value of the *intensity function of reflected waves* to the estimated noise level is much higher than in the case of method I.

In the literature many other methods of feature extraction can be found (wavelet analysis, pattern recognition, outlier analysis, etc.).

Localisation Estimation of the propagation time is a basis for adequate location of the damage. Techniques of time propagation estimation, based on the maximal signal envelope created using the Hilbert transform, may be unreliable if the level of noise in the signal is too high. Instead of the signal envelope, one may use, as suggested in section 5, the *intensity function of reflected waves.*

Estimation of the propagation time may be conducted on the signal received at the sensor number 2, when excitation occurs in transducer number 1 (t_{1-2}) and inversely (t_{2-1}). From the experimental simulation one obtains adequately $t_{1-2} = t_{2-1} = 0.0393$ ms, which gives, with the distance between sensors $L_2 - L_1 = 20.32$ cm, a propagation velocity of 5166 m/s, almost identical to the theoretically calculated velocity. Those values were obtained through application of the above-mentioned methods I and II. The time of propagation, calculated with the use of the envelope and Hilbert

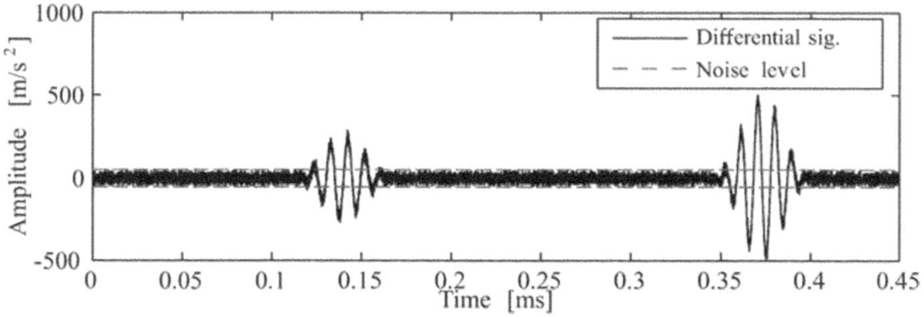

(a) Crack with a size of 15% of the height of the rod.

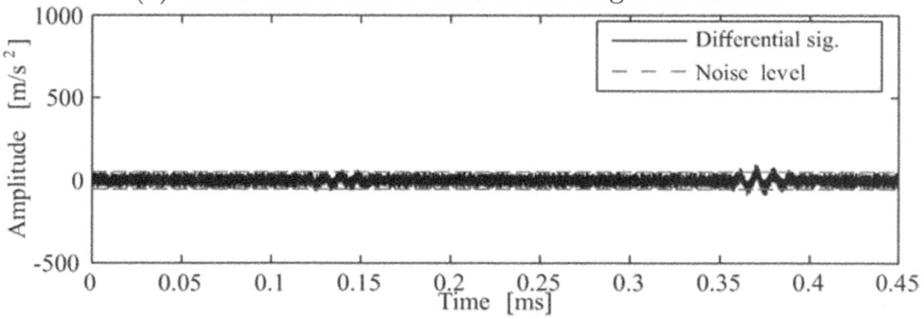

(b) Crack with a size of 5% of the height of the rod.

Figure 15. Difference in signals between non-damaged rod and rod with crack against estimated noise level background.

transform, adequately shows $t_{1-2} = 0.0391$ ms and $t_{2-1} = 0.0395$ ms. Such a discrepancy shows the lower precision of estimation.

Knowing the velocity of wave propagation and the location of the sensors, the *intensity function of reflected waves* can be transformed from the time domain to the distance domain in such a way that suitable superpositions of reflected waves occur,

$$E(x_j) = \sum_{i=1}^{N_S} e_i\left(t(x_j)\right), \qquad t(x_j) = \frac{d_{Tx}^j + d_{xS}^j}{c} \tag{45}$$

$$d_{Tx}^j = \sqrt{(x_j - x_T)^2}, \qquad d_{xS}^j = \sqrt{(x_j - x_S)^2} \tag{46}$$

where x_j is the presently considered rod coordinate, x_T stands for the excitation coordinate, x_S stands for the sensor coordinate, N_S denotes the total number of sensors, and c is the estimated velocity of wave propagation.

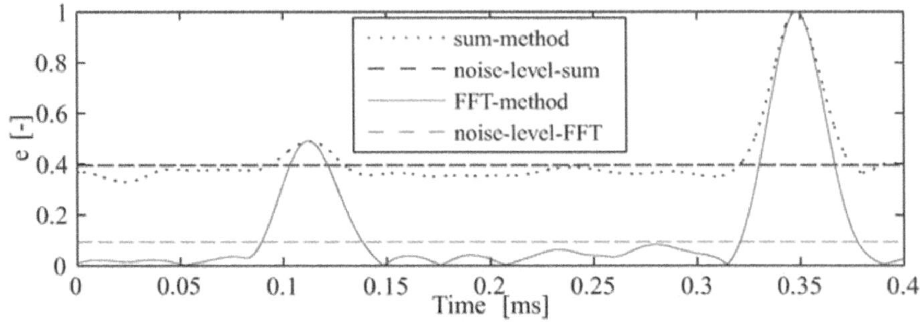

Figure 16. Differential signal filtered with the use of methods I and II, against estimated noise level background.

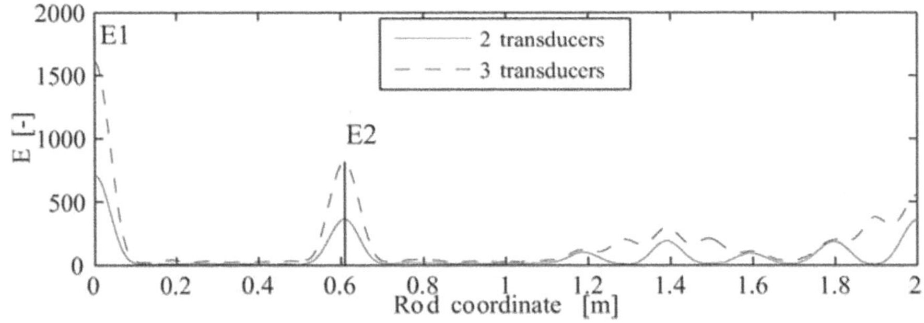

Figure 17. Amplification *intensity function of reflected waves* with application of two and three sensors.

The procedure presented in equations (45)-(46) was applied for the sensor configuration given in Figure 14 and for a configuration with additional sensors located between sensors number 1 and number 2. As a result, the amplified *intensity functions of reflected waves* presented in Fig. 17 were obtained. Amplification of the wave reflected from the crack occurs about 0.6 m from the left end of the rod (amplitude E_2). However, the wave reflected from the edge of the rod is also amplified (amplitude E_1). It can be observed, that the amplification is connected with the number of sensors and is equal to $E/e = N_s^2$.

Estimation of the damage size The size of the damage may be estimated by taking into account the relation of the amplitude of the wave reflected from the damage A_R with the amplitude of the excited wave A_T.

Because of the noise in the signal, the amplitudes A_R and A_T may be replaced with the amplitudes of the *intensity functions of reflected waves* e_R and e_T. However, it is necessary to familiarise oneself with the relation of e_R/e_T to crack depth. Such a relation may be defined in the process of experimental research or based on a numerical model (Fig. 18).

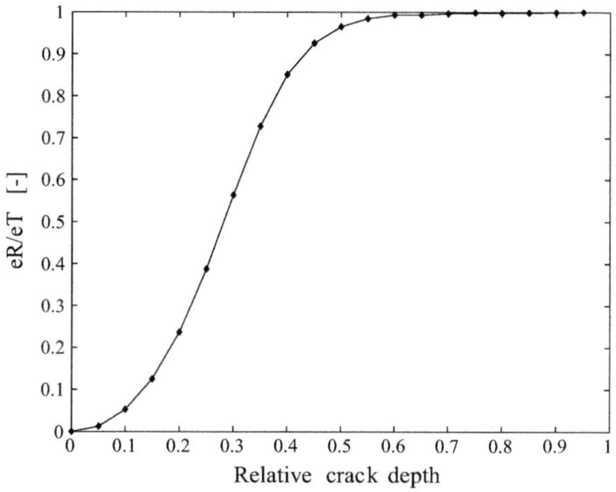

Figure 18. Relative crack influence on relation e_R/e_T with carrier frequency 100 kHz (numerical model).

Attention should be brought to the fact that, for very small cracks, it is difficult to calculate the amplitudes e_R from signals containing noise. Therefore, it is easier to calculate the relation e_R/e_T indirectly. Assuming amplitude symbols according to Figure 17 for the case with three sensors, one obtains $E_2 = 818.0$. Applying the procedures given in equation (44) for signals registered for a rod without damage, one obtains the amplitude of the *intensity function of reflected waves* directly at the location of excitation $e_{ref} = 728.25$. The relation e_R/e_T may be expressed by the equation,

$$\frac{e_R}{e_T} = \frac{E_2/N_s^2}{e_{ref}} \tag{47}$$

In the analysed example, $e_R/e_T = 0.1248$, and with the reference to the relation presented in Fig. 18, corresponds to a crack with a depth of 15% of the height in cross–section of the rod.

Genetic algorithms in the problem of identification Genetic algorithms are extremely suitable for the problem of optimisation of functions with multiple minima (maxima) or discontinuous functions (Goldberg, 1989). The genetic algorithm (GA) differs substantially from more traditional search and optimisation methods. The four most significant differences are:

- GAs search a population of points in parallel, not a single point.
- GAs do not require derivative information or other auxiliary knowledge; only the objective function and corresponding fitness levels influence the directions of search.
- GAs use probabilistic transition rules, not deterministic ones.
- GAs work on an encoding of the parameter set rather than the parameter set itself.

Those features predestine genetic algorithms to applications in problems regarding damage identification.

It is important to note that the GA provides a number of potential solutions to a given problem and the choice of final solution is left to the user. In cases where a particular problem does not have one individual solution (for example the solution is a family of crack locations) as in the case of multi-objective optimisation and scheduling problems, then the GA is potentially useful for identifying these alternative solutions simultaneously (Chipperfield et al., 1994).

Having the signals from the simulated experiment and numerical model of the rod presented in Section 4.2, an attempt was made to identify the location and size of the damage simultaneously. It was assumed that both decision variables, location of the crack m and depth of the crack n, would be encrypted in one chromosome divided into two parts,

$$\underbrace{1\,0\,0\,1\,0}_{m}\;\underbrace{0\,1\,1\,1}_{n}$$

The decision variables are represented by integers, ranging over $[0, 2^p - 1]$, where p denotes the number of bits in the chromosome. This enables easy modelling of the damage, because the locations agree with nodes of the mesh of the spectral finite elements. What is more, the following data were assumed:

- number of individuals 40,
- maximum number of generations 20,
- crossover probability 0.7,
- mutation probability 1/40,

The objective function, which is to be minimised, was suggested in the form of,

$$f(m, n) = \sum_{i=1}^{N} \sum_{j=1}^{4} |(R_{ij} - S_{ij}(m, n))| \tag{48}$$

where N denotes the number of points in the registered signal, R_{ij} stands for the j-th amplitude of the i-th signal from the simulated experiment, and $S(m,n)$ are the signal amplitudes obtained by using the numerical model, with the parameters specified by the decision variables m and n. The signals $S(m,n)$ are continuously calculated, because the number of possible locations and sizes of the crack may be relatively large (depending on the precision of the calculation). To shorten the calculation time, the problem was solved in two stages.

Stage I In this stage a 9-bit chromosome was assumed, where 5 bits fall to the decision variable m, and 4 bits fall to decision variable n. This assumption corresponds with ranges of representation, $m \in [0, 31]$ and $n \in [0, 15]$, when $m = 0$ and $n = 0$ denote that damage does not occur, and the remaining whole numbers correspond to positions of the crack location and size given by the number of divisions (Figure 19). In this way the crack cannot occur at the distance equal to 0 as well as equal the length of the rod, and the size of the crack cannot reach 100% of the cross–section height.

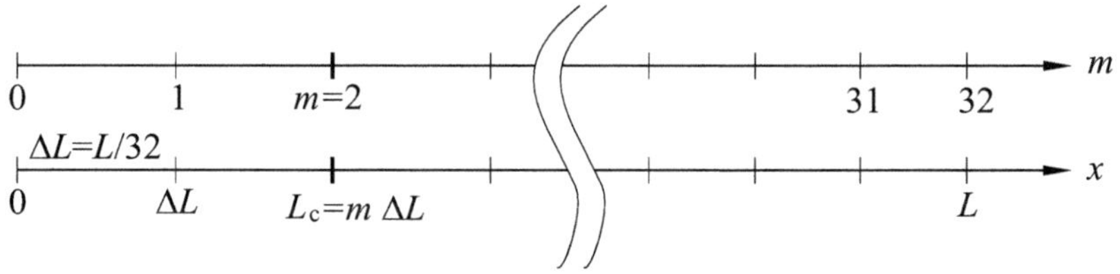

Figure 19. Schematic diagram of rod division in stage I.

In the analysed example, the genetic algorithm already converges to a result after two generations: $m^I = 10$, $n^I = 2$; this refers to a coordinate of crack location $L_c^I = 0.625$ and a size of crack $a^I = 0.125$. This solution is the first approximation, which is the starting point to stage II.

Stage II In this stage an 8-bit chromosome is assumed, where 4 bits fall to the decision variables m and n. The solution is searched for within a range situated to the left and to the right of the approximation found in stage I, thus,

$$L_c \in \left[\,(m^I - 1)\,\Delta L^I,\,(m^I + 1)\,\Delta L^I\,\right)$$

and similarly for the size of the crack a. The division resulting from the assumption of a four-bit representation of the decision variables, is presented in Figure 20. This time, the decision variables $m = 0$ and $n = 0$ correspond to the left ends of the ranges.

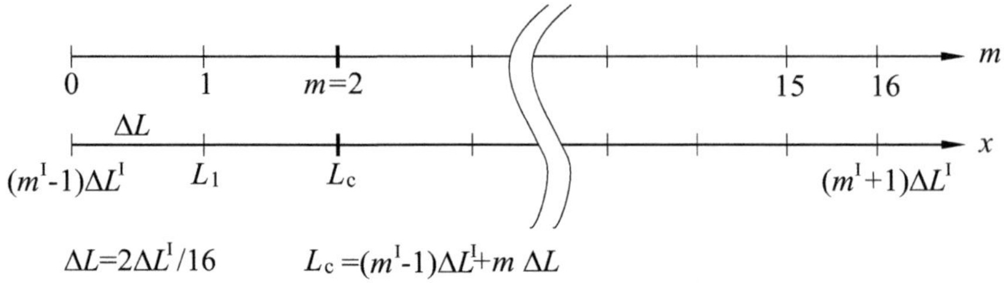

Figure 20. Schematic diagram of rod division in stage II.

In this example, after six generations the genetic algorithm gives a solution: $m^{II} = 5$, $n^{II} = 11$; this corresponds to coordinates of the crack location $L_c^I = 0.6016$ m and the size of the crack $a^{II} = 0.1484$. The obtained solution is exact as far as the location of the crack is concerned - this results from the assumed discretisation (255 elements, between which one may model the crack), and shows about 0.2% deviation as far as the depth of the crack is concerned.

Conclusions To summarise, the conception of a damage identification system based on knowledge of the field of wave propagation as well as the conception based on genetic algorithms, leads to considerable results. In both cases, the location and size of the damage were identified with high precision. On the other hand, the identification of damage in two–dimensional elements of a structure based on genetic algorithms may be too complex for contemporary computers.

6 Experimental Applications of Lamb Waves

6.1 Test stand profile

The test stand in the Department of Mechanics of Intelligent Structures in The Institute of Fluid-Flow Machinery of the Polish Academy of Sciences consists of piezoelectric transducers, measuring devices designed for wave generation and data acquisition, and a computer (Fig. 21). The measuring device is a prototype device, constantly developed and improved. In general, the device is superb for the needs of wave propagation analyses. An electronical system enables registration of signals from 12 measuring channels, while the 13th channel is used for wave generation. The device may be connected to the computer through a USB connector in order to control its parameters and transfer measured signals.

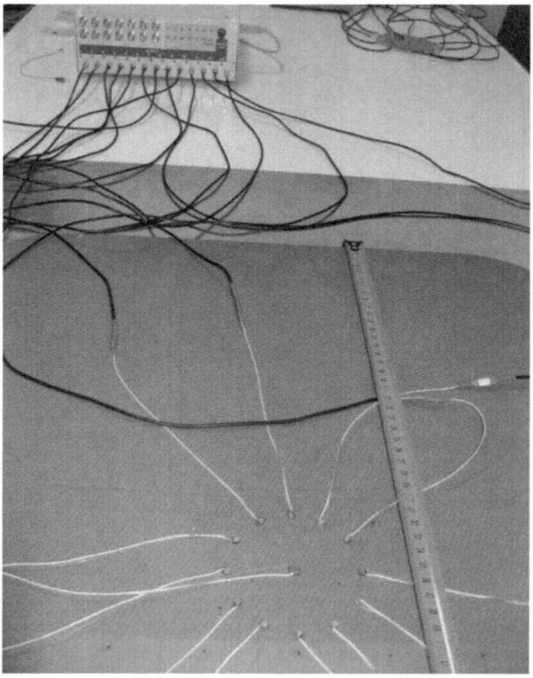

Figure 21. View of the fragment of a test stand.

The configuration of the matrix of piezoelectric transducers is in the shape of a clock with transducers placed on each "hour" of the round "clock face" and an additional transducer which is placed in the centre of the transducer configuration (Fig. 22). Elastic waves are generated by means of a central piezoelectric transducer CMAP11 (5 mm x 5 mm x 2 mm) or CMAP10 (3 mm x 3 mm x 2 mm) made by Noliac. Wave registration takes place at circumferential transducers CMAP10 (3 mm x 3 mm x 2 mm) also made by Noliac. However, it is also possible to generate the waves in the circumferential transducers. The limitation of the device is that the transducer generating the elastic waves cannot operate at the same time as the sensor registering the elastic waves.

The subject of research here is a composite panel, whose shape and dimensions are given in Figure 22. It is a part of a door from an Agusta AW-139 helicopter. The panel is made of six layers of carbon-epoxy laminate

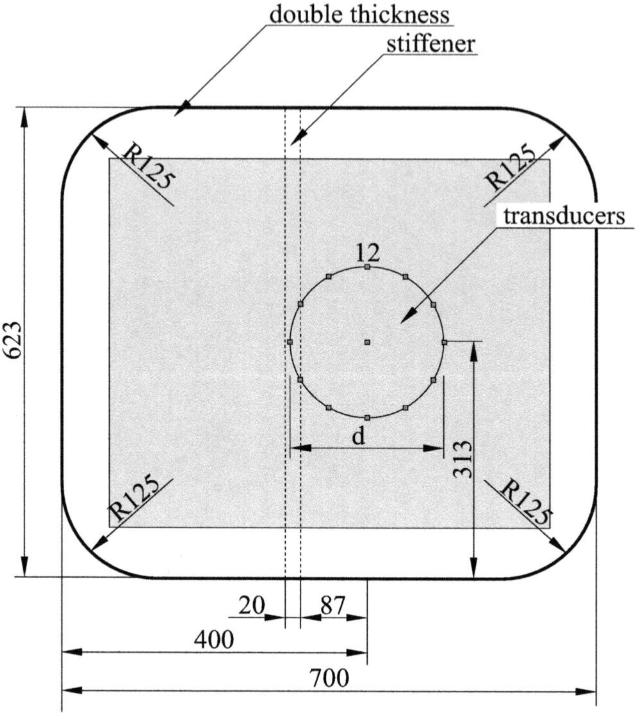

Figure 22. Geometry of the sample and distribution of sensors.

with the ply stacking sequence [+45/-45/0/90/+45/-45], and covered with a sealing compound with a thickness of about 0.14 mm, which makes up the face board of the door. The theoretical total thickness of the composite plate is equal to about 1.15 mm.

Transducers were attached to the studied specimen using a wax substance, the same that is used in assembling accelerometers. This enables easy assembly and non-destructive disassembly of transducers, which is important owing to the lowered costs of the experimental research. Connecting transducers to the surface of a structure using wax is a relatively cheap solution and at the same time practical. However, it is not an optimal solution because the wax strongly attenuates elastic waves and what is more, precise bonding of transducers is difficult. In practical monitoring systems, one should apply durable bonding e.g. using epoxy resin with mechanical properties similar to the surface, to which transducers are attached. The influence of the bonding layer thickness and its Young's modulus, on the glued

Table 2. Properties of materials (Żak et al., 2000; com)

	Sealing compound	Epoxy resin	Carbon fibre
Youngs Module, GPa	3.43	3.43	230
Poissons coefficient	0.35	0.35	0.2
Density, kg/m^3	1350	1250	1750

connection of the piezoelectric element, was studied in the work by Qing et al. (2006). Experimental results indicate that increasing the glue thickness changes the electromechanical impedance, the resonance frequency of piezoelectric element and the amplitude of the signal registered by the sensor. The influence of Young's modulus of the bonding layer on the signal amplitude is insignificant.

6.2 Theoretical dispersion curves

Because the mechanical properties of the specimen are unknown, for theoretical calculations the data given in the Table 2 were assumed. These values are verified according to group velocity profiles measured by experiment. With the use of a procedure described in work by Kudela et al. (2007b), dispersion curves were calculated (i.e. the group velocity dependence on frequency). It should be emphasised that this procedure was extended in a way that takes into account the asymmetrical ply stacking sequence and covering face lamina. Results are presented in Figure 23. Within the range of frequency around 1-600 kHz the longitudinal wave propagates the fastest (S0), next is the shear wave (SH0), and the slowest is the bending wave (A0). Above a frequency of about 600 kHz there occur other modes of Lamb waves.

Similarly, a group velocity distribution diagram was obtained depending on propagation angle (Figure 24) with frequency of 120 kHz. It is important that values of group velocities of the longitudinal wave (S0) and the shear wave (SH0) significantly differ from the average, i.e. the shape of the profile of group velocity significantly differs from a circle. The situation is different in the case of the bending wave (A0), where the mentioned differences are much smaller. Owing to this fact, in methods of damage localisation based on the time of flight of waves reflected from the damage, the most favourable modes for analysis are the bending waves. It is also important that in the analysed element of structure, the bending wave is characterised by minimal dispersion within the frequency range above 100 kHz (the course of the dispersion curve for the velocity of A0 mode for frequencies above 100 kHz in Figure 23 is similar to a horizontal line).

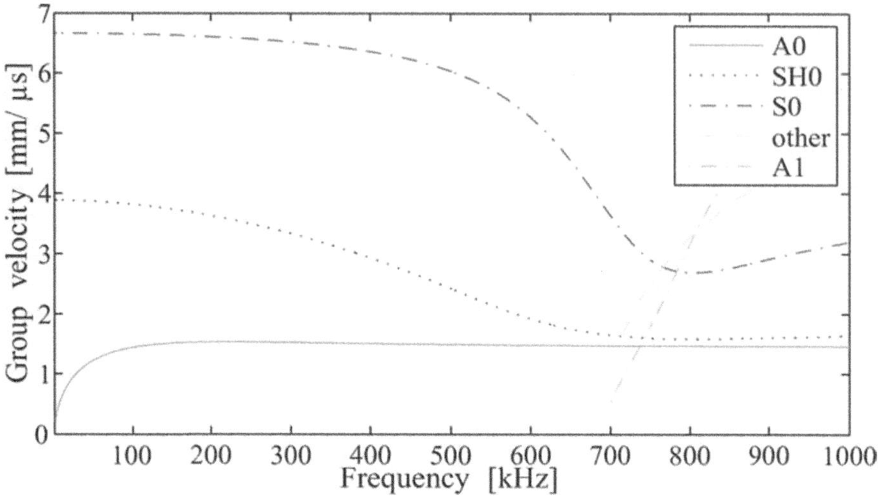

Figure 23. Theoretical dispersion curves obtained for the studied carbon-epoxy laminate.

6.3 Estimation and verification of wave group velocities

Measurements were conducted using a sinusoidal excitation modulated by a Hanning window within a frequency range of 50–150 kHz with a 10 kHz step. As a result of the modulation, signals with 3, 5, 7 and 10 cycles have been investigated. For the estimation of group velocity, a windowing method based on the signal spectrum energy was used. Selected results of the experiment are presented in Fig. 25.

Figure 25 presents the dependence of group velocities on the wave propagation direction, with given excitation frequencies for different number of cycles. Individual points are distributed with a 30 degree angle, corresponding to the distribution of sensors. Diagrams were also plotted with a dashed line, the theoretical curve of velocity profile for a composite with a fibre stacking sequence [+45/-45/0/90/+45/-45]. With frequencies above 100 kHz (the minimal dispersion area) the velocities estimated based on experimental signals almost agree regardless of the number of cycles, and their distribution depending on propagation angle to a large extent agrees with the theoretical profile. It should also be added that for a frequency range corresponding to the dispersion profile, velocities are the most similar to circles, which is beneficial in application to the methods of damage localisation.

A conclusion from this research is that, owing to high wave attenuation

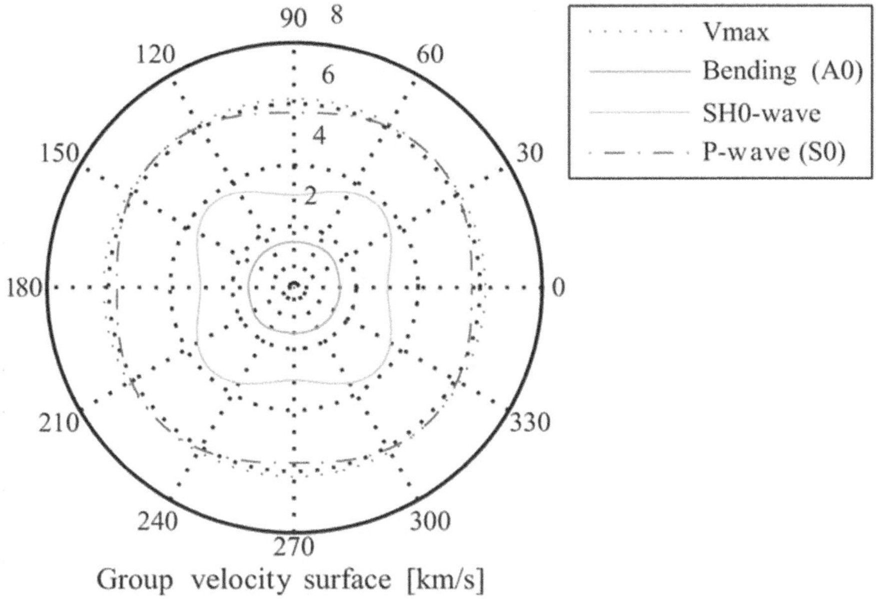

Figure 24. Theoretical graph of group velocity distribution depending on propagation angle, obtained for the studied laminate with a frequency of 120 kHz.

(due to composite material damping), inspection of a panel area is only possible within a radius of 0.5 m for a frequency of 50 kHz. With an increase of the frequency, wave attenuation also increases.

It should be emphasised that, in the registered signals, the predominant wave amplitudes are connected with bending waves. Shear and longitudinal waves are also present in the signals but their amplitudes are very small with reference to the amplitudes of the bending waves. For this reason, based on registered signals one cannot estimate the velocity of longitudinal and shear wave propagation. However, shear and longitudinal waves propagate with higher velocity than bending waves. Moreover, they undergo mode conversion, which causes additional signal disturbances.

6.4 Damage detection

Piezoelectric transducers were attached to the studied sample using wax. A "clock" configuration of sensors with a radius of about 4 cm was applied. A series of measurements for a non-damaged sample as well as for a sample

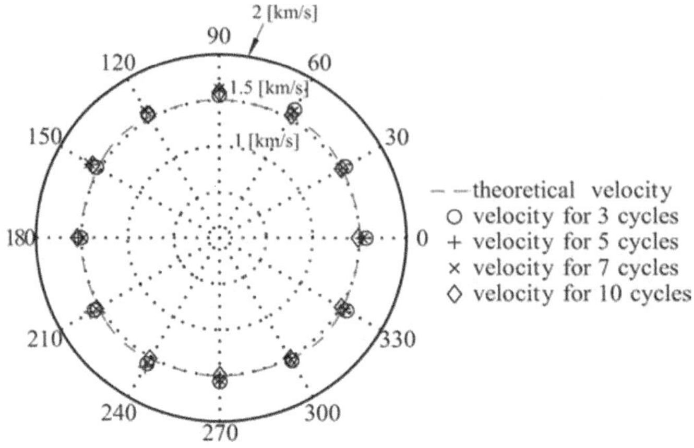

Figure 25. Group velocity dependence on the propagation angle with excitation frequency of 120 kHz for an even number of cycles.

with various artificially inserted damages were conducted. Signal processing was carried-out using an algorithm suggested by Kudela et al. (2008). The complicated geometry of the studied sample and the composite material with strong damping behaviour, of which the sample was made, mean that distinction between waves reflected from damage and reflected from features of the structure (edges, stiffener, depth change, etc.) is impossible. For this reason, it is necessary to use reference signals. That is why, after application of the signal processing procedure, the amplitudes of signals registered for the damaged sample s_r were linked to the amplitudes of signals registered for the non–damaged sample s_d using a dB scale:

$$E_{d-r} = 10 \log_{10}\left(\frac{E_d}{E_r}\right) \tag{49}$$

where E_d and E_r refer to superimposed *intensity functions of reflected waves* e_d and e_r given by Eq. (44) in the case of signals registered for the damaged (s_d) and reference (s_r) structure, respectively. The functions E_d and E_s in the case of 2D problems can be described by the formula:

$$E = \sum_k \int_S e_k(x,y)dS \approx \sum_k \sum_{i,j} e_k(x_i,y_j), \qquad k = 1,\ldots,12 \quad i = 1,\ldots,N \quad j = 1,\ldots,M \tag{50}$$

Table 3. Inserted cracks scenario

no	crack no 1		crack no 2	
	depth [mm]	length [mm]	depth [mm]	length [mm]
a	0.2	4	-	-
b	1	4	-	-
c	1	8	-	-
d	1	14	-	-
e	1	18	-	-
f	1	18	1	4
g	1	18	1	8
h	1	18	1	15

where S is the surface of the investigated structure, and N and M stand for the total number of nodes i and j, located on the surface of the investigated structure; k is the sensor number. This function gives a map which can be called a *damage influence map* or *damage intensity map*.

As an indicator of the damage level of the element of structure, a value described with the following formula was assumed:

$$D = \sum_{ij} |E_{d-r}(x_i, y_j)| \tag{51}$$

After a series of tests it was found that a frequency of about 120 kHz is the optimal frequency for signal excitation[3]. Further researches were conducted using only a frequency of excitation of 120 kHz.

6.5 Crack detection

Within the studied sample, cracks were inserted according to the scenario given in Table 3. As a result of the damage detection algorithm suggested in the work by Kudela et al. (2008), one finds maps of damage influence as presented in Figure 26. On the maps of damage influence, the real location of the crack for reference purposes was clearly marked with a white line.

While analyzing the maps of damage influence presented in Figures 26a–c one may notice an increase in the value of map amplitudes in the form of circumferential stripes, which the radius agrees with the radius where damage is located. By increasing the damage to a length of 8 mm, the map of damage influence (Figure 26c) reaches a maximal value in the close

[3] For a signal excitation frequency of 120 kHz the biggest differences were noted between signals in damaged and non-damaged sample, what manifested in the biggest values of damage level indicator D

neighbourhood of the damage (difference between centre of the crack and maximal value of damage influence map is equal to about 25 mm). Further increasing of the crack length brings about surprising results. Amplitudes of damage influence maps presented in Figures 26d–e undergo strengthening, not only in the place where the crack occurs but also in other areas. What is more, the maximal amplitude value of the damage influence map decreases with increase of the damage length. The observed anomalies may be explained by the effect of mode conversion. Shear and longitudinal waves converted from the bending mode propagate faster. Damage influence maps are made based on the bending wave velocity profile, therefore strengthening or attenuation of the amplitudes of the damage influence maps occur also in areas which do not agree with the location of the crack. What is more, a problem may be caused by noise (*coherent noise*) that results from inaccuracy in signal amplitude subtraction (in the dB scale of logarithmic relation between signal amplitudes), which for example are shifted at random against one another.

Insertion of the second crack causes significant changes in the damage influence maps (Figures 26f–h). One may observe that the maximal value of the damage influence maps increase along with the increase of the second crack's length. Despite the fact that the second crack is located closer to the sensor configuration, it is impossible to locate unambiguously the crack based on the damage influence map. One may only estimate the radius where the crack is located, and only in the case of the crack with a length over 10 mm (Figure 26h).

Despite difficulties in formulating a crack localisation method, experimentation in their detection proves to be relatively simple. Damage detection may occur based on the damage indicator values represented in formula (51). First, one must experimentally establish a threshold value which, once surpassed, will indicate damage of the structure. Figure 27 presents dependence of the damage level indicator on the damage scenarios put together in Table 3. Also, an arbitrary established threshold level of the value $1.75 \cdot 10^5$ was applied to the Figure 27. The damage level indicator takes a lower value than the threshold level only in the case of damage scenario "a", i.e. a crack within the face layer of the sample. Insertion of a crack with a depth of 1 mm into studied sample, causes exceedance of the threshold level. One may state that the indicator of damage level, shows an increasing tendency along with size and number of inserted cracks with the exception of damage scenario "d". This indicator may be effectively used for damage detection purposes in the early stage of development.

In the case of experimental research concerning delamination detection, the situation is similar to the case of matrix crack detection. However,

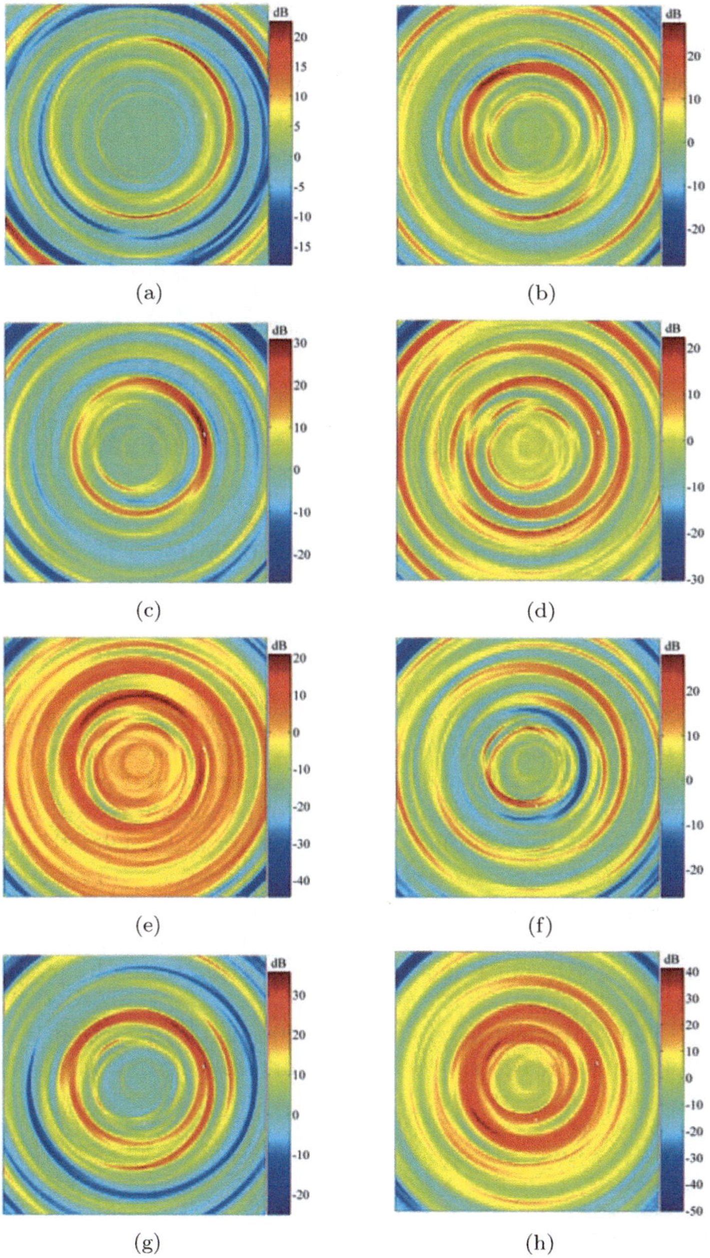

Figure 26. Experiment: damage influence maps.

location of the delamination is much more difficult. It derives from the fact that amplitudes of waves reflected from delamination are significantly lower than amplitudes of waves reflected from cracks.

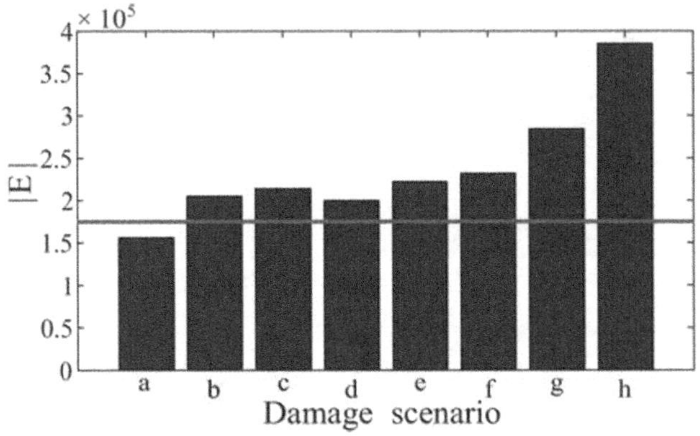

Figure 27. Damage size indicator.

6.6 Conclusions

Analyses of signals measured with the use of specialist testing equipment were conducted. Estimation of wave group velocities for various frequencies and numbers of cycles was made. Dispersion curves, and profiles of dependence of wave group velocities on the angle of propagation, were calculated. Experimental results were compared with theoretical results. High conformity between theory and experiment both in dispersion curves and profiles of velocity depending on wave propagation angle were observed.

A serious problem with the studied specimen of a composite structure is the high damping, which increases in proportion to the frequency of the excitation signal.

The developed damage detection algorithm enables detection of extremely small cracks. In some cases, it is also possible to locate the crack.

Bibliography

URL `http://www.performance-composites.com`.

D.N. Alleyne and P. Cawley. Optimization of Lamb wave inspection techniques. *NDT&E International*, 25:11–22, 1992.

K. Amaratunga and J.R. Williams. Time integration using wavelets. In *Proceedings of SPIE, Wavelet Application for Dual Use*, pages 894–902, 1995.

A. Bachschmid, G. Diana, and B. Pizzigoni. The influence of unbalance on cracked rotors. In *Vibrations of Rotating Machinery*, pages 193–198. Institution of Mechanical Engineers, 1984.

L.J. Bond. Numerical techniques and their use to study wave propagation and scattering: a review. In S.K. Datta, J.D. Achenbach, and Y.S. Rajapakse, editors, *Elastic Waves and Ultrasonic Non-destructive Evaluation*. North-Holland, 1990. The Proceedings of the IUTAM Symposium.

J.P. Boyd. *Chebyshev and Fourier spectral methods*. Springer, 1989.

C. Canuto, M.Y. Hussaini, A. Quarteroni, and T.A. Zang. *Spectral methods in fluid dynamics*. Springer, 3 edition, 1991.

P. Cawley and R.D. Adams. The location of defects in structures from measurements of natural frequencies. *Journal of Strain Analysis*, 14: 49–57, 1979.

A. Chakraborty and S. Gopalakrishnan. A spectrally formulated finite element for wave propagation analysis in layered composite media. *International Journal of Solids and Structures*, 41:5155–5183, 2004.

A. Chakraborty and S. Gopalakrishnan. A spectrally formulated plate element for wave propagation analysis in anisotropic material. *Computer Methods in Applied Mechanics and Engineering*, 194:4425–4446, 2005.

A. Chakraborty and S. Gopalakrishnan. A spectral finite element model for wave propagation analysis in laminated composite plate. *Journal of Vibration and Acoustics*, 128:477–488, 2006a.

A. Chakraborty and S. Gopalakrishnan. An approximate spectral element for the analysis of wave propagation in inhomogeneous layered media. *AIAA Journal*, 44:1676–1685, 2006b.

L.W. Chen and C.L. Chen. Vibration and stability of cracked thick rotating blades. *Computers and Structures*, 28:67–74, 1988.

Y.K. Cheung. *Finite Strip Method in Structural Mechanics*. Pergamon Press, Oxford, 1976.

A.J Chipperfield, P.J. Fleming, H. Pohlheim, and C.M. Fonseca. Genetic algorithm toolbox user's guide. ACSE Research Report 512, University of Sheffield, 1994.

Y. Cho and J.L. Rose. A boundary element solution for a mode conversion study on the edge reflection of lamb waves. *Journal of the Acoustical Society of America*, 99(4):2079–2109, 1996.

W. Dauksher and A.F. Emery. Accuracy in modeling the acoustic wave equation with Chebyshev spectral finite elements. *Finite Elements in Analysis and Design*, 26:115–128, 1997.

P.P. Delsanto, T. Whitecomb, H.H. Chaskelis, and R.B. Mignogna. Connection machine simulation of ultrasonic wave propagation in materials I: the one-dimensional case. *Wave Motion*, 16:65–80, 1992.

P.P. Delsanto, T. Whitecomb, H.H. Chaskelis, R.B. Mignogna, and R.B. Kline. Connection machine simulation of ultrasonic wave propagation in materials II: the two-dimensional case. *Wave Motion*, 20:295–314, 1994.

P.P. Delsanto, R.S Schechter, and R.B. Mignogna. Connection machine simulation of ultrasonic wave propagation in materials III: The three-dimensional case. *Wave Motion*, 26:329–339, 1997.

J.F. Doyle. *Wave Propagation in Structures.* Springer-Verlag, 2 edition, 1997.

D. E. Goldberg. *Genetic Algorithms in Search, Optimization and Machine Learning.* Kluwer Academic Publishers, Boston, 1989.

G. Gounaris and A.D. Dimarogonas. A finite element a cracked prismatic beam for structural analysis. *Computers and Structures*, 28:309–313, 1988.

B.S. Haisty and W.T. Springer. A general beam element for use in damage assessment of complex structures. *Journal of Vibration, Acoustics, Stress, and Reliability in Design*, 110:389–394, 1988.

Y. Kim, S. Ha, and F.-K. Chang. Time–domain spectral element method for built–in piezoelectric–actuator–induced Lamb wave propagation analysis. *AIAA Journal*, 46(3):591–600, 2008.

D. Komatitsch and J.P. Vilotte. The spectral element method: An efficient tool to simulate the seismic response of 2D and 3D geological structures. *Bulletin of the Seismological Society of America*, 88:368–392, 1998.

D. Komatitsch, Ch. Barnes, and J. Tromp. Simulation of anisotropic wave propagation based upon a spectral element method. *Geophisics*, 65(4): 1251–1260, 2000.

M. Koshiba, S. Karakida, and M. Suzuki. Finite element analysis of Lamb waves scattering in an elastic plate waveguide. *IEEE Transactions on Sonic and Ultrasonic*, 31:18–25, 1984.

M. Krawczuk. Modelling and identification of cracks in truss constructions. *Finite Element Analysis and Design*, 12:41–50, 1992.

M. Krawczuk. A rectangular plate finite element with an open crack. *Computers and Structures*, 46:487–493, 1993.

M. Krawczuk. Rectangular shell finite element with an open crack. *Finite Elements in Analysis and Design*, 15:233–253, 1994.

M. Krawczuk and W. Ostachowicz. Identification of delamination in composite beams by genetic algorithm. *Science and Engineering of Composite Materials*, 10(2):147–155, 2002.

M. Krawczuk and W. Ostachowicz. Hexahedral finite element with an open crack. *Finite Elements in Analysis and Design*, 13:225–235, 1993a.

M. Krawczuk and W. Ostachowicz. A finite plate element for dynamic analysis of a cracked plate. *Computer Methods in Applied Mechanics and Engineering*, 115:67–78, 1994.

M. Krawczuk and W. Ostachowicz. Journal of vibration and acoustics. *Natural vibrations of a clampedclamped arch with an open transverse crack*, 119:145–151, 1997.

M. Krawczuk and W.M. Ostachowicz. Transverse natural vibrations of a cracked beam loaded with a constant axial force. *Journal of Vibration and Acoustics*, 115:524–528, 1993b.

M. Krawczuk, W. Ostachowicz, and A. Zak. Modal analysis of cracked, unidirectional composite beam. *Composites Part B: Engineering*, 28B: 641–650, 1997.

M. Krawczuk, A. Zak, and W. Ostachowicz. Elastic beam finite element with a transverse elastoplastic crack. *Finite Elements in Analysis and Design*, 341(1):61–73, 2000.

M. Krawczuk, A. Zak, and W. Ostachowicz. Finite element model of plate with elastoplastic through crack. *Computers and Structures*, 79:515–532, 2001.

M. Krawczuk, M. Palacz, and W. Ostachowicz. The dynamic analysis of cracked Timoshenko beam by the spectral element method. *Journal of Sound and Vibration*, 264:11391153, 2003.

M. Krawczuk, M. Palacz, and W. Ostachowicz. Flexural-shear wave propagation in cracked composite beam. *Science and Engineering of Composite Materials*, 11:55–67, 2004.

P. Kudela. *Damage detection in selected composite elements of structures by the elastic wave propagation method.* PhD thesis, Polish Academy of Sciences, Institute of Fluid-Flow Machinery, 2008. in polish.

P. Kudela and W. Ostachowicz. Damage detection in composite rods by the elastic wave propagation method. In *Computer Methods in Mechanics*, 2007.

P. Kudela, W. Ostachowicz, and A. Żak. Influence of temperature fields on wave propagation in composite plates. In *Key Engineering Materials*, volume 347, pages 537–542, Switzerland, 2007a. Trans Tech Publications.

P. Kudela, A. Żak, M. Krawczuk, and W. Ostachowicz. Modelling of wave propagation in composite plates using the time domain spectral element method. *Journal of Sound and Vibration*, 302:728745, 2007b.

P. Kudela, W. Ostachowicz, and A. Żak. Damage detection in composite plates with embedded PZT transducers. *Mechanical Systems and Signal Processing*, 22:1327–1335, 2008.

B.C. Lee, M. Palacz, M. Krawczuk, W. Ostachowicz, and W.J. Staszewski. Wave propagation in a sensor/actuator diffusion bond model. *Journal of Sound and Vibration*, 276:671–687, 2004.

G.R. Liu and Z.C. Xi. *Elastic waves in anisotropic laminates.* CRC Press, 2002.

G.R. Liu, J. Tani, K. Watanabe, and T. Ohyoshi. Harmonic wave propagation in anisotropic laminated strips. *Journal of Sound and Vibration*, 139:313–330, 1990.

T. Liu, K. Liu, and J. Zhang. Unstructured grid method for stress wave propagation in elastic media. *Computer Methods in Applied Mechanics and Engineering*, 193:2427–2452, 2004.

T. Liu, K. Liu, and J. Zhang. Triangular grid method for stress-wave propagation in 2–D orthotropic materials. *Archive of Applied Mechanics*, 74: 477–488, 2005.

D.R. Mahapatra and S. Gopalakrishnan. A spectral finite element model for analysis of axialflexuralshear coupled wave propagation in laminated composite beams. *Computers and Structures*, 59:67–88, 2003.

M. Mitra and S. Gopalakrishnan. Wavelet based spectral finite element for analysis of coupled wave propagation in higher order composite beams. *Composite Structures*, 73:263–277, 2006.

W. Ostachowicz and M. Krawczuk. On modelling of structural stiffness loss due to damage. In *Key Engineering Materials*, volume 204-205, page 185200, 2001.

W. Ostachowicz and M. Krawczuk. Coupled torsional and bending vibrations of a rotor with an open crack. *Archive of applied mechanics*, 62: 191–201, 1992.

W. Ostachowicz, M. Krawczuk, and M.P. Cartmell. The location of a concentrated mass on rectangular plates from measurements of natural vibrations. *Computers and Structures*, 80:1419–1428, 2002.

W. Ostachowicz, M. Krawczuk, and M. Palacz. Detection of delamination in multilayer composite beams. In *Key Engineering Materials*, pages 245–246, 2003.

M. Palacz and M. Krawczuk. Analysis of longitudinal wave propagation in a cracked rod by the spectral element method. *Computers and Structures*, 80:1809–1816, 2002.

A.T. Patera. A spectral element method for fluid dynamics: Laminar flow in a channel expansion. *Journal of Computational Physics*, 54:468–488, 1984.

C. Pozrikidis. *Introduction to Finite and Spectral Element Methods using MATLAB(R).* Chapman & Hall/CRC, 2005.

J.S. Przemieniecki. *Theory of matrix structural analysis.* McGraw Hill, New York, 1968.

G.L. Qian, S.N. Gu, and J.S. Jiang. A finite element model of cracked plates and application to vibration problems. *Computers and Structures*, 39: 483–487, 1991.

X.P. Qing, H-L. Chan, S.J. Beard, T.K. Ooi, and S.A. Marotta. Effect of adhesive on the performance of piezoelectric elements used to monitor structural health. *International Journal of Adhesion & Adhesives*, 26: 622–628, 2006.

E. Ruffino and P.P Delsanto. Problems of accuracy and reliability in 2–D LISA simulations. *Computers and Mathematics with Applications*, 38: 89–97, 1999.

R.S. Schechter, H.H. Chaskelis, R.B. Mignogna, and P.P. Delsanto. Real-time parallel computation and visualization of ultrasonic pulses in solids using the connection machine. *Science*, 265:1188–1192, 1994.

G. Seriani. A parallel spectral element method for acoustic wave modeling. *Journal of Computational Acoustics*, 5(1):53–69, 1997.

G. Seriani. 3–D spectral element-by-element wave modelling on Cray T3E. *Physics and Chemistry of the Earth*, 24(3):241–249, 1999.

R. Spall. Spectral collocation methods for one dimensional phase change problems. *International Journal of Heat Mass Transfer*, 15:2743–2748, 1995.

R. Sridhar, A. Chakraborty, and S. Gopalakrishnan. Wave propagation in anisotropic and inhomogeneous uncracked and cracked structures using pseudospectral finite element method. *International Journal of Solids and Structures*, 43:4997–5031, 2006.

J.C. Strickwerda. *Finite Difference Schemes and Partial Differential Equations*. Wadsworth-Brooks, 1989.

H. Tada, P.C. Paris, and G.R. Irwin. *The stress analysis of cracks handbook*. Del Research Corporation, 1973.

G.S. Verdict, P.H. Gienand, and C.P. Burger. Finite element study of Lamb wave interactions with holes and through thickness defects in thin metal plates. *Review of Progress in Quantitative Non-destructive Evaluation*, 11:97–104, 1992.

J.R. Vinson and R.L. Sierakowski. *Behaviour of Stuctures Composed of Composite Materials*. Martinus Nijhoff, Dorchester, 3rd edition, 1989.

H. Yamawaki and T. Saito. Numerical calculation of surface waves using new nodal equation. *NDT&E International*, 8-9:379–389, 1992.

H. Yim and Y. Sohn. Numerical simulation and visualization of elastic waves using mass-spring lattice model. *IEEE Transaction on Ultrasonics, Ferroelectrics, and Frequency Control*, 47(3):549–558, 2000.

M.M.F. Yuen. A numerical study of the eigenparameters of damaged cantilever beam. *Journal of Sound and Vibration*, 103:301–310, 1985.

A. Żak, M. Krawczuk, and W. Ostachowicz. Numerical and experimental investigation of free vibration of multilayer delaminated composite beams and plates. *Computational Mechanics*, 26:309–315, 2000.

A. Zak, M. Krawczuk, and W. Ostachowicz. Vibration of a laminated composite plate with closing delamination. *Journal of Intelligent Material Systems and Structures*, 12(8):545–551, 2001.

A. Żak, M. Krawczuk, and W. Ostachowicz. Propagation of in-plane waves in an isotropic panel with a crack. *Finite Elements in Analysis and Design*, 42:929–941, 2006.

A. Żak, M. Krawczuk, W. Ostachowicz, P. Kudela, and M. Palacz. Elastic wave propagation in cracked isotropic plates. In *Proceedings of the Third European Workshop Structural Health Monitoring 2006*, pages 316–323, 2006.

A.J. Zak, M.P. Cartmell, and W.M. Ostachowicz. A sensitivity analysis of the dynamic performance of a composite plate with shape memory alloy wires. *Composite Structures*, 60:145–157, 2003.

GPSR Compliance

The European Union's (EU) General Product Safety Regulation (GPSR) is a set of rules that requires consumer products to be safe and our obligations to ensure this.

If you have any concerns about our products, you can contact us on ProductSafety@springernature.com

In case Publisher is established outside the EU, the EU authorized representative is:

Springer Nature Customer Service Center GmbH
Europaplatz 3
69115 Heidelberg, Germany

Batch number: 10732218

Printed by Printforce, the Netherlands